Forestry Economics

Forestry Economics introduces students and practitioners to all aspects of the management and economics of forestry. The book adopts the approach of managerial economics textbooks and applies this to the unique processes and problems faced by managers of forests. While most forestry economics books are written by economists for future economists, what many future forest and natural resource managers need to understand is what economic information is and how to use it to make better business and management decisions. John E. Wagner draws on his twenty years of experience teaching and working in the field of forest resource economics to present students with an accessible understanding of the unique production processes and problems faced by forest and other natural resource managers.

There are three unique features of this book:

- The first is its organization. The material is organized around two common economic models used in forest and natural resources management decision making.
- The second is the use of case studies from various disciplines: outdoor and commercial recreation, wood products engineering, forest products, and forestry. The purpose of these case studies is to provide students with applications of the concepts being discussed within the text.
- The third – at the end of each chapter – is revisiting the question of how to use economic information to make better business decisions. This ties each chapter to the preceding ones and reinforces the hypothesis that a solid working knowledge of these economic models and the information they contain is necessary for making better business decisions.

This textbook is an invaluable source of clear and accessible information on forestry economics and management for not only economics students, but also for students of other disciplines and those already working in forestry and natural resources.

John E. Wagner is Professor of Forest Resource Economics at the State University of New York College of Environmental Science and Forestry in Syracuse, USA.

Routledge Textbooks in Environmental and Agricultural Economics

Forestry Economics

A managerial approach

John E. Wagner

LONDON AND NEW YORK

First published 2012
by Routledge
2 Park Square, Milton Park, Abingdon, Oxon, OX14 4RN

Simultaneously published in the USA and Canada
by Routledge
711 Third Avenue, New York, NY 10017

*Routledge is an imprint of the Taylor & Francis Group, an informa
business*

British Library Cataloguing in Publication Data
A catalogue record for this book is available from the British Library

Library of Congress Cataloging in Publication Data
Wagner, John E.
    Forestry economics : a managerial approach/by John E. Wagner.
        p. cm.
    Includes bibliographical references and index.
    1. Forests and forestry–Economic aspects. 2. Forest management.
    I. Title.
SD393.W26 2011
634.9'2–dc22   2011002712

ISBN: 978-0-415-77440-6 (hbk)
ISBN: 978-0-415-77476-5 (pbk)
ISBN: 978-0-203-80802-3 (ebk)

Typeset in Times New Roman by Sunrise Setting Ltd, Torquay, UK

Contents

vi   *Contents*

List of figures

List of tables

Preface

Textbooks that I have used in the past with a focus on Forestry or Natural Resource Economics were written from the perspective of an economist to a future economist. However, most students who graduate from environmental, forestry, or natural resource management programs are not going to be economists: They are going to be managers. They will need to understand what economic information is and how they can use it to make better business and management decisions. In this regard, this book takes an approach similar to Managerial Economics textbooks that are used in business schools at the undergraduate and graduate levels. However, standard business school-type Managerial Economic textbooks do not address the unique production processes and problems faced by managers of forests and other natural resources. Those issues are addressed by this book.

There are three unique features of this book. The first feature is its organization. The introduction develops and describes two common economic models used in the decision making process: (1) maximization of net benefits and (2) least cost/cost effective. Chapters 2 through 4 break down each model into its components and identifies and critically analyzes the economic information each component contains. Chapter 5 reconstitutes the models combining the knowledge gained from the decomposition into a useful analysis tool. Chapters 6 and 7 expand the models to encompass the market place. Chapters 8 through 10 further expand the models to take into account compounding and discounting, capital budgeting, and risk, with Chapter 9 examining the forest rotation problem. Chapter 11 examines the impacts of forest taxes on the decision making models. Finally, Chapter 12 discusses techniques commonly used to estimate the demand for non-market goods.

The second feature is the use of case studies from various disciplines, such as outdoor and commercial recreation, wood products engineering, forest products, and forestry. The case studies come from a variety of sources and will be used to illustrate key points. The final feature is revisiting the question: "How to use economic information to make better business decisions?" at the end of each chapter. This ties each chapter to the preceding ones and reinforces the hypothesis that a solid working knowledge of these economic models and the information they contain are necessary for making better business decisions.

Acknowledgments

There are a few people who should be acknowledged, as without their help this book would have been much harder to complete. First, to my wife Janice and my parents Albert and Mirney. Second, to Mr Stephen Singer who provided the initial editorial review of the entire book. Third, to the following colleagues who provided professional reviews of various chapters:

Dr Douglas Carter	University of Florida
Dr Donald Grebner	Mississippi State University
Dr Donald Hodges	University of Tennessee
Dr David Newman	SUNY – College of Environmental Science and Forestry
Dr Erik Nordman	Grand Valley State University
Mr Patrick C. Penfield	Syracuse University
Dr Jeffery Prestemon	Forest Service – Southern Research Station
Dr Thomas Stevens	University of Massachusetts – Amherst

Fourth, the following agencies, professional societies, and businesses graciously allowed me copyrights to material that I used as my case studies: United States Department of Agricultural Forest Service, Canadian Farm Business Management Council, Ontario Ministry of Agriculture – Food and Rural Affairs, J.W. Sewall Company, Ohio Sea Grant – The Ohio State University, Society of American Foresters, Forest Products Society, Manomet Center of Conservation Sciences, and Micromill Systems, Inc. Finally, to the editorial staff at Routledge Press for their help and patience.

Abbreviations

AC	Average Cost
ac	Acres
AFC	Average Fixed Cost
AP	Average Product
AR	Average Revenue
ARP	Average Revenue Product
ATC	Average Total Cost
AVC	Average Variable Cost
BCR	Benefit Cost Ratio
bd.ft	Board feet
CAI	Current Annual Increment
CMAI	Culmination of Mean Annual Increment
cu.ft	Cubic feet
DBH	Diameter at Breast Height
FV	Future Value
irr	Internal Rate of Return
LEV	Land Expectation Value
MAI	Mean Annual Increment
MC	Marginal Cost
MFC	Marginal Factor Cost
MP	Marginal Product
MR	Marginal Revenue
MRP	Marginal Revenue Product
NPV, NPV_t, NPV_T	Net Present Value, Net Present Value at time t, $t = 0, 1, 2, 3, \ldots, T$
P, P_Q, P*	Market Price, Own-Price, Market Equilibrium Price
PAI	Periodic Annual Increment
PI	Profitability Index
PV	Present Value
Q, Q*	Market Quantity, Market Equilibrium Quantity
ROI	Return on Investment
SEV	Soil Expectation Value

TC	Total Cost
TFC	Total Fixed Costs
TR	Total Revenue
TRP	Total Revenue Product
TVC	Total Variable Costs,
USD	United States Dollars
$	Dollars

1 Introduction

In its most basic form, managing a forest entails measuring what is there, determining an appropriate management plan given what is there and the landowner's goals and objectives, physically manipulating the forest based on the management plan, and assessing the results. The merchantable trees are sold to a buyer.[1] This straightforward description leads to some very interesting questions. I will only highlight the following two: Where did the price the buyer offered come from? Who defined what is merchantable?

The answer to the first question is straightforward. The price offered by the buyer is based on what the primary processor (e.g. sawmill, pulp, or paper mill, etc.) is willing to paying. The answer to the second question is also straightforward. The primary processor defines what is merchantable.[2] Interestingly, a similar set of relationships can be described between secondary and final processors. Consequently, not only is your prescription dependent on "what is there" and the landowner's goals and objectives, but also on the supply and demand of the stumpage and wood fiber markets and the supply and demand of the secondary and primary markets.

What if the landowner's primary objective is not the production of timber or wood fiber? Again, in its most basic form, managing a forest entails measuring what is there, determining an appropriate management plan given what is there and the landowner's goals and objectives, and then physically manipulating the forest based on the management plan with an additional proviso; namely, the cost of implementing the plan does not exceed the landowner's defined budget. In this example, not only is your plan dependent on "what is there" and the landowner's goals and objectives, but also on the implementation costs and the landowner's budget. What if the landowner's primary objective is still not the production of timber or wood fiber and the management plan could be described as neglect, benign or otherwise? As forests grow without any human intervention, does this latter example imply there is no economic information the landowner can use to inform their choice to take no management action? Or, is the landowner ignoring or ignorant of the relevant economic information surrounding their choice to take no management action?

What if the example was of a sawmill or a charter boat fishing business? I would assert that, regardless of the business, using relevant economic information will lead to better business decisions more often than not.[3] Thus, the purpose of this book is to address the question: "How to use economic information to make better business decisions?". The introduction will lay the groundwork necessary to address this question. The remaining chapters will build upon this groundwork.

I will start with a story that may be more familiar to most of you than operating a commercial recreational enterprise or sawmill. It is a dark and stormy night, the semester is underway and you have four hours after supper to spend on one or more activities of your choice leading to the most satisfying outcome. You can (a) study, (b) spend time with friends, or (c) watch some television. With nothing much to watch on TV this evening, you decide the relevant choices would be either studying or spending time with friends. Weighing studying – you have some reading from two classes to finish – versus meeting friends, you decide that for this evening you will use the first two hours studying and spend the rest of the time with friends.

How did you go about making the decision to allocate those four hours among the different uses? What tradeoffs did you make? Studying implied that you gave up watching television; you could have studied with your friends but reading and studying school material is an activity best done with few distractions (Foerde *et al.* 2006).[4] You also know that reading school material usually takes a lot of concentration and after about two hours you need a rest. As a result, you decided to study for the first two hours (giving up spending the entire time with friends), then spend the remaining time with your friends (giving up two hours of study time). If it were a sunny warm afternoon rather than a stormy night, you probably would have opted to spend more than two hours outside with your friends because you could have then studied all evening. As the semester comes to a close there always seems to be more work due – term papers, projects, etc. – in addition to the normal reading and homework. You might have spent the entire four hours studying. Having a test tomorrow tends to focus your attention on preparing for the test, especially if you did not spend as much time earlier in the week preparing for it. Again, you might have spent the entire four hours studying. So while we are talking about a four-hour block of time, the context in which that four hour time block occurs also seems to affect your decision on how to use that time.

What, if anything, does this example have to do with economics, managerial economics, managing forests or other natural resources, or managing a business that uses natural resources as inputs? I will use the following sawmill example to address this question. In much of the United States there is an abundance of small-diameter trees (e.g. 6- to 9-inches diameter at breast height); unfortunately, there are very few markets for these trees. The most common uses are for pulp, oriented strandboard raw material, or firewood. While dimension lumber or beams made from small-diameter trees could provide the greatest profit to a mill owner, dimension lumber or beams made from this material often warp on drying (Patterson and Xie 1998). Patterson *et al.* (2002) and Patterson and Xie (1998) have proposed an alternative use from these small-diameter trees: creating 4 × 4 inch inside-out (ISO) beams. Figure 1.1 illustrates the procedure for manufacturing an ISO beam.

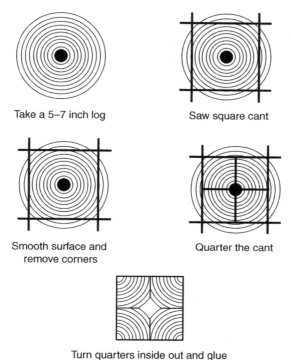

Take a 5–7 inch log

Saw square cant

Smooth surface and
remove corners

Quarter the cant

Turn quarters inside out and glue

Figure 1.1 Illustration of an inside-out beam.

(Used with permission from the *Forest Product Journal* and
Patterson *et al.* (2002))

Patterson and Xie (1998) examined if it is technically possible to produce ISO
beams and if the physical and mechanical properties of 4 × 4 inch ISO beams
compare favorably with 4 × 4 inch solid beams made from small-diameter hardwoods. They determined that the physical and mechanical "properties of the ISO
beams were equal to or superior to the sold beams tested" and that ISO beams
average warpage was only half of solid beams.[5] Patterson *et al.* (2002) used an
engineering economic analysis approach and showed that, for both softwood and
hardwood, producing ISO beams was profitable.

In terms of managerial economics, what do both of these examples have in
common? Both require identifying alternate ways to achieve given objectives, and
selecting the alternatives that accomplish those objects in, most often, the most
resource-efficient manner. The objective of the stormy-night example is the most
satisfying outcome; for the sawmill example it is maximizing profit. The input in
the stormy-night example is a four-hour block of time; for the sawmill example it
is small-diameter trees. In the stormy-night example, to analyze the tradeoffs
between studying and relaxing requires economic information concerning how
much studying or relaxing would add to your overall satisfaction in a given

context. In the sawmill example, the economic information required to analyze the tradeoffs is the physical and mechanical properties of ISO beams compared to solid beams and the profitability of producing ISO beams. In the stormy-night example, if you are not efficient with the time allotted to studying you will have to spend more time studying, have less time to relax, and reduce your overall satisfaction. In the sawmill example, if the owner is not efficient in using small-diameter trees to produce ISO beams, they have to use more input and this will reduce their profitability. While it could be reasonably argued that the level and detail of the economic information and the methodology used to analyze the tradeoffs in the sawmill example are more complex than in the stormy-night example, the process of analyzing the tradeoffs are the same. Although you would not consider your decision of how to use a four-hour time block a business decision, it can in fact be viewed as one.

What is economics?

Economics textbooks, some with greater effectiveness than others, regularly provided a definition of economics. As I believe it is necessary to define and describe important terms, I will also provide one. The following is based on a definition used by Samuelson (1976a):

> Economics is the study of how individuals choose, with or without money, to employ scarce productive resources that have alternative uses, to produce various commodities and services and distribute them for current or future consumption among various persons in society.

There are four key concepts in this definition that need further clarification. The first is scarcity. Pearce (1994) defines scarcity as "usually reserved for situations in which the resources available for producing outputs are insufficient to satisfy wants." Scarcity is often cited as the fundamental reason why economics exists. There are two types of scarcity: absolute and relative. Absolute scarcity implies that there is a finite quantity of a resource. For example, oil and coal are often referred to as absolutely scarce. Relative scarcity implies that sacrifices must be made to keep the resource in its current use or to obtain it, or they have alternative uses. A barrel of oil can be refined into gasoline or other petrochemicals. On the other hand, I have an obsolete calculus and analytical geometry textbook published in 1951. This textbook is for all practical purposes absolutely scarce. There exists a finite number of these books. However, it is not useful as a reference, while it is old it is not a collectable item, and I have not had to make any sacrifices to keep it in my bookshelf. There are, for all practical purposes, no alternative uses of this book. In fact, I have not made sacrifices of any kind with respect to this book. Consequently, it is not relatively scarce. The fact that the book is absolutely scarce is less important than the fact that it is not relatively scarce. Oil on the other hand is absolutely scarce, and also relatively scarce. I must make sacrifices to obtain it (now or in the future) and keep it in its current use.

Thus, in terms of the comparative importance, relative scarcity is more important than absolute scarcity.[6]

The second key concept is choice. Only individuals choose and, therefore, only the individual is responsible for the choices that they make. While it may appear that organizations and institutions choose, it is the individuals in the organizations and institutions who make the decisions (Heyne *et al*. 2006). The reason individuals must choose is because of scarcity; more specifically, relative scarcity. Relative scarcity implies choice or:

Relative Scarcity ↔ Choice

On that dark and stormy night, you had four hours to spend. Spending time with your friends was the alternative you gave up or sacrificed to study. Spending time studying was the alternative you gave up or sacrificed to spend time with your friends. The sawmill owner had to decide how to use the small-diameter tree. If they manufactured ISO beams the alternative given up was solid beams and vice versa. Stated differently, relative scarcity implies choice and choice implies an opportunity cost

Relative Scarcity ↔ Choice ↔ Opportunity Cost

The first two key concepts lead to one of the fundamental building blocks of economics and that is "opportunity cost." Opportunity cost is defined as the value of the next best alternative forgone. Returning to the examples, studying implied that you gave up watching TV and seeing your friends. You valued being with your friends more than watching TV. Thus the correct enumeration of opportunity cost was relaxing with your friends. While there are a number of, albeit limited, uses of small-diameter trees, the relevant choice or tradeoff for using small-diameter trees is producing either ISO or solid beams (Wagner *et al*. 2004; Patterson *et al*. 2002). Opportunity cost is the "What if" question: What am I giving up to continue what I am doing? Learning to ask and answer this question explicitly is vital to comprehending the economic way of reasoning about scarcity. Ignoring the choices associated with scarcity does not make the opportunity cost of those choices go away. You can either be ignorant or informed.

The third key concept is "with or without money." There are basically three functions of money: (1) medium of exchange, (2) unit of account, and (3) store of value. As a medium of exchange, money is a convenient way to exchange goods that allow individuals to gain the advantages of geographic and human specialization (Heyne *et al*. 2006; McConnell and Brue 2005). As a unit of account, it is a standard for measuring relative value; it aids an individual by making it easy to compare the prices of goods, services, and resources (Heyne *et al*. 2006; McConnell and Brue 2005). Finally, as a store of value it enables individuals to transfer purchasing power from the present to the future (Heyne *et al*. 2006; McConnell and Brue 2005). Anything widely accepted as a medium of exchange, unit of account, and store of value can serve as money. The US dollar can serve as a medium of exchange, but salt was once used as a medium of exchange, most notably in the Roman Empire. If you do not like dollars, change to some other medium of exchange. However, changing to some other medium of exchange, even a barter system, does not make relative scarcity or the opportunity costs of choices associated with relative scarcity go away.

The fourth key concept is consumption of the resource now as opposed to in the future. Just as there is an opportunity cost of using a resource now (for example, the choice of using a small-diameter tree to produce either an ISO beam or solid beam), there is also the choice of not cutting the tree and letting it grow to produce a more valuable product at a future date. This is known as the opportunity cost of time. Your use of a credit card to purchase a commodity is also an illustration of this. You buy the commodity using the credit card company's money with the promise of paying back the amount of the purchase, and, if necessary, a premium, at a later date.

What is the economy?

As with the definition of economics, there are many definitions of the economy. The definition I prefer is:

> The economy is a collection of technological, legal, and social arrangements through which individuals in society seek to increase their material and spiritual well-being.
>
> (Field and Field 2002: 23)

This definition highlights two important concepts. The first concept is the importance of "technical, legal, and social arrangements" or institutions necessary to allow individuals to realize gains from trading. These institutions include: (1) Property rights – without a system of property rights gains from trading are difficult to realize.[7] (2) A legal system – without a legal system there are no established rule for contracts, and property rights, etc.; nor is a means of redress available if a wrong has been committed. (3) Money – without money, transactions are more expensive. To serve as a medium of exchange, unit of account, and store of value, money must be stable. (4) Governments (local, state, and national) – Without governments the conditions for these institutions to function and evolve will not exist (Heyne *et al.* 2006).

The second concept is individuals seeking to "increase their material and spiritual well-being"; other authors use the terminology "increase their wealth" (Heyne *et al.* 2006; Silberberg and Suen 2001). The economic concept of wealth is simply whatever people value (Heyne *et al.* 2006; Klemperer 1996; Pearce 1994). People value material goods, and they also value mental and spiritual well-being. Thus, the goal of the economy is to increase individual wealth. One way in which wealth can increase is by the production of physical commodities. However, it does not follow logically that only the production of physical commodities increases wealth.[8] Trade, a voluntary exchange of goods or services, increases the wealth of those involved in the exchange.[9] For example, Kyle McDonald started on 12 July 2005, with one red paper clip and 14 trades later on 15 July 2006 traded for a house on 503 Main Street in the town of Kipling, Saskatchewan

Canada (http://oneredpaperclip.blogspot.com/2005/07/about-one-red-paperclip.html, accessed on 1 March 2007).

Economic reasoning

Economics, like physics or biology, is a science. More specifically, economics is a social science. As a science, economic theories can be used to hypothesize cause-and-effect relationships that can be tested empirically (Silberberg and Suen 2001; Friedman 1953). Facts without a theory are useless, because one set of facts cannot be shown to cause another set of facts. There are no means of separating the irrelevant from relevant facts (Heyne *et al.* 2006; Silberberg and Suen 2001). For example, if you throw a ball straight up in the air, it will rise rapidly, then less rapidly, and finally reaching a point where it stops rising, stops, and descends toward you, slowly at first but gaining speed. Fact A: You toss a ball in the air nightly. Fact B: A week before your economics exam you study economics for at least two hours a night. Fact C: You receive an A on your economics test. There are no theories that can be identified to show a cause-and-effect relationship between Fact A and Fact C. Stated differently, Fact A is irrelevant in explaining Fact C as Fact C cannot be inferred from Fact A. There are theories that can be identified to show a cause-and-effect relationship between Fact B and Fact C. Thus, Fact B is relevant in explaining Fact C as Fact C can be inferred from Fact B.

Economic theories can help systematically analyze the costs and benefits of different resource allocation decisions made by individuals. That is, economic theories will not identify what to choose, but examine the tradeoffs or opportunity costs associated with the choices. If the context in which the analysis was done holds in the future, it allows the development of testable statements about the individual's future choices. Economic reasoning is thus about systematically examining the choices or tradeoffs individuals make in a given context. This requires developing a consistent set of terminology, models, and methods (Silberberg and Suen 2001; Friedman 1953). The rest of the chapter will be devoted to introducing the minimum set of terminology, models, and methods necessary to reason economically. Subsequent chapters will build on these.

Individual decision behavioral assertions

In a nutshell, economics is the study of individual choice given scarcity and the resulting opportunity costs of those choices. Because the principal agent of interest is the individual, if we are to study individual choice we need to develop two assertions or postulates concerning their decision behavior.[10] First, individuals are asserted to be maximizers; in terms of voluntary exchanges, individuals will seek to maximize their satisfaction, net benefit, utility, profit, or smiles-per-minute, etc. Consistent with relative scarcity, net benefits are defined as benefits associated with the choice minus opportunity costs of the choice. Second, individuals have preferences for goods and services. They can order these preferences, and they are consistent in that ordering. For example, if an individual prefers Coke® to Pepsi®, and Pepsi® to all other carbonated cola beverages, and if given a choice of Coke® or Pepsi®, they will always choose Coke® voluntarily. Further, if given a choice between Pepsi® and all other carbonated cola beverages, they will always choose Pepsi® voluntarily. Finally, if given a choice between Coke® and all other carbonated cola beverages, they will always choose Coke® voluntarily.

The assertions that individuals are maximizers of net benefits (i.e. benefits minus opportunity costs) and can order their preferences are often not well understood and lead to confusion. McKenzie and Lee (2006: 108–109) provide the clearest and concise discussion of what these assertions imply that I will paraphrase. Being a maximizer does not imply an individual is selfish. People receive pleasure from others' happiness; however, if the goal is to improve their or others' well-being, people are motivated to make decisions that do the most to accomplish this objective. These assertions do not mean that individuals never make mistakes. Individuals do not have perfect knowledge or can fully control the future, thus they base their choices on what they expect to happen, not on what does happen. Given the plethora of possible choices (e.g. breakfast cereal), individuals may make choices based on habit. While a better choice may be made if individuals collect and examine all information, the opportunity cost (time) is too high relative to the benefit. Individuals do maximize their net benefits and order their preferences; however, anyone who equates rational behavior with what *they* would do will have no trouble concluding that others are irrational. Do not make the assumption that everyone has the same set of preferences. Do not make the assumption that everyone has *your* set of preferences.

While the principal agent of interest is the individual, economic theories are not designed or intended to develop hypotheses about an identified individual's choices; for example, the person sitting next to you. The individual referenced is a representative or average individual in a given context.

Value versus price

Value can have many different meanings, as economics is a social science and the principal agent of interest is the individual, I will be concerned with preference-related concepts of value. According to a seminal work on the economic concept of value by Brown (1984), an assigned value is the expressed relative importance or worth of an object to an individual in a given context. Therefore, an assigned value is observable based on the choices an individual makes. Assigned values are relative; it is not the intrinsic nature of the object but the object relative to all other objects that gives rise to an assigned value (Brown 1984). Assigned values are relative, not absolute; assigned values depend on the context surrounding an individual's choice. If the context changes, the value an individual assigns to an object will also change. As can be seen the preference-related concept of value is tied to the assertions concerning individual decision behavior.

Price is defined as a per-unit measure of assigned value and thus a measure of relative scarcity. As the price of a good or service increases relative to other goods and services, this particular good or service is scarcer relative to the other goods and services. It bears repeating that the concept of relative scarcity is a function of an individual's preference for goods and services (Silberberg and Suen 2001). It is left to the reader to draw the relationship between price and opportunity cost through answering the following question: If the price of a resource is zero, what is its opportunity cost?

Market price is economic information that consumers use in examining their opportunity cost of purchasing a good or service. That same market price is economic information that suppliers use in examining their opportunity cost of producing or providing a good or service. As with value, price is not a measure of the intrinsic value of the good; it is information individuals employ to allocate a scarce good or service to a specific use. If a person purchases a good, we know the minimum assigned value they place on that good, other things being equal. Although price is usually denoted in monetary terms, it need not be.

To illustrate the relationship between value and price, let the market price of an ISO beam be $7.76 per beam (Patterson *et al.* 2002).[11] If you bought one beam, then the observed assigned value you place on that beam would be

$$\$7.76 = 7.76 \frac{\$}{beam} \cdot 1 \text{ beam}$$

as you could have spent that $7.76 on another item. If you bought two beams, the observed assigned value you place on those beams would be

$$\$15.52 = 7.76 \frac{\$}{beam} \cdot 2 \text{ beams}$$

as you could have taken that $15.52 and purchased something else.

Assumptions concerning the market and workable competition

The economic concepts of value and price are based on a given context. The market provides a place where information concerning the relative scarcity of a good or service is revealed and the choices individuals make are observed. Given the information generated by the market, an individual can choose to provide or produce a particular good or service and another individual can choose to purchase that particular good or service. A market can be a physical place like a grocery store or local farmers' market, or it can be place like eBay® where individuals offer to sell Gibson Les Paul guitars, bamboo flyrods, or one red paperclip, and others can choose to purchase the item or make a counter-offer.

Price is a key piece of information generated by the markets and used by buyers and sellers. Generally speaking, the more sellers of a good or service in a market, the greater the inability of a single seller to have control over the price. Similarly, the more buyers of a good or service in the market, the greater the inability of a single buyer to have control over price. Thus, if the market price is to only reflect how scarce a good or service is relative to other goods and services, no one buyer or group of buyers and no one seller or group of sellers can influence price. This is the characteristic of workable competition. Deviations from workable competition imply that the resulting price not only reflects some concept of relative scarcity but also an individual's market power. In the terminology of economic reasoning, deviations from workable competition imply a market imperfection.

Economics versus accounting

When an accountant discusses costs it is in terms of actual dollar outlays. For example, if you look at the debt column of your checkbook, you can tell me the amount of money you spent last month and what you purchased. These are explicit costs. When an economist discusses costs, it is in terms of opportunity cost: What am I giving up to continue what I am doing? Opportunity costs obviously include explicit costs or actual dollar outlays. In addition, there are other costs that may not be included in explicit costs.

Table 1.1 gives the annual costs and revenues associated with owning and operating a small hardwood sawmill. Total receipts of $1,100,000 are the sum of selling dimension lumber and chips. The accounting or explicit costs of $1,050,000 include employee wages, employee benefits, taxes, utilities, purchasing delivered logs, and fixed costs. At the end of the year net receipts total $50,000. For the sawmill's owner, net receipts are salary. Accounting profits are $000.00.

In order to own and operate this sawmill, you are giving up working as the manager of another sawmill in a different county. The owner of that sawmill

Table 1.1 Economics versus accounting – hardwood sawmill example

Sale: dimension lumber	$1,000,000	
Sale: chips	$100,000	
Total Revenue		$1,100,000
Cost: employee wages	$120,000	
Cost: employee benefits	$60,000	
Cost: delivered logs	$320,000	
Cost: fuel, oil, etc.	$190,000	
Cost: utilities and maintenance	$170,000	
Cost: fixed costs (i.e. depreciation, interest, rent, taxes, insurance, etc.)	$190,000	
Total (Explicit) Costs		$1,050,000
Net Receipts		$50,000
Cost: salary sawmill owner	$50,000	
Accounting Profit		$000
Forgone: salary ($60,000 you could have earned elsewhere; $60,000 – $50,000 = $10,000)	$10,000	
Forgone: interest on capital	$5,000	
Total (Implicit) Costs		$15,000
Economic Profit		–$15,000

would have paid you $60,000 per year as a manager. You are forgoing or giving up $10,000 in salary to own and operate your own mill. This implicit cost cannot be found on any accounting sheet; nonetheless it is a cost to you. To operate this mill, you have used some of your savings as a down payment to buy equipment. You are forgoing or giving up interest that you could have earned by investing those savings in a financial instrument such as a mutual fund. Again this implicit cost cannot be found on any accounting sheet but is a cost to you. The sum of these forgone alternatives is $15,000. The opportunity cost of owning and operating the sawmill is then $1,050,000 + $15,000 = 1,065,000 or the sum of explicit and implicit costs.[12] If explicit and implicit costs are included in the analysis, the economic profit of owning and operating your sawmill is −$15,000.

Efficiency versus equity

Efficiency is the technical relationship between inputs and outputs: the greater the output relative to input, the greater the efficiency. This is an objective analysis. Equity denotes the concept of a just or fair (not necessarily equal) distribution of goods, services, output, income, etc., among all consumers. This is a subjective analysis. In this book, I will concentrate on efficiency, but in a broader context, efficiency objectives cannot be pursued independently of equity objectives. It is important to note that the objectives of efficiency and equity often conflict and it may become necessary to compromise one for another. For example, measures that could expand output relative to the amount of input used (i.e. increased efficiency) might create unwanted changes in the distribution of income (i.e. decreased equity), and vice versa, illustrating the tradeoff between improvements in equity and efficiency. The relative weights placed on equity versus efficiency and the appropriate compromises are not matters that can be solved by economic analysis. Political and electoral processes must be used to reconcile divergent opinion about equity and efficiency. Economics and managerial economics are intended to address questions of efficiency; namely, the optimal allocation of resources that accomplish an entrepreneur's objective such as maximizing profits. Economics and managerial economics are not intended to address questions of equity. Consequently, the focus of this book will be on efficiency.

Economic models

Reality is complex and trying to model this complexity could lead to intractable models. Economic models, which are succinct representations of complex systems, focus on describing the underlying fundamental characteristics of the system that are critical for systematically examining individual choices within a given context (Field and Field 2002; Kay 1996; Pearce 1994). Thus, developing an economic model requires identifying the underlying fundamental characteristics. Following the framework provided by Silberberg and Suen (2001) and Friedman (1953), I will develop two common economic models: profit maximization and least cost/ cost effective.

Profit model

Individuals, regardless of whether they are buyers or sellers, seek to increase, improve, or maximize their net benefits or profits in the context of a market and assumptions concerning competition in that market. Thus, given the assertion that individuals are maximizers and the assumption of workable competition, the following economic model can be used to systematically examine and develop testable statements about individual choice:

$$Max\ NB = B - C$$

or

$$Max\ \Pi = TR - TC$$
$$= P \cdot Q - TVC - TFC \tag{1.1}$$

where

Max NB denotes maximization of net benefit (*NB*),
  B denotes benefits,
  C denotes costs;
Max Π denotes maximization of profit (Π);
  TR denotes total revenue, $TR = P \cdot Q$,
    P denotes market price,
    Q denotes market quantity;
  TC denotes total cost, $TC = TVC + TFC$
    TVC denotes total variable costs,
    TFC denotes total fixed costs.

Basically, equation (1.1) compares what you gave up to obtain a good or service (its opportunity cost) with the benefits received from having it. Simply, individuals search to find the greatest positive difference between benefits and costs or total revenue and total cost. The solution to equation (1.1) is often referred to as economically efficient.

While individuals provide some of what they need with their own labor, individuals do not generally increase their net benefit or profit in isolation: as was stated previously, net benefit or profit is increased by trade. This has been recognized by many economists from Adam Smith (1776) to the present. Figure 1.2 illustrates this using equation (1.1).

Buyer: max net benefit = benefit – cost

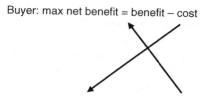

Seller: max profit = total revenue – total cost

Figure 1.2 Increase in net benefit or profit
through trade.

The buyer's objective is to maximize their net benefit. For example, one potential use of an ISO beam is for lighting as the wires can be run through the hollow center (Patterson *et al.* 2002). Not wanting to buy material, machinery, and other inputs necessary to produce ISO beams, the buyer will instead seek a seller who is offering to provide them. The seller's objective is to maximize their profit by offering to sell ISO beams. From the buyer's perspective, the buyer's cost of the ISO beams is the total revenue to the seller. The benefit to the buyer is using the ISO beams as lighting standards in an open beam ceiling of their house. From the seller's perspective, the seller's total revenue is the buyer's cost, or what the buyer exchanged for the ISO beams. And the total cost to the seller is the opportunity cost of not being able to sell the ISO beams to someone else.

Least cost/cost effective models

Equation (1.1) also implies that maximizing profit or net benefit requires decisions that make total revenue as large as possible while simultaneously making total costs as small as possible. Therefore, if net benefits or profits are to increase toward their maximum, then costs must be minimized. Becker *et al.* (2004) recognized this in their economic assessment of using a Mobile Micromill® for processing small-diameter ponderosa pine:

> The ability to develop value-added products by using small-diameter ponderosa pine is in part contingent on the ability of finished products to compete in the market place. Transportation and processing costs need to be minimized to make processing small-diameter logs more competitive.
>
> (Becker *et al.* 2004: 4)

This leads to the second economic model, given that individuals are maximizers and the assumption of workable competition:

$$Min\ Cost = TVC$$
$$\text{s.t.} \tag{1.2}$$
$$Q(x) = Q^0$$

where
  Min Cost denotes the objective of minimizing costs;
  TVC denotes total variable costs (e.g. transportation and processing costs);
  s.t. denotes "subject to" and what follows is a constraint placed on the objective;
  $Q(x) = Q^0$ denotes producing or providing a good or service, $Q(x)$, using inputs, x, at a given quantity or quality, Q^0.

Equation (1.2) is described as a least cost model; in other words, determining the least cost means of producing or providing a given level of a good or service. Using the stormy-night example, studying is one of the jobs you must perform while at school. You use various inputs when studying, but I will focus on your time. The more time you spend studying, the less time you can spend on doing anything else (the opportunity cost of your time). Therefore, you want to use as little time as possible studying (minimizing cost) while maintaining a given quality of understanding. If you do not maintain the quality, that means you have wasted your time because you will have to spend additional time to redo what you should have done in the first place. Using the ISO sawmill example, if the small-diameter trees are not sawn square and glued correctly (see Figure 1.1), then the ISO beams will not have physical and mechanical properties equal to or superior to solid beams and the sawmill owner will have wasted those inputs. The sawmill owner will want to use as few inputs (minimizing costs) while maintaining the quality and quantity of the ISO beams produced.

An alternative form of a least cost model is the cost effective model:

$$Max \ Q(x)$$
$$s.t. \quad\quad\quad\quad\quad\quad\quad\quad\quad\quad\quad\quad\quad\quad\quad\quad\quad (1.3)$$
$$TVC = C^0$$

where
   Max Q(x) denotes the objective of providing or producing the maximizing quantity or quality of a good or service using inputs, x;
   s.t. denotes subject to;
   $TVC = C^0$ denotes that the total variable costs (TVC) of using inputs x to produce or provide the good or service cannot exceed a given budget, C^0.

Thus, equation (1.3) states for any given budget level, your goal is to produce the most output. Using the stormy-night example, you have a given amount of time you want to spend studying on any given day. For that given amount of time you want to be as productive (or efficient) as possible in terms of quality of studying. If you are not, then you have wasted the time budgeted to this activity. Using the ISO sawmill example, purchasing small-diameter trees is one of many variable costs the sawmill owner must budget for. Given this budget, the owner wants the sawmill to be as productive (or efficient) as possible in converting small-diameter trees into ISO beams.

From the example, it can be seen that the least cost model and the cost effective model are heads and tails of the same coin.[13] The solution as determined by the least cost model is also the solution for the cost effective model. And, the solution as determined by the cost effective model is also the solution for the least cost model. The solutions to equations (1.2) and (1.3) can be referred to as being production cost efficient.

The logic of the least cost or cost effective models was based on the concept that maximizing profit or net benefit requires decisions that make total revenue as large as possible while simultaneously making total costs as small as possible. Thus, a solution to the profit maximization model is also a solution for the least cost or cost effective models (Silberberg and Suen 2001). If you have no control over your costs, you will have less control over your profits. However, it is up to the reader to determine if a solution to the least cost or cost effective model does or does not imply a profit maximization solution.

Marginal analysis

A margin denotes an incremental change from a starting point. Marginal analysis is a systematic examination of observable choices by individuals resulting from observable market conditions. For example, the amount of output you produce or provide is a variable you control. If I assert that you maximize your profits and are consistent in your behavior with respect to profit maximization, then you will increase the amount of output you provide as long as the incremental or marginal revenue of increasing the amount of output you provide is greater than the incremental or marginal cost of providing it. Stated differently, profits are increasing as long as marginal revenue is greater than marginal cost. If the incremental or marginal cost is greater than the incremental or marginal revenue, you will decrease the level of output you will provide; or, profits are decreasing when marginal cost is greater than marginal revenue.

Using the stormy-night example, in deciding how to spend a four-hour time block, you determined that you would study for two hours. You have some reading in two classes to finish; from past experience you know that for this material one hour is not long enough and three hours are too long. The incremental benefit from studying for the initial hour is greater than the incremental cost. The incremental benefit from studying for the second hour is also greater than the incremental cost. However, the incremental cost of studying for the third hour is greater than the incremental benefit, because this type of reading usually takes a lot of concentration and after about two hours you need a rest. As long as the incremental benefit is greater than the incremental cost, you will continue to study because your net benefits from studying are increasing.

Using the ISO sawmill example, given the production limitations of fixed equipment (e.g. quartering saw), the incremental costs of manufacturing an ISO beam will decrease at first but ultimately increase as the number of small-diameter trees put through the sawmill increases. Each small-diameter tree will produce an ISO beam that will sell for $7.76. The sawmill owner will use additional small-diameter trees if the incremental cost of milling it is less than the incremental benefit of selling the ISO beam it produces for $7.76. If the incremental cost of milling the additional small-diameter tree is greater than the incremental benefit of selling the ISO beam it produces for $7.76, then the owner will not mill that additional small-diameter tree. As long as the incremental production costs of milling a small-diameter tree are less than the incremental benefit of selling the ISO beam it produces, the sawmill owner's incremental profits will increase and the owner will increase production by milling additional small-diameter trees. If the incremental costs are greater than the incremental benefits, the sawmill owner will decrease production of ISO beams.

Marginal analysis will not prove if in fact you are maximizing profits nor prove if the output level you are producing is "optimal." Marginal analysis allows for statements concerning a direction of change; for example, increasing or decreasing the amount of output you provide. Continuing with the profit maximization model, marginal analysis will also allow for a testable statement concerning whether you will use more or less of an input if the price you pay for that input increases. As a manager you will always seek ways to improve or increase your profits. Marginal analysis provides you with a systematic methodology to examine choices that are consistent with the assertion of profit maximization.

The Architectural Plan for Profit

Maital (1994) defines the three Pillars of Profit as Price, Value, and Cost. While I agree with his assessment in terms of the three pillars:

> Business decisions are built on three pillars – cost, value and price. Cost is what businesses pay out to their workers and suppliers in order to make and market goods and services. Value is the degree to which buyers think those goods and services make them better off, than if they did without. And Price is what buyers pay. Those are the three essential elements in the day-to-day choices managers make. Juggling those three elements is what managers are paid for.
>
> Managers who know what their products cost and what they are worth to customers – and who also know the costs, values, and prices of competing products – will build good businesses, because their decisions will rest on sound foundations. Businesses run by managers who have only fuzzy knowledge of one of those three pillars will eventually stumble. It is deceptively difficult to build them, precisely because they have to be built – the information required is often incomplete or not readily at hand
>
> (Maital 1994: 6–7)

I would include an additional component: the Production System. While Maital implies that managers must have knowledge of production systems – managers should know "their production costs" – I believe this component should be explicit.

Knowing production costs rests on knowledge of the production system. Searching for ways to minimize these costs, and increasing profit, rests on knowledge of the production systems. This holds whether you are producing maple syrup, dimension lumber, or furniture.

Figure 1.3 illustrates that the Production System is the foundation on which the three Pillars of Profit are set. On top of the three pillars rests Profit. Profits cannot be held up if the foundation or one or more of the three pillars are weak. Figure 1.3 also provides a road map for the first four chapters of this book, starting with building the foundation, moving onto the pillars that represent total cost and total revenue, and ending with profits.

How to use economic information to make better business decisions?

The purpose of this book is to answer the question: "How to use economic information to make better business decisions?" This question will be revisited at the

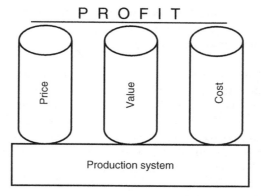

Figure 1.3 The Architectural Plan For Profit.

end of each chapter. To answer this question, I need to first define what economic information is: to that end, I will re-examine the two economic models discussed earlier and the Architectural Plan for Profit. Equation (1.1) defines the net benefit maximization or profit maximization model. The net benefit model is composed of two variables: benefits and costs. Benefits reflect the value that the purchaser places on the good or service. Costs represent the opportunity cost to the purchaser of buying the good or service. I would also argue that this opportunity cost represents a value concept. I will leave it to the reader to puzzle out this argument, which I will address in Chapter 3. The purchaser could be a consumer buying a final product or a producer buying an input. Thus, the economic information contained in the net benefit model is net benefit, value, and cost. The economic information contained in the net benefit model is illustrated in the Architectural Plan for Profit.

In the profit maximization version of equation (1.1), there are two variables: total revenue and total cost. Total revenue is composed of three pieces of economic information: price (P), quantity (Q), and $P \cdot Q$. Price is a per unit measure of value, a measure of relative scarcity, and what buyers use to help determine how to use a good or service. Price as a measure of relative scarcity depends on the assumption concerning the market and competition. Quantity is whatever the purchaser is buying. Goods and services do not appear magically, but must be produced or provided to the purchaser. This implies some knowledge of the production system used to produce or provide the goods or services. $P \cdot Q$ represents the value to the purchaser of buying the good or service.

Total cost is composed of two pieces of economic information: total variable and total fixed costs. As with the net benefit model, these costs represent the opportunity cost to the purchaser of buying the good or service as well as a value concept. However, not enough detail is given in the current version of the profit maximization model to explain exactly what is included in total variable and total fixed costs and how they differ. This detail will be provided in Chapter 3.

Thus, the economic information contained in the profit maximization model is the production system, value, price, cost, and profit. The economic information contained in the profit maximization model is illustrated in the Architectural Plan for Profit.

Equations (1.2) and (1.3) define the least cost and cost effective models, respectively. The variables included in the two models are total variable cost and the production system. The concept of total variable cost in equations (1.2) and (1.3) is the same as in equation (1.1) and represents the opportunity cost to the purchaser of buying the good or service and a value concept. In addition, the production system defined in equations (1.2) or (1.3) is the same concept as that given in equation (1.1). This implies some knowledge of the production system used to produce or provide the goods or services. Thus, the economic information contained in the least cost and cost effective models are the production system, value, and cost. The economic information contained in these two models is illustrated in the Architectural Plan for Profit.

After examining equation (1.1), the economic information it contains is identified as the production system, price, value, cost, and profit. After a similar examination of equations (1.2) and (1.3), the economic information they contain are identified as the production system, value, and cost. This is the same economic information that is displayed in Architectural Plan for Profit (Figure 1.3). The quote from Maital (1994: 7) is important to repeat here: "It is deceptively difficult to build them [the three pillars of profit], precisely because they have to be built – the information required is often incomplete or not readily at hand." Thus, to answer the question posed, I first need to identify the required economic information, then obtain it at the appropriate level of detail. I have identified broad categories of economic information required; the detail will be discussed in the following chapters.

The case studies

I will use a number of case studies to illustrate the concepts discussed, drawn from various disciplines such as outdoor and commercial recreation, wood products engineering, forest products, and forestry. Below is a list of the case studies. They can be found on the Routledge website for this book. While the case studies used will be summarized briefly in various chapters, you should read and familiarize yourself with each one.

References

Amateis, R.L. and Burkhart, H.E. (1985) Site Index Curves for Loblolly Pine Plantations on Cutover, Site-Prepared Lands, *Southern Journal of Applied Forestry*, 9(3):166–169.

Becker, D.R., Hjerpe, E.E. and Lowell, E.C. (2004) Economic assessment of using a Mobile Micromill® for processing small-diameter ponderosa pine, Gen. Tech. Rpt. PNW-GTR-623, Portland, OR: US Dept. of Agriculture, Forest Service, Pacific Northwest Research Station.

Burdurlu, E., Ciritcioğlu, H.H., Bakir, K. and Özdemir, M. (2006) Analysis of the most suitable fitting type for the assembly of knockdown panel furniture, *Forest Products Journal*, 56(1):46–52.

Burkhart, H.E., Cloeren, D.C. and Amateis, R.L. (1985) Yield Relationships in Unthinned Loblolly Pine Plantations on Cutover, Site-Prepared Lands, *Southern Journal of Applied Forestry*, 9(2):84–91.

Graham, W.G., Goebel, P.C., Heiligmann, R.B. and Bumgardner, M.S. (2006) Maple syrup in Ohio and the impact of Ohio State University (OSU) Extension Program, *Journal of Forestry*, 104(2):94–100.

Hagan, J.M., Irland, L.C. and Whitman, A.A. (2005) Changing timberland ownership in the Northern Forest and implications for biodiversity. Report # MCCS-FCP-2005-1 Brunswick, ME: Manomet Center for Conservation Science.

Hahn, J.T. and Hansen, M.H. (1991) Cubic and Board Foot Volume Models for the Central States, *Northern Journal of Applied Forestry*, 8(2):47–57.

Huyler, N.K. (2000) Cost of maple map production for various size tubing operations, Research Paper NE-RP-712, Newtown Square, PA: US Dept. of Agriculture, Forest Service, Northeastern Research Station.

Lichtkoppler, F.R. and Kuehn, D. (2003) New York's Great Lakes charter fishing industry in 2002, Ohio Sea Grant College Program OHSU-TS-039, Columbus, OH: Sea Grant Great Lakes Network, The Ohio State University.

Patterson, D.W., Kluender, R.A. and Granskog, J.E. (2002) Economic feasibility of producing inside-out beams from small-diameter logs, *Forest Products Journal*, 52(1):23–26.

Patterson, D.W. and Xie, X. (1998) Inside-out beams from small-diameter Appalachian hardwood logs, *Forest Products Journal*, 48(1):76–80.

Final thoughts

While this work is my own (as well as the errors), I would be truly negligent if I did not acknowledge those books and authors that provided the foundation upon which I have built my theoretical background: Buongiorno and Gilles (2003), Davis *et al.* (2001 and earlier additions), Rideout and Hesseln (2001), Klemperer (1996), Pearce (1990), Gregory (1987), Johansson and Löfgren (1985), Leuschner (1984), Clutter *et al.* (1983), and Duerr (1960 and later editions).

2 Production systems

Why is knowledge of the production system important?

The production of goods and services may take many forms. It could be the manufacturing of one or more outputs by combining inputs such as labor, capital, and other goods. It could be described as providing a service such as a forestry consultant, charter boat fishing captain, medical doctor, tax accountant, or lawyer. Whether it is the manufacturing of a good or the process of providing a service, the production system is the conversion of a set of inputs into outputs. A competent manager, or for that matter a competent employee, of *any* business – for profit or not-for-profit, forestry consultant or outdoor recreation, etc. – will search for ways to do this in generally the most resource efficient manner.

A good understanding of the production system is necessary for seeking opportunities to minimize costs and ultimately maximize profits. Thus the production system is the foundation upon which profits rest. This is illustrated in Figure 2.1.

Huyler (2000) confirms this observation:

> The key to a successful maple syrup operation is controlling production to maintain an acceptable profit margin. It is important that sugarbush operators keep accurate records so that areas of high cost can be identified and steps taken to reduce them.

> (Huyler 2000: 5)

As production systems range from simple to complex, how can we incorporate production systems into the economic models described in Chapter 1? And how can this be done so the manager will have the economic information useful to search for ways to be resource efficient? The next two sections will address the first question and I will return to the second question at the end of this chapter.

Profit model[1]

As the focus is on the producers or providers of goods and services, the profit model will be used.[2]

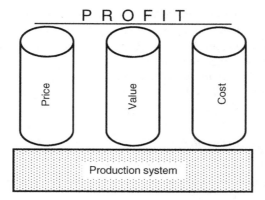

P R O F I T

Figure 2.1 The Architectural Plan for Profit.

$$Max\ \Pi = TR - TC$$
$$= P \cdot Q(x_1, x_2, ..., x_j) - TVC - TFC \qquad (2.1)$$
$$= P \cdot Q(\bullet) - \sum_j w_j \cdot x_j - TFC$$

where

Max Π denotes maximization of profit (Π);

TR denotes total revenue, $TR = P \cdot Q(x_1, x_2, ..., x_j)$;

P denotes the market price of the good or service,

$Q(x_1, x_2, ..., x_j) = Q(\bullet)$ denotes the system of producing or providing a good or service, Q, using inputs that the manager has direct control over, $x_1, x_2, ..., x_j$,

TC denotes total cost, $TC = TVC + TFC$;

TVC denotes total variable costs, $TVC = \sum_j w_j \cdot x_j$,

$\sum_j$ denotes the summation operator,

w_j denotes the *j*th wage or price paid for *j*th input x_j,

x_j denotes the *j*th input; for example, labor,

TFC denotes total fixed costs.

As before, profit is equal to total revenue minus total cost, and total revenue is equal to market price, P, times quantity, Q. However, an individual must produce or provide the good or service, which requires the use of one or more inputs. The way inputs are combined to produce or provide one or more outputs describes the production system. Thus, if an individual is going to make a good or service available to be sold, they must have knowledge of its production system. To reflect this, it is important to examine the quantity term in the total revenue function more closely.

Total cost is equal to total variable and fixed costs. While more detail will be provided in Chapter 3, total variable costs are basically the cost of using variable inputs, such as labor, to produce or provide a good or service. Just as there is a market for the output there are corresponding markets for inputs. Just as a market price is paid by the buyer for purchasing a good or service, a market price or wage must be paid by the seller for inputs used in producing or providing an output.

A goal of entrepreneurs that produce or provide an output is to seek ways to improve profits. As stated previously searching for opportunities to minimize costs and ultimately maximize profits requires a superior understanding of the production system, which is the foundation for examining costs (see Figure 3.1 in Chapter 3).

Least cost/cost effective models[3]

Similar arguments can be made for the least cost or cost effective models. Searching for opportunities to minimize costs is the direct goal of the least cost model given in equation (2.2)

$$Min\ TVC = \sum_j w_j \cdot x_j$$

s.t. $\qquad\qquad\qquad\qquad\qquad\qquad\qquad\qquad\qquad$ (2.2)

$$Q(x_1, x_2, ..., x_j) = Q^0$$

where

Min TVC denotes minimizing total variable costs, $TVC = \sum_j w_j \cdot x_j$;

$\sum_j$ denotes the summation operator,

w_j denotes the *j*th wage or price paid for *j*th input x_j,

x_j denotes the *j*th input,

s.t. denotes "subject to" and what follows is a constraint placed on the system;

$Q(x_1, x_2, ..., x_j) = Q^0$ denotes the system of producing or providing a good or service, *Q*, using inputs the manager has direct control over, $x_1, x_2, ..., x_j$, at a given quantity or quality, Q^0.

As was described in Chapter 1, the least cost and cost effective model are two sides of the same coin. The cost effective model is given in equation (2.3):

$$Max\ Q(x_1, x_2, ..., x_j)$$

s.t. $\qquad\qquad\qquad\qquad\qquad\qquad\qquad\qquad\qquad$ (2.3)

$$\sum_j w_j \cdot x_j = C^0$$

where

Max Q($x_1, x_2, ..., x_j$) denotes the system of providing or producing the maximizing quantity or quality of a good or service, *Q*, using inputs that the manager has direct control over, $x_1, x_2, ..., x_j$;

s.t. denotes subject to;

$\sum_j w_j \cdot x_j = C^0$ denotes that the total variable costs of using inputs $x_1, x_2, ..., x_j$ to produce or provide the good or service cannot exceed a given budget, C^0;

$\sum_j$ denotes the summation operator,

w_j denotes the *j*th wage or price paid for *j*th input x_j,
x_j denotes the *j*th input.

In the least cost or cost effective model, the production system is important. In the least cost model, equation (2.2), the entrepreneur is searching for ways to minimize the total variable costs of producing or providing a defined level of output. In the cost effective model, equation (2.3), the entrepreneur is searching for ways to produce or provide the most output for a set budget. In either case, the entrepreneur must have an excellent understanding of the production systems.

Case studies

I will summarize briefly the case studies used in this chapter to represent various production systems. They can be found on the Routledge website for this book. However, you should read the full case studies.

Select Red Oak (Hahn and Hansen 1991)

This production system estimates the gross and net board foot volume measured in International ¼ log rule, Vol_{gross} and Vol_{net} respectively, of an individual red oak tree using the following two equations:

$$Vol_{gross} = 744.2 \cdot S^{0.2627} \cdot \left(1 - e^{-0.00003487 \cdot D^{2.804}}\right) \tag{2.4}$$

$$Vol_{net} = Vol_{gross} \cdot \left[1 - \left(\frac{-6.696 + 1.077D}{100}\right)\right] \tag{2.5}$$

where
S = site index ($25 \leq S \leq 99$, with the average S being 66.5),[4] and
D = diameter at breast height (maximum D is 51.2 inches for sawtimber)

The difference between gross and net board foot volume is a deduction for cull or non-merchantable volume.

Loblolly Pine Plantation (Amateis and Burkhart 1985 and Burkhart et al. 1985)

This production system is more complex, using a series of equations to estimate the total cubic-foot volume, outside bark, per acre of planted loblolly pine (Yld) using a number of different inputs:

$$\ln(\text{Yld}) = -1.00184 + \frac{0.97745}{A} + 2.14146 \cdot \ln(H_d) + 0.00105 \cdot N \tag{2.6}$$

where
A = plantation age (years since planting) with $0 \leq A \leq 40$,
H_d = average height of dominant and codominant planted loblolly pines (feet),
N = number of planted loblolly pines surviving (per acre), and

ln denotes the natural logarithm.

$$\log\left(\frac{N_p}{N}\right) = A\left[0.013480 \cdot \log(N_p) + 0.00060783 \cdot H_d - 0.0084124 \cdot \sqrt{H_d}\right]$$

(2.10)

where
N_p = the number of seedlings planted per acre, and
log denotes the base 10 logarithm.

$$\ln(H_d) = \ln(S) \cdot \left(\frac{A_I}{A}\right)^{-0.10283} \cdot e^{-2.1676(A^{-1} - A_I^{-1})}$$

(2.8)

where
S = site index in feet ($45 \leq S \leq 87$ with the average S being 66),
A_I = some index age (years from planting), and
e denotes the exponential function.

Inside-out (ISO) Beams (Patterson et al. 2002 and Patterson and Xie 1998)

This production system uses small-diameter trees (e.g. 6- to 9-inches diameter at breast height) to manufacture 8-foot, 10-foot, 12-foot, 14-foot, or 16-foot 4 × 4 inch ISO beams depending on the length of the small-diameter tree input. Figure 2.2 illustrates the procedure for manufacturing an ISO beam.

Mobile Micromill® (Becker et al. 2004)

The Micromill will produce four-sided cants, rough-cut lumber (i.e. 1 × 4, 1 × 6, 2 × 4, 2 × 6, 4 × 4, 6 × 6), and chips using a single-pass system. The mill can process 4- to 20-foot logs with a maximum large-end diameter (outside bark) of 13 inches and a minimum small-end diameter (outside bark) of 4 inches. Figure 2.3 illustrates the single-pass system.

The production system

As was shown in Figure 2.1 and equations (2.1), (2.2), and (2.3), the production system is the foundation on which costs are minimized and profits ultimately maximized. Becker *et al.* (2004) recognized the relationship between the knowledge of the production system and minimizing costs and maximizing profit with respect to the Mobile Micromill:

> The number of productive machine hours will not only significantly affect operation and maintenance costs but will also determine the annual investment costs and ultimate mill profitability. To maximize the number of productive hours for each scheduled period of operation, it is imperative that mill owners and operators have a good understanding of how to process different tree

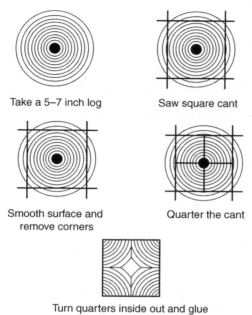

Take a 5–7 inch log   Saw square cant

Smooth surface and   Quarter the cant
remove corners

Turn quarters inside out and glue

Figure 2.2 Illustration of the steps to manufacture
an inside-out beam.

(Used with permission from the *Forest Product Journal*
and Patterson *et al.* (2002))

species and have the ability to maximize the product potential from each size
of log with given characteristics such as taper or amount of defect.

(Becker *et al.* 2004: 14)

Production systems can be described from many different points of view – for
example, an engineer to an ecologist – and can take many forms.

Figure 2.4 illustrates a description continuum, from depictive using flow charts
or graphics to numerical using mathematical models. For example, the ISO Beam
and the Mobile Micromill case studies use graphics to describe the relationship
between inputs and outputs and the production process in Figures 2.2 and 2.3,
respectively. The Select Red Oak and Loblolly Pine Plantation case studies use
sets of explicit equations to describe the relationship between inputs and outputs
and the production process, equations (2.4), (2.5) and equations (2.6), (2.7), (2.8),
respectively. As a manager, you will need a description of the production system
that focuses on identifying the underlying fundamental characteristics of the system that are critical to following the Architectural Plan for Profit (Figure 2.1).
Figure 2.4 also leads to the three fundamental questions needed to systematically
examine any production system: (1) What are the inputs? (2) What are the outputs?
and (3) How do I describe the production process?

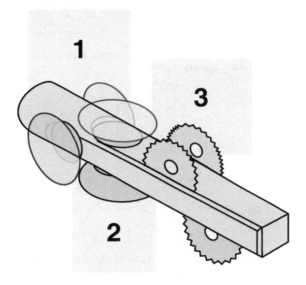

Figure 2.3 Single-pass production system used by the Mobile Micromill.

(Used with permission from the Micromill Systems, Inc.)

What are the inputs?

Identifying *all* the inputs of producing or providing an output could create a very extensive list. In creating this list, you must decide which inputs depict the underlying fundamental characteristics of the production system that will be critical for systematically following the Architectural Plan for Profit (Figure 2.1). In short, relevant inputs are those that will be critical for systematically following the Architectural Plan for Profit; they are those which you, as manager, have direct control over. In the case of the Select Red Oak case study, the explicit mathematical model, equations (2.4) and (2.5), define two inputs. For the Loblolly Pine Plantation case study, six inputs are used in the explicit mathematical model, equations (2.6), (2.7), and (2.8).[5] For the ISO Beams and the Mobile Micromill case studies, the graphics given in Figures 2.2 and 2.3 illustrate the processes

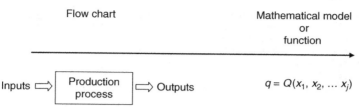

Figure 2.4 Production systems description continuum.

associated with each production system. As compared to the mathematical models, these graphics do not provide an explicit list of the inputs. Table 2.1 gives a list of inputs for the ISO Beams and the Mobile Micromill case studies.

As you will notice, the level of detail is greater for the ISO Beams than for the Mobile Micromill. The level of detail for the ISO Beams will make constructing the Pillar denoting Cost much easier.

Mathematical models used to depict the production system such as those from the Select Red Oak and Loblolly Pine Plantation case studies are probably intimidating to most readers (the Loblolly Pine Plantation mathematical model more so). However, does the input list given by these mathematical models describe the underlying fundamental characteristics of the production system that you, as a manager, can directly control? Are these inputs critical for systematically following the Architectural Plan for Profit (Figure 2.1) – specifically constructing the Pillars of Cost and Value? In the case of the Select Red Oak, the simple answer is no. You cannot directly control the inputs – diameter at breast height (D) and site index (S) – used in equations (2.4) and (2.5). However, you can indirectly control diameter at breast height by an action such as thinning. Equations (2.4) and (2.5) were developed by biometricians to estimate volume. These biometricians did not have the Architectural Plan for Profit in mind when they developed these mathematical models.[6] The answer is not as simple for the Loblolly Pine Plantation case study. The plantation age (A) in equations (2.6), (2.7), (2.8) and the number of seedlings planted per acre (N_P) in equation (2.7), are inputs over which you have direct control.[7] The index age (A_I) and site index (S) in equation (2.8) are inputs over which you have indirect control. For example, the site index can be increased through management actions such as fertilization and better drainage. In contrast, while no mathematical models were given in the ISO Beam and the Mobile Micromill case studies, the authors had the Architectural Plan for Profit in mind when developing the input list.

It bears repeating that in creating the list of inputs, you must decide which inputs depict the underlying fundamental characteristics of the production system that will be critical for systematically following the Architectural Plan for Profit (Figure 2.1). Comparing and contrasting the inputs described in the Select Red Oak and Loblolly Pine Plantation case studies with those in the ISO Beam and Mobile Micromill case studies also identifies an additional difference. No explicit fixed inputs are described in the Select Red Oak and Loblolly Pine

Table 2.1 Input lists for the Inside-Out Beams and Mobile Micromill case studies

Inside-Out Beams (Patterson et al. *2002)*	*Mobile Micromill® (Becker* et al. *2004)*
Land, buildings, scale	Mobile Micromill
Lift trucks (2)	Support equipment
Log deck and conveyor	Tractor – loader
Debarker	Forklift
Cut-off saw	Maintenance vehicle
Chipper	Chainsaw
Skrag mill	Mobile Micromill siting
Moulder (2)	Operation and maintenance – Micromill
Quartering saw	Operation and maintenance – support
	equipment
ISO clamping	Labor
Boiler and kiln(s)	Delivered raw logs
Pre-dryer (hardwoods)	Inventory
Trim saw	Relocation and equipment setup
Residue to boiler	
Utilities/office	
Salaries/wages	
Logs	

Plantation case studies, while there are in the ISO Beam and Mobile Micromill case studies. The Select Red Oak and Loblolly Pine Plantation case studies only describe variable inputs or those inputs that change as the level of output produced or provided changes.

The ISO Beam and Mobile Micromill case studies describe variable and fixed inputs. For example, logs delivered to both mills are variable inputs; if no logs are available to be run through either mill, then no beams or dimension lumber can be produced. However, the production of beams or dimension lumber requires fixed inputs such as capital equipment (e.g. the machines listed in Table 2.1). These inputs are required in the production process, but there is a period of time in which they cannot be varied. For example, the owner of an ISO beam mill cannot buy and install a new quartering saw in the same manner as they can vary the daily number of logs run through the mill.

What are the outputs?

Identifying *all* the outputs of a production system seems deceptively simple. For example, in the Select Red Oak and Loblolly Pine Plantation case studies, the output is volume. However, this is not the best measurement of volume. In the case of the Select Red Oak, the measure of volume is the gross and net board foot volume (International ¼ log rule) of an *individual* tree and the measure of volume for the Loblolly Pine Plantation is the total cubic-foot volume, outside bark, per acre of planted loblolly pine. While this is a better answer, it still is not very good. As with identifying "What are the inputs?", this list must also depict the underlying fundamental characteristics of the production system that will be critical for systematically following the Architectural Plan for Profit (Figure 2.1). Specifically, this list should make constructing the Pillar denoting Value much easier. Many hardwood trees, such as red oak, (and some softwood trees) are characterized by an index of quality. A tree designated as a Grade 1 denotes the highest quality, Grade 2 is next in quality followed by Grade 3 and so on. Diameter at breast height is one – of many – attribute used to distinguish between tree grades. Other things being equal, the larger the diameter the better the tree grade.[8] Individuals who purchase red oak place a higher value on Grade 1 trees, then Grade 2 and so on. The difference between the stumpage price of a Grade 1 tree versus a Grade 2 tree versus a Grade 3 tree can be substantial (Stier 2003). Thus, the output for red oak should include board foot as well as quality as defined by Grades 1, 2, or 3. In this example, the variable diameter at breast height does not provide information to help in constructing the Pillar denoting Cost, but provides information to help construct the Pillar denoting Value.

For the Loblolly Pine Plantation case study, the trees at any given stand age will be fairly uniform (as is the nature of trees in a plantation). Smaller loblolly pines are used for pulpwood and as the trees increase in diameter and height, they can be used as Chip-n-Saw, sawtimber, plywood logs, or power poles. The value of the output increases with greater diameter and height. The lowest valued output is pulpwood and the highest valued output is power poles. Thus, the output for the Loblolly Pine Plantation should include total cubic-foot volume, outside bark, per acre as well if the volume represents pulpwood, Chip-n-Saw, sawtimber, plywood logs, or power poles.

The output from the ISO Beams case study was previously defined as an 8-foot, 10-foot, 12-foot, 14-foot, or 16-foot 4 × 4 inch ISO beam depending on the length of the small-diameter tree input. Based on prices for solid wood 4 × 4 beams, the value of an ISO beam is estimated to increase with its length (Patterson *et al.* 2002). Thus, the output for the ISO Beams case study should be described as the number and length of ISO beams manufactured for a given period of time.

The output from the Mobile Micromill case study is the most variable depending on the inputs available. For example, Table 2.2 shows the percentage of lumber size and grade given inputs available in the study area.

In the case of the Mobile Micromill case study, it bears repeating that the list of outputs should depict the underlying fundamental characteristics of the production system that will be critical for systematically following the Architectural Plan for Profit (Figure 2.1). This list should make it much easier to construct the Pillar denoting Value. Other things being equal, lumber of the same size (e.g. 1 × 4) but of better quality is valued more by the consumer. In addition, kiln drying adds value to lumber. This is illustrated by Table 2.3.

The trick is to saw the most valuable lumber from any given input. Thus output for the Mobile Micromill case study should be described in terms of quality and quantity of the various different outputs for a given period of time.

Finally, one output that is rarely discussed is waste from a production process. This is often a more acute issue for manufacturing processes similar to those

Table 2.2 Percentage of lumber size and grade recovery for ponderosa pine

Lumber size	Lumber grade	Log small-end diameter (in)				
		3	4	5	6	7
Inches		Percent				
1 by 4	Common mid better	0	0	0	38	62
1 by 4	3, 4, and 5 Common	0	0	2	71	2
3 by 3	Standard and better	1	46	52	1	0
3 by 3	Economy	6	64	28	2	0
4 by 4	Standard and better	0	2	42	51	5
4 by 4	Economy			51	30	8

Data from Lowell.
(Used with permission from the USDA Forest Service)

Table 2.3 Output price by lumber size and grade for ponderosa pine

Product size	Finished product grade	Market scenario[a]			
		Observed	Low	Average	High
In		Dollars per thousand board feet			
Kiln dried, surfaced:[b]					
1 by 4, 6	2 Common and better	530	453	500	548
1 by 4, 6	3 and 4 Common	275	201	245	278
4 by 4	Standard	350	—[c]	—[c]	—[c]
4 by 4	Economy	300	—[c]	—[c]	—[c]
Green, rough sawn:					
4 by 4	Standard, Economy	200	—[c]	—[c]	—[c]
1 by 4	3 Common, Utility	220	164	190	219
1 by 4	5 Common, Economy	150	110	134	155
Byproducts[d]					
Wood chips	Dirty, including bark	4-20/BDT	—[c]	—[c]	—[c]

Notes
a Observed markets are for ponderosa pine products sold in the Four-Corners region in October 2003. Low, average, and high market prices were compiled by the Western Wood Products Association and reported by Haynes (2003).
b Kiln-dried, planed products will require additional manufacturing and transportation to a secondary processing facility independent of mobile Micromill operations.
c The Western Wood Products Association does not compile price data for this category.
d Chip byproducts are "dirty" owing to bark and other impurities. Chip prices are based on biomass purchases for bioenergy powerplants and pellet manufacturing in the Southwest (per bone dry ton [BDT], zero percent moisture). Prices will differ significantly based on distance to market.

(Used with permission from the USDA Forest Service)

associated with the ISO Beam and the Mobile Micromill case studies. For example, sawdust and other residue from sawing is an output that must be used in some manner or disposed of. If the sawdust and other residue can be used to generate electricity for the mill, then the ash must be disposed of. All production process will create waste and disposing of waste can be costly, depending on its amount, type, and toxicity.

How do I describe the production process?

Referring back to Figure 2.4, production system descriptions may take many forms ranging from the numerical approach of the classic production function (Carlson 1974; Henderson and Quandt 1980; Silberberg and Suen 2001) to a combination of verbal and graphics. The description of the production process must satisfy two key conditions. First, the description must depict the underlying fundamental characteristics of the production system that will be critical for systematically following the Architectural Plan for Profit (Figure 2.1). This condition has been satisfied if the answers to "What are the inputs?" and "What are the outputs?" are complete and precise. Second, the description must define the technical and physical relationships between combinations of inputs and the maximum outputs that each combination can produce.[9] In terms of economic information, this description must allow identifying input–output combinations that are technically efficient. A technically efficient input–output combination is when it is not possible to increase output without increasing input or when the maximum output is produced with the least amount of input. Maital (1994) describes the difference between technical inefficiency and technical efficiency as what is produced relative to what could be produced feasibly given the existing resources, knowledge, and ability. Technical efficiency is measured based only on inputs that the manager can control.

The Select Red Oak and the Loblolly Pine Plantation case studies describe the production process using mathematical equations (2.4), (2.5) and (2.6), (2.7), (2.8), respectively. Basically, these equations summarize a complex biological process using two variable inputs for the Select Red Oak and six variable inputs for the Loblolly Pine Plantation case studies. The ISO Beam and the Mobile Micromill case studies provide written descriptions and a graphical illustration of the production process. For the ISO Beams case study, step-by-step directions to manufacture ISO beams are given in the Procedures section of Patterson and Xie (1998) and briefly summarized by Patterson *et al.* (2002) and illustrated in Figure 2.2:

> Small trees (6- to 9-in. diameter at breast height) were cut into 8-foot logs. These logs were cut into 4-1/4-inch square cants with sides parallel to the pith. The sides of the cants were smoothed on a moulder and the corners chamfered. The cants were quartered, the exterior surfaces heat treated, resorcinol adhesive was applied, and the quarters turned inside-out and clamped together. After the adhesive cured, the new beams were kiln-dried and molded to the final size of 3-1/2 by 3-1/2 inches.
>
> (Patterson *et al.* 2002: 24)

Patterson *et al.* (2002) examines three different daily production levels given the
list of inputs in Table 2.1.

For the Mobile Micromill case study, descriptions of the production process are
scattered throughout Becker *et al.*'s document. However, a brief description of the
single-pass production process is given by Becker *et al.* (2004) and illustrated in
Figure 2.3:

> The Micromill is a 300-horsepower dimension sawmill capable of processing
> 4- to 20-ft logs with a maximum large-end diameter (outside bark) of 13 in
> and a minimum small-end diameter (outside bark) of 4 in. Designed to cut
> about 850,000 to 3,500,000 ft³ per year, depending on tree species and desired
> product, the mill produces four-sided cants, rough-cut lumber, and chips in a
> single-pass, automatic-feed system.
>
> (Becker *et al.* 2004: 1)

Table 2.4 gives observed and published production rates for the Mobile Micromill.

To compare the observed and published production rates given the difference
in the logs used as inputs described in the footnotes of Table 2.4, Becker *et al.*
(2004) recommends using hourly production rates for a 4-inch square cant.

Do the production process descriptions in the case studies satisfy the second
key condition given above? In terms of the Select Red Oak and the Loblolly Pine
Plantation case studies, the mathematical equations allow generating yield tables
given various levels of the inputs. For example, Tables 2.5 and 2.6 show the yields
of the average select red oak tree and the average acre of a loblolly pine plantation
given defined input levels.

Table 2.4 Observed and published production rates for the Micromill SLP5000D

Specification	Production rate					
	Observed[a]	Published[b]				
Average butt-end log diameter (in)	8	12	10	8	6	4
Square cant size (in)	4	8	7	6	4	3
Estimated feed rate (ft/minute)	55	50	55	60	65	70
Hourly production (board feet)	3,000	12,250	10,375	6,500	3,250	2,125
Monthly production (thousand board feet)	504	2,050	1,750	1,100	550	350

Notes
a Based on 8-ft log length, 21 working days/month, and one 8-hour shift per day. Figures are
based on a continuous run of presorted ponderosa pine with an average butt-end diameter
of 8 in.
b Based on 12-ft log length, 21 working days/month, and one 8-hour shift per day. All figures
are based on a continuous run of presorted, uniform-sized spruce, pine, or fir logs.

(Used with permission from the USDA Forest Service and Micromill Systems, Inc.)

Table 2.5 Select Red Oak board foot volume per tree (International ¼)

D	Site index 25		Site index 66		Site index 99	
	Volume gross	Volume net	Volume gross	Volume net	Volume gross	Volume net
1	0	0	0	0	0	0
5	6	6	7	7	8	8
10	38	37	49	47	55	52
15	116	105	150	135	166	151
20	249	212	321	274	357	304
25	436	348	563	449	626	499
30	664	494	857	638	954	709
35	910	628	1,175	810	1,307	902
40	1,146	729	1,479	941	1,646	1,047
45	1,349	786	1,741	1,014	1,937	1,128
50	1,504	795	1,941	1,026	2,159	1,141
55	1,610	764	2,078	986	2,311	1,097

D denotes diameter at breast height measured in inches.
Gross and net volume are measured as board foot volume (International ¼) per tree.

Table 2.6 Loblolly Pine Plantation yield table

Plantation age	Site index 45 yield	Site index 66 yield	Site index 87 yield
1	7	7	7
5	160	266	385
10	661	1,321	2,164
15	1,281	2,757	4,678
20	1,897	4,183	7,068
25	2,462	5,427	9,022
30	2,964	6,470	10,644
35	3,408	7,370	12,169
40	3,806	8,206	13,798

Yield is measured as cubic feet per acre and is calculated given the following information: $A_I = 25$; $N_p = 1250$.

These yield tables also define the efficiency frontier for the identified input levels. These yield tables allow you as a manager to develop economic information on efficiency. For example, you have a select red oak with a diameter at breast height of 25 inches growing on a site with a site index of 66. If the net volume of your tree is greater than (less than) 554 board feet (International ¼ log rule) then your tree is more productive than the average red oak on a similar site with the same diameter at breast height. A similar example can be described for an acre of planted loblolly pine using Table 2.6.[10]

Technical efficiency cannot be measured directly for the ISO Beams case study because no information relating inputs to outputs directly is given in either Patterson *et al.* (2002) or Patterson and Xie (1998). However, Patterson *et al.*

(2002) provide information about three daily production levels (i.e. 464 ISO beams per day, 696 ISO beams per day, and 928 ISO beams per day), a list of the machines required (Table 2.1), and the purchase costs and number of machines for each daily production level. Pre-drying of each quarter's exterior surfaces is required before the adhesive is applied, according to the data provided by Patterson *et al.* (2002). The quarters are turned inside-out and clamped, and the new beams are dried in kilns. This would lead me to infer that while most of the machines listed in Table 2.1 (e.g. the cut-off saw) are not running at capacity nor are they constraining the product system in any manner, the capacity of the kiln(s) and pre-dryers are the limiting capital inputs in this production process. In other words, to produce more than 464 ISO beams per day more capital equipment – larger capacity kilns and pre-dryers – are required. The same argument can be made for producing more than 696 ISO beams per day. I would conclude that at each different daily production level given, the production system is technically efficient.

As part of the economic assessment of using a Mobile Micromill, Becker *et al.* (2004) assessed the Micromill's productivity by comparing observed and published production rates. Table 2.4 shows the observed production rates for a 4-inch square cant are lower than the published rates for a 4-inch square cant. The authors' stated differences between observed and published production rates "will result from variability in operator experience, average log diameter, and cant size." Based on the information provided in Table 2.4, I would conclude that there are some technical inefficiencies that have the potential of being reduced as labor becomes more familiar with the production process.

Economic descriptors of the production system

The three classic economic descriptors of the production system – total product, average product, and marginal product – provide a different piece of economic information about the production system. However, the economic information they provide should not be viewed in isolation from the Architectural Plan for Profit. The relative value of the economic information is its use in helping to construct the Pillars of Cost and Value.

Total product (TP)

Total product defines the most efficient level of production given the production system. Stated differently, total product defines the greatest combination of outputs for a given combination of inputs or the least amount of inputs needed to produce a given level of outputs. There are three means of depicting total product: tables, graphs, or mathematical equations. For the Mobile Micromill case study, total product is depicted in Table 2.4 which compares observed production with published production rates. For the Select Red Oak case study, total product is calculated using equations (2.4) and (2.5).[11] Given the input ranges (i.e. $25 \leq S \leq 99$ and $1 \leq D \leq 51.2$ inches), Table 2.5 depicts the gross and net volume (board feet per tree) at three levels of site index. For the Loblolly Pine Plantation case study, total product is calculated using equations (2.6), (2.7), and (2.8). A similar yield table (Table 2.6) can be developed for the Loblolly Pine Plantation case study using the following input information: $A_I = 25$; $N_p = 1250$, $S = 45$, 66, 87, and $1 \leq A \leq 40$.

The information in Tables 2.5 and 2.6 can also be displayed graphically. Figures 2.5 and 2.6 are a two-dimensional view of total product or in this case yield of Select Red Oak and the Loblolly Pine Plantation, respectively. In these figures, all the inputs are held constant except one. In Figure 2.5, site index is held constant at 66 and diameter is allowed to vary. In Figure 2.6, site index, trees planted per acre, and the index age are held constant and plantation age is allowed to vary. Even though these are vastly different species from different geographic areas, the units that describe the output are different scales (i.e. board feet per tree (bd.ft/tree) vs cubic feet per acre (cu.ft/ac)), and different inputs (diameter vs plantation age), the shapes of the curves shown in Figures 2.5 and 2.6 are similar.

A three-dimensional view of total product can also be generated given the information in Tables 2.5 and 2.6. Figure 2.7 shows total product or yield of select red oak as a production surface allowing both diameter and site index to vary. Figure 2.8 shows the yield of the Loblolly Pine Plantation holding the index age and trees planted per acre constant (i.e. $A_I = 25$; $N_p = 1210$) and allowing plantation age and site index to vary.

As with Figures 2.5 and 2.6, even though these are vastly different species from different geographic areas, the units that describe the output are different scales (i.e. bd.ft/tree vs cu.ft/acre), and different inputs (diameter and site index vs plantation age and site index), the shapes of the curves shown in Figures 2.7 and 2.8 are comparable.

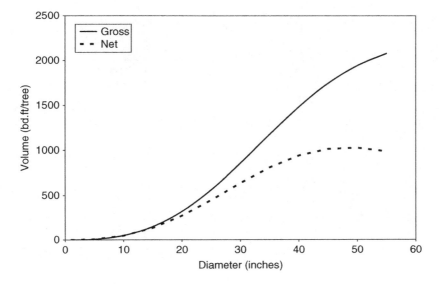

Figure 2.5 Gross and net yields for Select Red Oak.

Volume is measured as board foot, International ¼, per tree (bd.ft/tree) given S = 66.

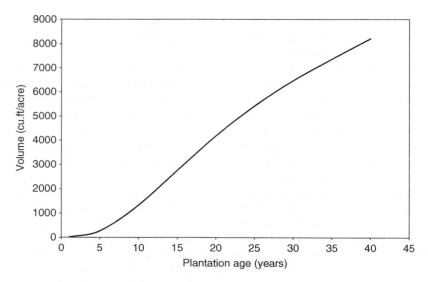

Figure 2.6 Yield of Loblolly Pine Plantation.

Volume is measured as cubic feet per acre (cu.ft/ac) given $A_I = 25$, $S = 66$, $N_p = 1210$.

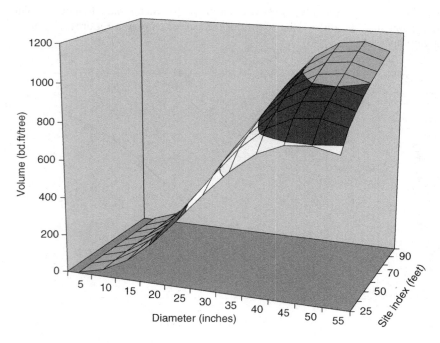

Figure 2.7 Net volume for Select Red Oak.

Volume is measured as board feet per tree (bd.ft/tree).

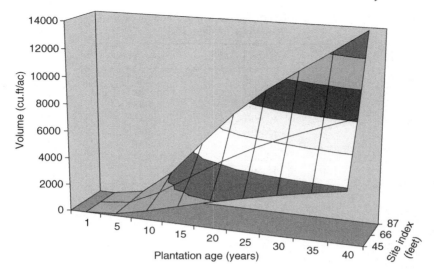

Figure 2.8 Loblolly Pine Plantation volume.

Volume is measured as cubic feet per acre (cu.ft/ac) given $A_I = 25$ and $N_p = 1210$.

To illustrate why the shapes in Figures 2.5 and 2.6 and Figures 2.7 and 2.8 are similar I will start with what Albert Einstein called a thought experiment, then use empirical data to examine the conclusions from the thought experiment. Take as given a fixed area of land; for example, an acre. This acre starts with no vegetation and is seeded, relatively uniformly, by the surrounding trees. The seeds germinate within a reasonable period of time and the seedlings can be considered all of the same age or even-aged. At this young age, there is very little if any competition for the fixed quantities of area, sunlight, water, and nutrients. And while there may be some mortality, there is positive net volume growth per acre for the surviving trees. This positive net volume growth per acre for the surviving trees will probably increase for a number of years and the measure of volume per acre will increase. Competition for the fixed quantities of sunlight, water, and nutrients will eventually cause an increase in mortality resulting in a decline in positive net volume growth per acre for the surviving trees. But the measure of volume per acre will still increase. In time, the volume loss due to mortality will be greater than the volume growth of any surviving trees and the measure of volume per acre will decrease. The testable conclusion of our thought experiment is that volume per acre will increase with time until mortality is greater than growth of the surviving trees and volume per acre will decline.

Because most forest businesses harvest their forests long before they reach the age where mortality is greater than growth, empirical data are very rare. Gevorkiantz and Scholz (1948) sampled 385 plots in southwestern Wisconsin. These samples were used to develop the gross and net volumes shown in Table 2.7.

Table 2.7 Unthinned oak stand in Southwestern
Wisconsin (medium site)

Age	Gross volume	Net volume
40	300	250
60	4,700	4,050
80	10,200	8,700
100	13,600	11,000
120	15,800	11,800
140	17,300	11,600
160	18,500	10,900
180	18,500	

Gross and net volumes are measured as board foot
(International ¼) per acre.

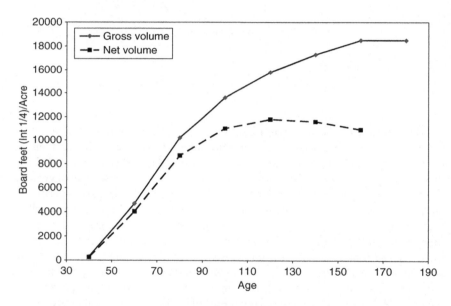

Figure 2.9 Unthinned oak stand in Southwestern Wisconsin (medium site).

The difference between gross and net volumes is determined by subtracting
mortality and cull losses. Thus, net volumes represent merchantable volumes.
Figure 2.9 is a graphic of the gross and net volume per acre vs stand age. Both
Table 2.7 and Figure 2.9 show that in terms of gross volume, the stand has reached
a physiological maximum at an age of 160. In terms of net volume, the stand
reaches a maximum at age 120. Gevorkiantz and Scholz (1948) provide a graphic
that shows that after age 160, losses due to mortality and cull are greater than
growth. This corroborates the conclusion of our thought experiment empirically.
We could repeat this thought experiment with any tree species, arrive at a similar

set of conclusions that could be examined empirically, and the same result obtained. The reason for this is the Law of Diminishing Returns:[12]

> When increasing quantities of a given input are added to fixed quantities of some other factors, first the change in output per unit change of the identified input and then the output per input for the identified input will eventually decrease.

In fact *any* production process must follow the Law of Diminishing Returns.

The reason that the graphs of the Select Red Oak and Loblolly Pine Plantation have similar shapes is due to the Law of Diminishing Returns. In the Select Red Oak and the Loblolly Pine Plantation, Figures 2.7 and 2.8, site index is identified as one of the two inputs. Site index is an input that can be indirectly controlled by you as manager. For example, it can be changed by fertilizing the site. However, while the right amount of fertilizer can increase the quality of the site, applying too much fertilizer can kill the trees. In the case of the Select Red Oak, diameter is the other identified input. Diameter is a proxy for time or stand age. According to Burns and Honkala (1990), red oak will grow on average about 0.2 to 0.4 inches in diameter per year. Thus, even at the scale of an individual tree, volume – especially net volume – will decrease as time increases.

The definition of the Law of Diminishing Returns includes two descriptors of the production process: (1) the change in output per unit change of the identified input and (2) output per input for the identified input. The first descriptor defines marginal product and the second defines average product. Each will be discussed, starting with average product.

Average product (AP)

Average product is calculated by dividing total output by the level of input necessary to produce that level of output:

$$AP = \frac{Q(x_j)}{x_j} \tag{2.9}$$

where

$Q(x_j)$ denotes the production system holding all other inputs and factors constant or fixed except the *j*th input, x_j.

Table 2.8 shows the AP for Select Red Oak. For example, the AP for a 30-inch diameter select red oak is

$$\frac{638 \text{ bd.ft/tree}}{30 \text{ inch}} = 21.3 \text{ bd.ft/tree/inch}$$

There are three points that are important about this calculation. First, the input site index is held constant – in this case site index is equal to 66. Second, the units given to AP must be determined correctly or the interpretation of AP will be meaningless in terms of economic information. The units on output are defined as net board feet per tree – International ¼ (bd.ft/tree). The units on the input diameter are measured in inches. Therefore, the units on any AP given in Table 2.8 are bd.ft/tree/inch. Third, the interpretation of the above AP calculation is that for every inch in diameter the tree grew between 0 and 30 inches, the tree gained 21.3 net board feet of volume.

Table 2.9 shows the AP calculations for the Loblolly Pine Plantation case study. For example, the AP for a 20-year-old loblolly pine plantation is

$$\frac{4183 \text{ cu.ft/acre}}{20 \text{ years}} = 209.2 \text{ cu.ft/acre/year}$$

As with AP for Select Red Oak, three points are important about this calculation. First, the input site index is held constant – in this case site index is equal to 66. Second, the units given to AP must be determined correctly or the interpretation of AP will be meaningless in terms of economic information. The units on output are defined as cubic feet per acre (cu.ft/acre). The units on the input plantation age are measured in years. Therefore, the units on any AP given in Table 2.9 are cu.ft/acre/year. Third, the interpretation of the above AP calculation is that between the plantation ages of 0 and 20, the plantation grew 209.2 cubic feet per acre per year.

Given the information in Table 2.5 and equation (2.9), calculate AP for the Select Red Oak holding diameter constant at 25 inches and allowing site index to vary. What are the units on the results? What is the interpretation of AP? Given the information in Table 2.6 and equation (2.9), calculate AP for the Loblolly Pine Plantation holding stand age constant at 20 years and allowing site index to vary. What are the units on the results? What is the interpretation of AP?

Table 2.8 Select Red Oak net volume, average, and marginal product

D	Net volume	Average product	Marginal product
1	0		
5	7	1.4	1.8
10	47	4.7	8.0
15	135	9.0	17.7
20	274	13.7	27.6
25	449	18.0	35.1
30	638	21.3	37.7
35	810	23.2	34.5
40	941	23.5	26.1
45	1,014	22.5	14.5
50	1,026	20.5	2.4
55	986	17.9	−8.0

D denotes diameter at breast height measured in inches.
Volume is measured as net board foot volume per tree – International ¼ (bd.ft/tree) given a site index of 66.
Average product is measured as bd.ft/tree/inch.
Marginal product is measured as bd.ft/tree/inch.

Table 2.9 Loblolly Pine volume, average product, and marginal product

Plantation age	Volume	Average product	Marginal product
1	7	7.1	
5	266	53.2	64.7
10	1,321	132.1	211.0
15	2,757	183.8	287.2
20	4,183	209.1	285.2
25	5,427	217.1	248.8
30	6,470	215.7	208.6
35	7,370	210.6	180.1
40	8,206	205.2	167.2

Volume is calculated in terms of cubic feet per acre (cu.ft/acre) given the following information: $S = 66$; $A_I = 25$; $N_p = 1250$.
Plantation age is measured in years since planting.
Average product is measured as cu.ft/acre/year.
Marginal product is measured as cu.ft/acre/year.

A measure of AP that is unique to forestry is called Mean Annual Increment (MAI). MAI is

$$MAI = \frac{Volume}{year} \tag{2.10}$$

Just like AP, MAI is calculated with respect to stand age. For the Loblolly Pine Plantation, the MAI at plantation age 20 is 209.2 cu.ft/acre/year. The interpretation of the MAI calculation is the same as the AP calculation: between the plantation ages of 0 and 20, the plantation grew on average 209.1 cubic feet per acre per year. Thus, in Table 2.9, the column titled Average product is equivalent to a MAI. However, the column entitled Average product in Table 2.8 is *not* equivalent to MAI as the measure of volume was *not* divided by time or year.

The Law of Diminishing Returns states that AP will decline at some point. This can be drawn out of the data in Tables 2.8 and 2.9; however, the graphs of AP, Figures 2.10 and 2.11, show this decline readily. For Select Red Oak the decline occurs at a diameter of 45 inches and for the Loblolly Pine Plantation the decline occurs at a plantation age of 25.

Marginal product (MP)

Marginal product is defined as the change in output per unit change of the identified input

$$MP = \frac{\Delta Q(x_j)}{\Delta x_j} = \frac{Q(x_j) - Q(x_{j-1})}{x_j - x_{j-1}} \tag{2.11}$$

where

$Q(x_j)$ denotes the production system holding all other inputs and factors constant or fixed except the *j*th input, x_j;

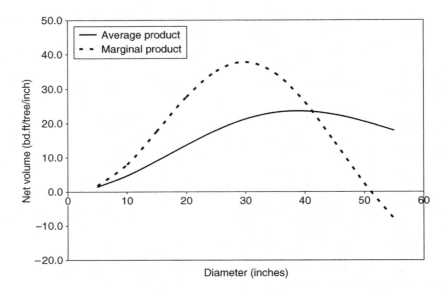

Figure 2.10 Select Red Oak average and marginal product.
Given S = 66.

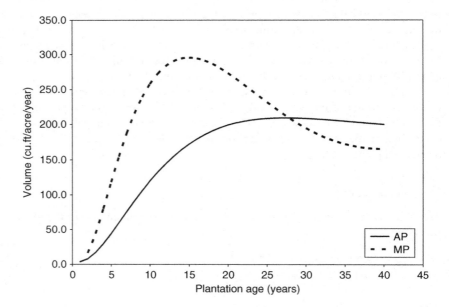

Figure 2.11 Loblolly Pine Plantation average and marginal product.
Given $A_I = 25$, $S = 66$, $N_p = 1210$.

$\Delta Q(x_j) = Q(x_j) - Q(x_{j-1})$ denotes change in the output produced or provided given a per unit change in the identified input x_j; and

$\Delta x_j = x_j - x_{j-1}$ denotes the per unit change in the identified input x_j.

The fourth column of Table 2.8 shows the MP of diameter on volume per tree given site index set at 66 for Select Red Oak. For example, the MP of a change in volume resulting from changing diameter from 20 to 25 inches is:

$$\frac{(554 - 318) \text{ bd.ft/tree}}{(25 - 20) \text{ inches}} = 47.2 \text{ bd.ft/tree/inch}$$

Three points are important about this calculation. First, the input site index is held constant – in this case site index is equal to 66. Second, the units given to MP must be determined correctly or the interpretation of MP will be meaningless in terms of economic information. The units on output are defined as net board feet per tree – International ¼ (bd.ft/tree). The units on the input diameter are measured in inches. Therefore, the units on any MP given in Table 2.8 are bd.ft/tree/inch. Third, the interpretation of the above MP calculation is that for every inch in diameter the tree grew between 20 and 25 inches, the tree gained 47.2 net board feet of volume.

Table 2.9 shows the MP calculations for the Loblolly Pine Plantation case study. For example, the MP for a loblolly pine plantation that is between 25 and 20 years old is

$$\frac{(5427 - 4183) \text{ cu.ft/acre}}{(25 - 20) \text{ years}} = 248.8 \text{ cu.ft/acre/year}$$

As with MP for Select Red Oak, three points are important about this calculation. First, the input site index is held constant, with the site index in this case equal to 66. Second, the units given to MP must be determined correctly or the interpretation of MP will be meaningless in terms of economic information. The units on output are defined as cubic feet per acre (cu.ft/acre). The units on the input plantation age are measured in years. Therefore, the units on any MP given in Table 2.9 are cu.ft/acre/year. Third, the interpretation of the above MP calculation is that between the plantation ages of 25 and 20, the plantation grew 248.8 cubic feet per acre per year.

Given the information in Table 2.5 and equation (2.11), calculate MP for the Select Red Oak holding diameter constant at 25 inches and allowing site index to vary. What are the units on the results? What is the interpretation of MP? Given the information in Table 2.6 and equation (2.11), calculate the MP for the Loblolly Pine Plantation holding stand age constant at 20 years and allowing site index to vary. What are the units on the results? What is the interpretation of MP?

A measure of MP that is unique to forestry is called periodic annual increment (PAI). PAI is

$$PAI = \frac{\Delta Volume}{\Delta year} \tag{2.12}$$

Just like MP, PAI is calculated with respect to stand age. For the Loblolly Pine Plantation, the PAI between the plantation ages of 25 and 20 is 248.8 cu.ft/acre/year. The interpretation of the PAI calculation is the same as the MP calculation: between the plantation ages of 25 and 20, the plantation grew on average 248.8 cubic feet per acre per year. Thus, in Table 2.9, the column titled Marginal product is equivalent to a PAI. However, the column titled Marginal product in Table 2.8 is *not* equivalent to PAI as the measure of volume was *not* divided by time or year.

The Law of Diminishing Returns states that MP product will decline at some point. This can be drawn out of the data in Tables 2.8 and 2.9. The graphs of MP, Figures 2.10 and 2.11, show this decline readily. For Select Red Oak the decline occurs at a diameter of 35 inches and for the Loblolly Pine Plantation the decline occurs at a plantation age of 15.

Total versus average versus marginal product

Comparing and contrasting the graphs of TP versus AP and MP is instructive.[13] Figures 2.12 and 2.13 show the graphs of TP, AP, and MP for Select Red Oak and Loblolly Pine Plantation, respectively.[14] Four points follow from this analysis.

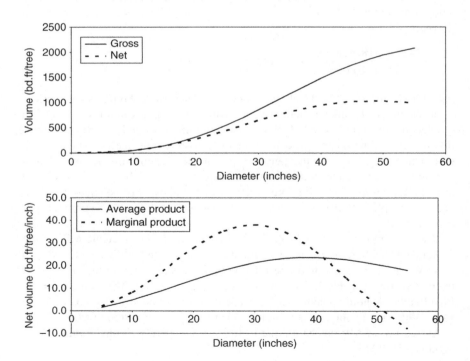

Figure 2.12 Select Red Oak total, average, and marginal product.

Given $S = 66$.

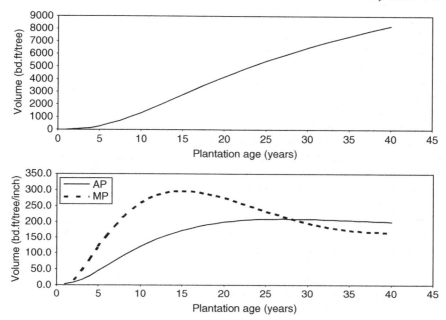

Figure 2.13 Loblolly Pine Plantation total, average, and marginal product.
Given $S = 66$, $A_I = 25$, and $N_p = 1210$.

First, the horizontal axes on the graphs of TP and AP and MP have the same scale. For example, in Figure 2.12 the scale is Diameter (inches) and in Figure 2.13 the scale is Plantation Age (years). The vertical axes on the graphs of TP and AP and MP have different scales. In Figure 2.12, the vertical axis of the TP graph has a scale of bd.ft/tree while the vertical axis of the AP and MP graph has a scale of bd.ft/tree/inch. In Figure 2.13, the vertical axis of the TP graph has a scale of cu.ft/ acre while the vertical axis of AP and MP graph has a scale of cu.ft/acre/year. This difference in scale is important in interpreting each graph.

Second, according to the Law of Diminishing Returns, "first the change in output per unit change of the identified input and then the output per input for the identified input will eventually decrease." In other words, MP will be the first to decrease followed by AP. This can be seen on Figures 2.12 and 2.13, and on Figures 2.10 and 2.11. However, the direct tie back to the TP graph is missing in Figures 2.10 and 2.11. In the thought experiment, at a young age when there was little competition for area, sunlight, water, and nutrients the trees grew rapidly and volume per acre increased quickly. This is illustrated in Figure 2.13 by examining the MP and TP curve. Between stand ages of 0 and 15, the Loblolly Pine Plantation is growing rapidly as shown by the MP curve and volume also increased quickly as shown by the TP curve. After stand age 15, mortality due to competition increases and while there is still positive net growth, the net growth has slowed as has the increase in total volume. Again, by examining the MP and TP curves in Figure 2.13 you can see that after stand age 15 the net growth is still positive, though it is decreasing, as illustrated by the MP curve; and while volume is increasing it is increasing more slowly, illustrated by the TP curve. It is left to the reader to develop a similar argument for describing the relationship between TP and MP for Select Red Oak given in Figure 2.12.

Third, examining Figures 2.12 and 2.13 shows that whenever MP is greater (less) than AP, AP is increasing (decreasing). In the case of the Loblolly Pine Plantation, MP describes the growth of trees at a given point in time and AP describes the growth of the trees from when they were planted to a point in time. If the current growth of the trees is greater than the average growth, then the average growth will increase. If the current growth of the trees is less than the average growth, then the average growth will decrease.[15] It is left to the reader to develop a similar argument for describing the relationship between MP and AP for Select Red Oak given in Figure 2.12. An additional relationship between MP and AP is also apparent by examining Figures 2.12 and 2.13. When AP is at its maximum value, MP equals AP.

Fourth, while is seems obvious, every point on the TP curve has a corresponding point on the AP and MP curves. To illustrate this concept I will use the Wisconsin oak data from Gevorkiantz and Scholz (1948). The net yields (TP), mean annual increment (AP), and periodic annual increment (MP) are given in Table 2.10 and graphically in Figure 2.14.

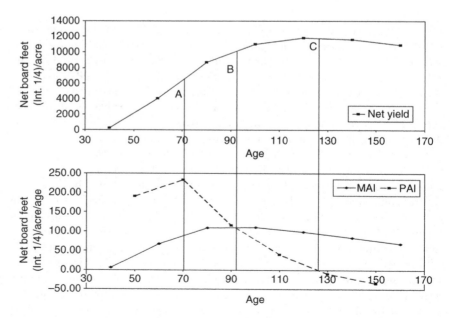

Figure 2.14 Unthinned oak stand in Southwestern Wisconsin (medium site).

Table 2.10 Unthinned oak stand in Southwestern Wisconsin (medium site)

Age	Net volume Bd.Ft/Ac	MAI Bd.Ft/Ac/Age	PAI age	PAI Bd.Ft/Ac/Age
40	250	6.25		
60	4,050	67.50	50	190.00
80	8,700	108.75	70	232.50
100	11,000	110.00	90	115.00
120	11,800	98.33	110	40.00
140	11,600	82.86	130	−10.00
160	10,900	68.13	150	−35.00

Volumes are measured board foot volume (International ¼) per acre.

As both the TP graph and the AP and MP graph have the same vertical axis, a vertical line can be drawn between the two graphs to illustrate this point. I will highlight three such points. The first point is where MP is maximized. This is shown by the vertical line labeled *A*. To the left of this line net volume is increasing at an increasing rate. The second point is where AP is maximized and *AP = MP*. This is shown by the vertical line labeled *B*. The third point is where TP is maximized and where *MP = 0*. This is shown by the vertical line labeled *C*. A final observation possible with Figure 2.14 is that between lines *A* and *C* net volume is increasing but at a decreasing rate. To the right of line *C* net volume is decreasing at an increasing rate.

Non-continuous production systems

Not every production process can be depicted with smooth graphs as those developed in the Select Red Oak and the Loblolly Pine Plantation case studies. For example, the Micromill and ISO Beams case studies describe production processes that require a production line or a series of sequential steps to convert a log into a final product. Each step must be completed before the input can move to the next step. I will concentrate on the ISO Beams case study and the production process described by Patterson *et al.* (2002). Given the fixed nature of the existing capital equipment (e.g. the machines listed in Table 2.1), labor, and the sequential steps required by the production process as defined by Patterson *et al.* (2002), the ISO Beam mill's daily total product with respect to the variable input stems per day is limited to three feasible output levels. This is shown in Figure 2.15.

While it would be possible to draw a line between the points shown in Figure 2.15, this line would imply that technically efficient production levels between the two points are possible by varying the input level. The production levels are either 464, 696, or 928 beams per day not 580 or 812 ISO beams per day.

There are three possible reasons for a non-continuous production process. First, the binary nature of the technology of one or more machines used in the production process means that it is either on or off. If it is on, the technically efficient point is to run the machine at its physical capacity due to the high fixed resource costs. Zudak (1970) uses the example of an open hearth furnace in steel fabrication to illustrate this idea:

> They cannot in general be started and stopped at random because the cost is too high. Failure to cool or heat the furnaces very gradually (over a period of several weeks) causes severe damage to their linings. Even after being correctly cooled, several days are required to rebrick certain parts of the system damaged by cooling. During a production period, furnaces are run at capacity or not used at all.
>
> (Zudak 1970: 257–258)

A kiln is similar, in some respects, to an open hearth furnace. Drying schedules prescribe the time required to safely dry commercial wood (e.g. Boone *et al.* 1988). These schedules should be followed regardless of whether the kiln is full or not. Thus, for example, splitting a load in half does not reduce the drying time by half, but doubles the drying time for the whole load. While splitting the load in half will reduce the amount of energy used per half load, the total energy required to dry both halves is greater than if the load is dried as a single unit. Consequently, it is more efficient to run kilns at full capacity. This implies that total product is a point as shown in Figure 2.15.

Second, the production process requires a number of steps to produce the finished product. Figure 2.2 illustrates the steps of the production process for

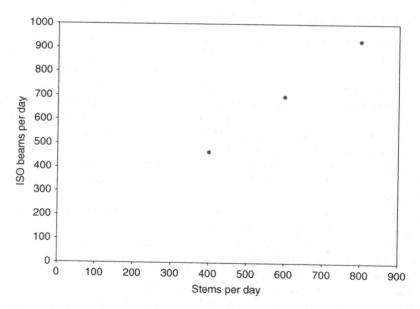

Figure 2.15 Inside-out beam total product.

Based on data from Patterson *et al.* (2002).

manufacturing an ISO beam. Each step must be completed before the next step can be started. Some machines (e.g. cut-off saw, molder, quarter, and trim saws) will take a minimum amount of time to complete a step. For example, a quarter saw has a maximum feed rate defined by its physical characteristics (e.g. motor power and speed, band saw, or circular saw, etc.). These physical characteristics define the minimum amount of time required to, for example, quarter a cant of a given species and length. Thus total product is always a point defining the minimum amount of time required to complete a step.[16]

Third, some inputs (e.g. labor) are not homogenous in the production process. Each step may require a different set of skills and/or a minimum number of employees per step. In this case, increasing the input labor by a single homogenous employee will not increase output. This outcome is similar to that discussed by Zudak (1970). According to Patterson *et al.* (2002), new capital equipment (i.e. boiler, kilns, and pre-dryers) is required to increase output. Again, total product is defined as a point. In addition, average and marginal product have no meaning. While it is true that adding a single homogenous employee will decrease average product – from equation (2.9), the numerator is constant while the denominator increases – the decrease is due to an increase in technical inefficiency not the Law of Diminishing Returns. Thus, due to the non-continuous nature of the production process, marginal produce may also not reflect the Law of Diminishing Returns accurately.[17]

As a result of the above three reasons, the total product for the ISO Beams case study is defined as three input–output combinations; i.e. 400, 600, and 800 stems per day producing 464, 696, and 928 beams per day, respectively (see Figure 2.15). Given the series of sequential steps required in the production process, the fixed nature of the capital equipment, the kilns, and non-homogenous labor, total product is increasing as the number of stems per day increases until it is constrained physically by the capacities of the kiln and pre-dryer. Thus, technical efficiency would logically be described as a point and marginal and average product have no meaning. The only way to increase production to greater than 464 beams per day, for example, is to purchase more capital equipment in terms of boilers, kilns, and pre-dryers. As a manager, you will need to recognize a non-continuous production process and what is causing the process to be non-continuous. It is the causes of the non-continuity that the manager should focus their attention on initially.

How to use economic information – *production systems* – to make better business decisions?

The following non-forestry example taken from the news illustrates the important relationship between measurement and management. In December 1998, the National Aeronautics and Space Administration (NASA) launched the Mars Climate Orbiter. The purpose of the orbiter was to collect information about the atmospheric temperature on Mars, dust, water vapor, clouds, and the amount of carbon dioxide (CO_2) that is added and removed from the poles each Martian year. In addition, the orbiter was to assist in data transmission to and from the

Mars Polar Lander. On 23 September 1999, the orbiter burned up in the Mars atmosphere. A review of the incident showed that one team of scientists used English units (e.g. inches, feet, and pounds) while another team used metric units (e.g. centimeters, meters, and kilograms). Incorrect information was given to the orbiter during the critical maneuvers required to place it in the proper Mars orbit. As a result, NASA lost the $125 million Mars Climate Orbiter (http://www.cnn.com/TECH/space/9909/30/mars.metric/ accessed on 11 September 2007; http://mars.jpl.nasa.gov/msp98/news/mco990930.html accessed on 11 September 2007).

Lord William Thomas Kelvin (1825–1907) recognized the relationship between measurement and management: "If you cannot measure it, you cannot improve it." Maital (1994: 8) built on this idea by stating: "What you cannot measure, you cannot effectively understand, control, or alter." A slight modification of this quote is often given as: "If you cannot measure it, you cannot manage it." If your measurements are not correct (e.g. incorrect units), your management decision based on this information may turn out to be disastrous. So, don't crash the Mars Orbiter.

To answer the question asked in the above heading, I will first need to identify the economic information obtained from the production system. Simply put, the production system contains the physical input–output relationships necessary to maximize profits. These are determined by answering the following three questions: (1) What are the inputs? (2) What are the outputs? (3) How do you describe the production process? The relevant inputs are those that you the manager have direct control over. Answering "What are the inputs?" must help you build the Pillar of Cost given in the Architectural Plan for Profit (Figure 2.1). The concept of output has at least two dimensions: quantity and quality. Both dimensions have a direct correlation with how the consumer values the output. Thus, answering the question "What are the outputs?" must help you build the Pillar of Value given in the Architectural Plan for Profit (Figure 2.1).[18] As value and price are important to the consumer, answering this question will help you as the manager in your decisions in pricing your outputs given your markets. This will be discussed in Chapters 3, 5, and 6. As will be discussed in the next chapter, answering the question "How do you describe the production process?" will have a direct bearing on estimating cost.

In summary, the foundation of the Architectural Plan for Profit must be built so that it will help to provide answers to future questions dealing with cost, value, price, and profit.

3 Costs

Why are costs important?

Simply put, if you do not manage costs you will lose control of profits. The importance of cost management is illustrated by the authors of the Mobile Micromill case study:

> The ability to develop value-added products by using small-diameter ponderosa pine is in part contingent on the ability of finished products to compete in the market place. Transportation and processing costs need to be minimized to make processing small-diameter logs more competitive.
>
> (Becker *et al.* 2004: 3)

Three new case studies will be introduced in this chapter and will be described in more detail in the next section. However, as with the Mobile Micromill, the author of the Maple Syrup Operation case study also identified the importance of cost management:

> The key to a successful maple syrup operation is controlling production to maintain an acceptable profit margin. It is important that sugarbush operators keep accurate records so that areas of high cost can be identified and steps taken to reduce them.
>
> (Huyler 2000: 5)

Thus, Cost is the first pillar of the Architectural Plan for Profit that I will examine (see Figure 3.1).

Figure 3.1 shows that the Pillar of Cost is built on the foundation of the Production System. As described in Chapter 2, answering the question "What are the inputs?" must help you build this pillar. The simple answer to this question is that the only relevant inputs are those that you as the manager control directly. Given this simple answer, the rest of the chapter will build this pillar. First however, I will discuss how costs are part of the two economic models given in Chapters 1 and 2 and describe the case studies used in this chapter.

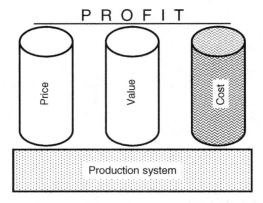

Figure 3.1 The Architectural Plan for Profit.

Profit model

The profit model is given by equation (3.1).

$$Max \; \Pi = TR - TC$$
$$= P \cdot Q(x_1, x_2, ..., x_j) - TVC - TFC \tag{3.1}$$
$$= P \cdot Q(\cdot) - \sum_j w_j \cdot x_j - TFC$$

where

Max Π denotes maximization of profit (Π);

TR denotes total revenue, $TR = P \cdot Q(x_1, x_2, ..., x_j)$;

P denotes the market price of the good or service,

$Q(x_1, x_2, ..., x_j) = Q(\cdot)$ denotes the system of producing or providing a good or service, *Q*, using inputs that the manager controls directly, $x_1, x_2, ..., x_j$,

TC denotes total cost, $TC = TVC + TFC$;

TVC denotes total variable costs, $TVC = \sum_j w_j \cdot x_j$,

$\sum_j$ denotes the summation operator,

w_j denotes the *j*th wage or price paid for *j*th input x_j,

x_j denotes the *j*th input; for example, labor,

TFC denotes total fixed costs.

Given equation (3.1), profit is equal to total revenue minus total cost. Maximizing profit requires decisions that make total revenue as large as possible while simultaneously making total costs as small as possible for any given input(s)–output(s) combination. Therefore, if profits are to increase towards their maximum, then costs must be managed and minimized.

Least cost/cost effective models

In the least cost model, equation (3.2), the entrepreneur's objective is to find ways to minimize the total variable costs of producing or providing a given level of output.

$$Min\ TVC = \sum_j w_j \cdot x_j$$

s.t.   (3.2)

$$Q(x_1, x_2, ..., x_j) = Q^0$$

where

Min TVC denotes the objective of minimizing total variable costs, $TVC = \sum_j w_j \cdot x_j$,

$\sum_j$ denotes the summation operator,

w_j denotes the *j*th wage or price paid for *j*th input x_j,

x_j denotes the *j*th input,

s.t. denotes "subject to" and what follows is a constraint placed on the objective;

$Q(x_1, x_2, ..., x_j) = Q^0$ denotes the system of producing or providing a good or service, *Q*, using inputs the manager controls directly, $x_1, x_2, ..., x_j$, at a given quantity or quality, Q^0.

In the cost effective model, equation (3.3), the entrepreneur's objective is to find ways to produce or provide the most output for a given budget.

$$Max\ Q(x_1, x_2, ..., x_j)$$

s.t.   (3.3)

$$\sum_j w_j \cdot x_j = C^0$$

where

Max Q($x_1, x_2, ..., x_j$) denotes the system of providing or producing the maximizing quantity or quality of a good or service, *Q*, using inputs that the manager controls directly, $x_1, x_2, ..., x_j$;

s.t. denotes subject to;

$\sum_j w_j \cdot x_j = C^0$ denotes that the TVCs of using inputs $x_1, x_2, ..., x_j$ to produce or provide the good or service cannot exceed a given budget, C^0;

$\sum_j$ denotes the summation operator,

w_j denotes the *j*th wage or price paid for *j*th input x_j,

x_j denotes the *j*th input.

Both equation (3.1) and equations (3.2) and (3.3) contain the same concept of costs. As was stated in the previous section and Chapter 1, optimizing equation (3.1) requires managing and minimizing costs. The optimal solution for equations (3.2) and (3.3) also requires managing and minimizing costs. Thus, the optimal solution to equation (3.1) is also an optimal solution for equations (3.2) and (3.3). The optimal input–output combinations resulting from solving equations (3.2) and (3.3) for different levels of output or budgets are defined as production cost efficient. Using the Rules of Inference (i.e. Hypothetical Syllogism, see Copi and Cohen 1998), the optimal input–output combination resulting from equation (3.1) must also be defined as production cost efficient.[1]

Case studies

The case studies can be found on the Routledge website for this book. After reading the brief summaries you should read and familiarize yourself with each one. Two of the five case studies used in this chapter, ISO Beams (Patterson *et al.* 2002 and Patterson and Xie 1998) and the Mobile Micromill (Becker *et al.* 2004), were used in Chapter 2. Their summaries will not be repeated here.

Maple Syrup Operation (Huyler 2000)

The primary focus of this case study will be on the sap collection component of a maple syrup operation as described by Huyler (2000). The secondary focus will be on the processing of sap into syrup and other maple sugar products as described by Heiligmann *et al.* (2006) and CFBMC (2000).

How do I describe the production process?

Maple syrup is made primarily from the sap of sugar maples (*Acer saccharum* Marsh.). Other sources of sap are black maple (*Acer nigrum* Michx. f.), red maple (*Acer rubrum* L.) and silver maple (*Acer saccharinum* L.) (Graham *et al.* 2006). The stand of maple trees where the sap is collected is called a sugarbush. Sap is collected in a short four- to six-week period in late winter/early spring by tapping the trees (Chapeskie and Koelling 2006). The taps are connected to a storage tank via vacuum tubing.[2] One tap produces approximately six to ten gallons of sap per season. The number of gallons of sap required to produce one gallon of syrup depends on the sap's percent sugar content as measured by a Brix scale and can be determined by dividing 86 by the percent sap's Brix sugar content (Stowe *et al.* 2006).[3] The normal sugar content of sap is 1.5 percent to 3.5 percent measured using the Brix scale (Sendak and Bennink 1985). In the sugarhouse, water is evaporated from the sap using either a "batch type" or "continuous-flow" evaporator concentrating the sugar into syrup (Graham *et al.* 2006; Stowe *et al.* 2006).

This leads to two observations. First, the higher the sugar content, the fewer gallons of sap are required to produce one gallon of syrup. For example, 57 gallons of sap with a sugar content of 1.5 percent would be required to produce one gallon of syrup while only 25 gallons of sap with a sugar content of 3.5 percent would be required to produce one gallon of syrup. Thus, the higher the sugar content of the sap, the less water has to be evaporated to concentrate the sugar into syrup. Second, for a given quantity of sap, the higher the sugar content, the more syrup can be produced using approximately the same inputs (Sendak and Bennink 1985).

What are the inputs?

Besides the sugarbush, Table 3.1 provides a list of the inputs for collecting sap.

A similar list can be found in CFBMC (2000); in addition, the authors of this study provide a list of inputs for further processing the sap. According to Huyler

Table 3.1 Standard equipment list for a vacuum tubing sap collection system

Item	Quantity
Nylon sap spout	1 per tap
5/16-inch sap tubing	15 feet per tap
1/2-inch mainline tubing	2 feet per tap
3/4-inch mainline tubing	1.2 feet per tap
1-inch mainline tubing	0.7 feet per tap
5/16-inch connector	0.05 per tap
1/2-inch connector	0.02 per tap
3/4-inch connector	0.012 per tap
1-inch connector	0.007 per tap
5/16-inch end cap	0.04 per tap
5/16-inch tee	1 per tap
4-way wye	0.02 per tap
1- × 3/4-inch reducer	0.002 per tap
3/4 × 1/2-inch reducer	0.004 per tap
Quick clamp	0.082 per tap
Aluminum fence wire	0.7 foot per tap
Quick clamp pliers	1 per operation
Wire ties	1 per operation
Wire tier	1 per operation
Fence wire stretcher	1 per operation
Spout puller	1 per operation
Sap vacuum pump	1 per operation
50-gallon vacuum storage tank	1 per operation
Snowshoes (pair)	1 per operation
Power tree tapper with battery pack	1 per operation
Tapping bit, bit file, and spark plug	1 per operation
Hand tool set	1 per operation

(Used with permission from the USDA Forest Service)

(2000) and CFBMC (2000) there is a fixed proportional relationship between the inputs and the number of taps used to collect the sap (see Table 3.1).

CFBMC (2000) estimates the labor input for collecting sap, maintaining the lines etc. at 4.08 minutes per tap and Huyler (2000) estimates the labor input between 2.92 and 6.93 minutes per tap with an average of 4.74 minutes per tap. In addition to collecting sap, there are the inputs of converting sap into syrup and retail sales. These include, but are not limited to, an evaporator, energy, and a building. CFBMC (2000) estimate the labor input from collecting the sap, converting the sap into final products, and selling the products to range from 8.78 to 27.12 minutes per tap depending on the number of taps and type of evaporator.

What are the outputs?

The output can be defined as maple sap, syrup, sugar, cream, and candies. For the purposes of the case study, the output will be defined as either the number of taps,

Table 3.2 Charter boat capital and operating inputs

	Capital inputs
Boat	Tackle and other equipment
	Operating inputs
Fuel/oil	Boat maintenance and repair
Dockage	Office and communications
Labor(hired)	Boat storage fees
Equipment repair	Boat repair not covered by insurance
Advertising	License and fees
Miscellaneous	Drug testing/professional dues
Insurance	Boat launch fees

(Based on information given by Lichtkoppler and Kuehn (2003))

which is consistent with Huyler (2000) and CFBMC (2000), or gallons of syrup. Huyler (2000) and CFBMC (2000) use the following conversion factor: 1 tap produces approximately 10 gallons of sap with a sugar content of 2.15 percent and 0.25 gallons or 1 liter of syrup. As a comparison, NEAS (2008) estimated, for the New England States (Connecticut, Maine, Massachusetts, New Hampshire, and Vermont), an average production ranged of 0.187 to 0.211 gallons of syrup per tap for the period 2006 to 2008.

Great Lakes Charter Boat Fishing (Lichtkoppler and Kuehn 2003)

How do I describe the production process?

Charter boat fishing is a commercial recreation enterprise. One or more anglers hire or charter a boat, captain/guide, and may be a crew for a half or full day fishing for walleye, lake trout and salmon, steelhead, smallmouth bass, or yellow perch.

What are the inputs?

Table 3.2 lists the inputs as either operating or capital.[4]

What are the outputs?

The outputs are either a half or full day of a recreational fishing experience.

Knockdown Panel Furniture Fittings (Burdurlu et al. 2006)

Knockdown Panel Furniture (KPF) or "ready-to-assemble" furniture requires no glue or other permanent fastening systems to assemble. KPF is easy to transport, store, and can be assembled with relative ease by the consumer. Furniture manufacturers have a choice among a number of different specialized fittings they can use:

Each fitting requires different machinery and worker-supported processes for the connection of the fitting to the panel and the connection of the panels with fittings to each other. The differences affect processing time, and consequently, the production costs. Processing time could be negligible for a single fitting: however, it could constitute significant annual or long-term cost elements for mass or lot production of large numbers of furniture. Manufacturers should make cost analyses to determine which fitting should be used to lower costs of assembling the panels to each other with the fittings available in the market.

(Burdurlu *et al.* 2006: 47)

Figure 3.2 illustrates the six different types of fittings used.

What are the outputs?

The outputs are not KPF. The focus of this study is to determine the least cost fitting to be used in producing a piece of KPF. According to Burdurlu *et al.* (2006: 48): "The least cost analysis was made using the data obtained to find the least cost fitting." Thus, the outputs for this case study are the installed fittings.

As this is a least cost analysis, equation (3.2) can be used to express the economic model used by Burdurlu *et al.* (2006). The Q^0 of equation (3.2) would be equal to 1,000 installed fittings for each of the six different types of fittings examined. The fittings are illustrated in Figure 3.2.

What are the inputs?

Burdurlu *et al.* (2006) identified the relevant inputs as: labor, energy, and materials. In terms of equation (3.2), labor, energy, and materials are the inputs for determining total variable cost.

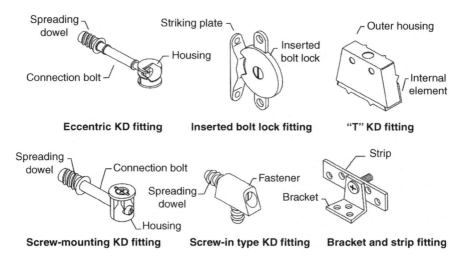

Figure 3.2 Knockdown Panel Furniture specialized fittings.

(Used with permission of the *Forest Products Journal* and Burdurlu *et al.* (2006))

How do I describe the production process?

The production process of interest describes the steps used to install each of the six fittings. These steps are described in Figure 3.3

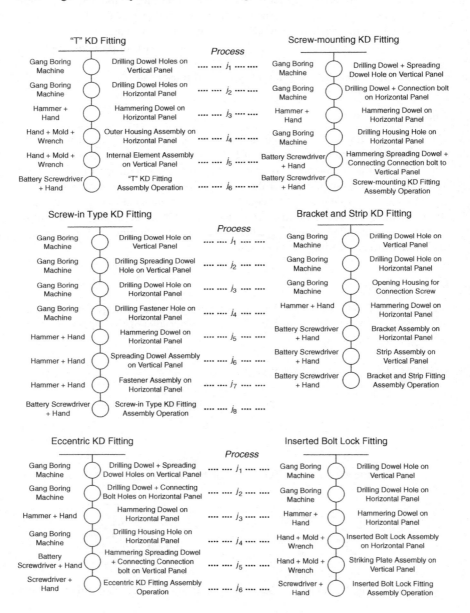

Figure 3.3 Steps to install Knockdown Panel Furniture fittings.

(Used with permission of the *Forest Products Journal* and Burdurlu *et al.* (2006))

Review: Economics versus accounting

Chapter 1 provides an example of the distinction between economic costs and accounting costs. The distinction between an accounting and economic cost is important and I will only summarize the main points of that discussion (I recommend you re-read that section in Chapter 1). Table 1.1 lists the annual costs and revenues associated with owning and operating a small hardwood sawmill. The costs are specified as explicit and implicit costs. Explicit accounting costs are those cash outlays that can be found on the debit/credit sheet. The implicit costs are opportunity costs of the time and other resources a sawmill's owner makes available for production with no direct cash outlay (Field and Field 2002; Henderson and Quandt 1980). Implicit costs are not included in an accounting debit/credit sheet. Nonetheless, they are a cost to you. The economic or opportunity cost – the value of the next best alternative forgone – is defined as the sum of explicit and implicit costs. Jensen (1982) reiterates the importance of opportunity costs with respect to management decisions. He defines opportunity costs with respect to production as the sum of explicit cash outlays plus the implicit costs of inputs that might be used in other ways to produce more or different outputs. These define the relevant production costs.

The disciplines of accounting and economics both use some of the same terminology with respect to costs. As shown by Table 1.1, explicit costs by themselves do not represent all that is forgone or given up to continue an action. Thus, the relevant cost information needed to make better decisions with respect to the Architectural Plan for Profit must include both explicit and implicit costs.

Economic descriptors of cost

There are three main descriptors of cost: total cost, average cost, and marginal cost. I will address each in turn.

Total cost (TC)

Examining equation (3.1) reveals that Total Cost is the sum of Total Variable Cost (TVC) plus Total Fixed Cost (TFC) or $TC = TVC + TFC$.

Total fixed costs (TFC)

Total fixed costs are costs associated with inputs that are constant, for various reasons, during the production process. In other words, these costs are constant no matter how much is produced or provided in the short-run.[5] TFC can be classified as the "DIRTI-5":[6]

> **Depreciation** – Capital equipment wears out as the result of producing or providing an output, age, and obsolescence (Pearce 1994). If allowed by law, depreciation is a "noncash expense" that reduces taxable net revenue. A purpose

of depreciation is to allow the owners of capital the opportunity to save funds to replace the capital.

Interest – Payments made to banks or other lenders for the use of money to purchase capital equipment.

Rent/Repairs – Payments made to the owners of land and capital equipment that are used by, but not owned by, the business, and annual maintenance and repair costs for the upkeep of capital equipment and buildings.

Taxes – Tax payments that are usually incurred on capital equipment and buildings.

Insurance – Costs of protecting the business against fire, weather, theft, etc.

For the Maple Syrup Operation case study, "the annual fixed cost for sap equipment and other fixed costs such as taxes and insurance were included in total fixed costs" (Huyler 2000: 1). The annual equipment fixed costs included a straight-line deprecation schedule and 8.5 percent interest charge on capital equipment and insurance payments. These costs are reported on an annualized basis.[7] Tables 3.3a and 3.3b give the total fixed costs.

Table 3.3a shows the total fixed costs for different-sized operations. Table 3.3b shows the total fixed costs for a 12,000-tap operation. According to the definition of fixed costs, they should be constant. However, in Table 3.3a, they appear to be increasing. Due to the fixed proportional relationships described in Table 3.1 as the size of the operation increases, the fixed costs will also increase. Table 3.3b shows

Table 3.3a Annual sap collection costs by size of operation – Maple Syrup case study (1998 USD[§])

Number of taps	Gallons of syrup	Total variable costs		Total fixed costs		Total cost ($)
		Labor ($)	Material ($)	Equipment (annual) ($)	Tax ($)	
500	127.5	260	260	1,750	50	2,320
1,000	255	520	370	2,170	100	3,160
2,000	510	1,040	600	3,020	200	4,860
3,000	765	1,560	840	3,870	300	6,570
4,000	1,020	2,080	1,040	4,720	400	8,240
5,000	1,275	2,600	1,300	5,350	500	9,750
6,000	1,530	3,120	1,500	6,420	600	11,640
7,000	1,785	3,640	1,750	7,210	700	13,300
8,000	2,040	4,160	1,920	8,080	800	14,960
9,000	2,295	4,680	2,160	8,910	900	16,650
10,000	2,550	5,200	2,400	9,800	1,000	18,400
11,000	2,805	5,720	3,080	12,650	1,100	22,550
12,000	3,060	6,240	3,360	13,680	1,200	24,480

[§]1998 USD denotes 1998 US dollars.

(Used with permission from the USDA Forest Service)

Table 3.3b Annual sap collection costs for a 12,000 tap operation – Maple Syrup case study (1998 USD[§])

Number of taps	Gallons of syrup	Total variable costs		Total fixed costs		Total cost ($)
		Labor ($)	Material ($)	Equipment (annual) ($)	Tax ($)	
500	127.5	260	260	13,680	1,200	15,400
1,000	255	520	370	13,680	1,200	15,770
2,000	510	1,040	600	13,680	1,200	16,520
3,000	765	1,560	840	13,680	1,200	17,280
4,000	1,020	2,080	1,040	13,680	1,200	18,000
5,000	1,275	2,600	1,300	13,680	1,200	18,780
6,000	1,530	3,120	1,500	13,680	1,200	19,500
7,000	1,785	3,640	1,750	13,680	1,200	20,270
8,000	2,040	4,160	1,920	13,680	1,200	20,960
9,000	2,295	4,680	2,160	13,680	1,200	21,720
10,000	2,550	5,200	2,400	13,680	1,200	22,480
11,000	2,805	5,720	3,080	13,680	1,200	23,680
12,000	3,060	6,240	3,360	13,680	1,200	24,480

[§]Cost data based on information from Huyler (2000); 1998 USD denotes 1998 US dollars.

that if the sugarbush has a 12,000-tap capacity (or producing 3,060 gallons of syrup) and you have the collection equipment necessary for this production level, the fixed costs are constant whether you produce 0, 6,000, or 12,000 taps. Thus, for a given sugarbush size the fixed costs of collecting the sap are constant. The CFMBC (2000) report shows the same pattern for the total fixed costs associated with the processing of the sap into syrup and retailing or wholesaling of the final products.

In the Mobile Micromill case study, the fixed costs were made up of financing the Mobile Micromill and the supporting equipment and the fixed costs associated with siting the mill (Becker *et al.* 2004).[8]

Table 3.4 and Figure 3.4 show that the fixed costs for the first year included the down payments and the financing costs on the Mobile Micromill and the supporting equipment and the siting costs. For the next four years, the fixed costs included the finance costs of the Mobile Micromill and the supporting equipment and the siting costs. In the final two years of the analysis, the fixed costs included the financing costs of the Mobile Micromill and the siting costs.[9] Again these costs should be constant. While it is true that there are three levels of fixed costs during an seven-year period, these costs are invariant with respect to how much output is produced or sold. The costs change due to factors not associated with production; namely, contracts signed with the banks for financing capital equipment. In the case of the Mobile Micromill, the loan period was for seven years while the support equipment was for five years.

Table 3.5 gives the annual operating cost for the Great Lakes Charter Boat Fishing (Lichtkoppler and Kuehn 2003) case study.

The cost data provided by Lichtkoppler and Kuehn (2003) do not distinguish explicitly between operating expenses that are fixed versus variable. In the ISO

Table 3.4 Fixed costs for the Mobile Micromill

Item	Year						
	1	*2*	*3*	*4*	*5*	*6*	*7*
Down payments							
Mobile MicroMill	129,956.00						
Tractor-loader	6,450.00						
Forklift	12,900.00						
Maintance vehicle	4,838.00						
Financing							
Mobile MicroMill	70,609.00	70,609.00	70,609.00	70,609.00	70,609.00	70,609.00	70,609.00
Tractor-loader	4,267.00	4,267.00	4,267.00	4,267.00	4,267.00		
Forklift	8,534.00	8,534.00	8,534.00	8,534.00	8,534.00		
Maintance vehicle	3,200.00	3,200.00	3,200.00	3,200.00	3,200.00		
Site lease and preparation	3,000.00	3,000.00	3,000.00	3,000.00	3,000.00	3,000.00	3,000.00
	243,754.00	89,610.00	89,610.00	89,610.00	89,610.00	73,609.00	73,609.00

Annual production level of 3600 thousand board feet (MBF) of product.

(Used with permission from the USDA Forest Service)

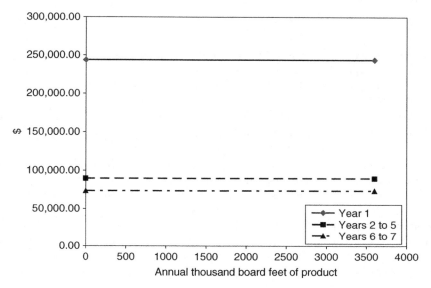

Figure 3.4 Fixed costs for the Mobile Micromill.

(Used with permission from the USDA Forest Service)

Table 3.5 Average annual operating costs for New York
boat-owing captains (2002 USD[§])

Item	Cost ($)
Fuel/oil	1,895
Advertising	1,200
Labor (hired)	1,168
Equipment repair	1,115
Dockage	1,096
Miscellaneous	901
Insurance	831
Boat repair and maintenance	717
Office and communications	531
Boat storage fees	429
Boat repair not covered by insurance	276
Drug testing/Professional dues	92
License fees	91
Boat launch fees	33

[§]2002 USD denotes 2002 US dollars.

(Based on information given by Lichtkoppler and Kuehn (2003))

Beams case study, the purchase price of the capital equipment are given, but costs are not distinguished as fixed costs or variable costs per se (Patterson *et al.* 2002; Patterson and Xie 1998). The primary fixed costs are interest payments on the capital equipment. It is left to the reader to determine why the Knockdown Panel Furniture Fittings (KPFF) case study does not include fixed costs.

Total variable costs (TVC)

Total variable costs are associated with inputs that the manager can manipulate during the production process. In other words, variable costs change with the level of output produced or provided. As with fixed costs, there is a direct relationship between the answer to the "What are the inputs?" question from Chapter 2 and TVC. This relationship is illustrated by equations (3.1), (3.2), and (3.3). Specifically, the description of the production system, $Q(x_1, x_2,..., x_j)$, is based on inputs, $x_1, x_2,..., x_j$, the manager controls directly: those inputs that are relevant to systematically following the Architectural Plan for Profit. These same inputs are used in defining TVC or wage (w) times input (x), $\sum_j w_j \cdot x_j$. Given this, it is possible to show this direct relationship between the production system and TVC graphically. Figure 3.5(a) depicts a generic production system using only one input to produce one output; its shape is dictated by the Law of Diminishing Returns as defined in Chapter 2. Figure 3.5(b) is the same production system with the axes flipped; namely, the Output, $Q(x)$, is now on the horizontal axis and the Input, x, is now on the vertical axis. Finally, as defined $TVC = w \cdot x$, multiplying the vertical axis Input, x, by its wage, w, converts the production system graph into a TVC graph as shown in Figure 3.5(c).

The Knockdown Panel Furniture Fitting case study focused solely on the variable production costs. The three inputs under direct control of the manager were labor, energy, and materials. The unit labor costs were $4.20 per hour. The eccentric fitting used the least hours of labor while the screw-in fitting used the most. The unit energy costs were $0.12 per kilo-watt hour. Two machines – a gang boring machine and chargeable drill/drivers – were used to install the fittings. In terms of energy consumption, inserted bolt lock and T fittings used the least

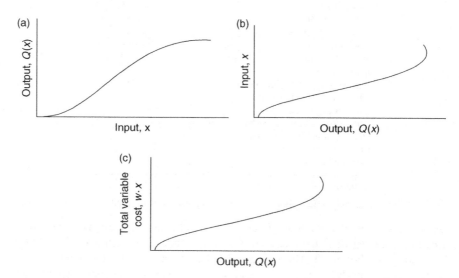

Figure 3.5 The relationship between the production system and total variable cost.

electricity and the screw-in fitting used the most. The material costs varied with fitting type. The least expensive was the screw-in at $0.022 per fitting and the most expensive was the screw-mounting at $0.299 per fitting. Total variable costs ranged from a high of $0.387 per fitting for the screw-mounting to a low of $0.146 per fitting for the screw-in type.[10]

For the Maple Syrup Operation case study, the "labor, materials, tap rental, and miscellaneous expenses were combined for each operation to determine the variable annual operating cost per tap" (Huyler 2000: 2). Tables 3.3a and 3.3b give the total variable costs of sap collection. In the sap collection component, labor performs four tasks: (1) preparing, (2) tapping, (3) tub checking and repair, and (4) cleanup and storage. According to Huyler (2000), the average labor requirement to perform these four tasks is 4.74 minutes per tap. Assuming a wage of $6.40 per hour as the variable labor costs (Table 3.3a), the variable labor costs for a 12,000-tap operation given in Table 3.3b are the same as those in Table 3.3a. This is due to the fact that the relationship between the time to complete the four tasks does not vary with respect to the number of taps. Thus, if the sugarbush has a 12,000-tap capacity (or producing 3,060 gallons of syrup) and only produces 2,000 taps (or 510 gallons of syrup), the variable labor costs are $1,040. The materials costs (e.g. paint, wire, coding tags, gas, oil, etc.) also vary with the level of output, but not in same fixed manner as with labor. The variable costs are shown graphically in Figure 3.6.

The variable costs of the Mobile Micromill were operation and maintenance costs of the mill (e.g. sharpening saw knives and blades, maintaining the sawdust blower, in-feed and out-feed decks, and diesel engine) and support equipment, labor (e.g. one operator, one log sorter/loader, two laborers on the green chain

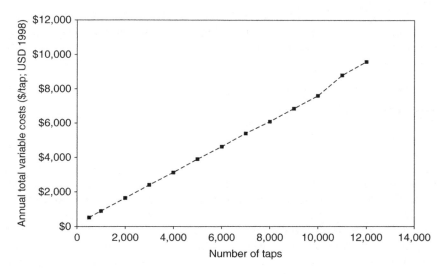

Figure 3.6 Annual total variable sap collection costs by size of operation – Maple Syrup Operation case study.

handling the outputs), fuel and oil for the Mobile Micromill and support equipment, and delivering and sorting raw logs. Variable costs of operation and maintenance for the Micromill and the support equipment depend directly on the number of productive machine hours (Becker *et al.* 2004). Productive machine hours are defined as the time actually spent sawing logs and do not include any downtime for maintenance, labor breaks, cleaning up, resetting saws, etc.

Average cost (AC)

Average cost is defined generically as some measure of production cost divided by the amount of output produced, or cost per unit of output. Let's explore briefly the economic information in the general concept of AC before we analyze the specific types of ACs. First, AC is a ratio or fraction. The numerator denotes some measure of production costs and the denominator is the amount of output produced. If a manager's goal is to maximize profits then the cost per unit should be minimized or production cost efficient. Thus, the concept of AC gives the manager information as to whether these costs are increasing or decreasing and to search for ways to make the costs as small as possible per unit output. Second, the units on AC are defined as $/Q. From Chapter 1 and examining equation (3.1), the units on price, P, must also be $/Q or value per unit output. Thus, if P is greater than AC, sales revenue will cover some measure of production costs.

The three main descriptors of AC are: (1) average fixed cost (AFC), (2) average variable cost (AVC), and (3) average total costs (ATC). I will discuss each in turn using the Maple Syrup Operation case study to illustrate the descriptors of AC. Tables 3.6a and 3.6b give the AFC, AVC, and ATC associated with the Maple Syrup Operation case study.

Table 3.6a gives the average annual sap collection costs by size of operation and Table 3.6b gives the average annual sap collection costs for a 12,000-tap operation. These same data are illustrated in Figures 3.7a and 3.7b, respectively.

Average fixed cost (AFC)

Average fixed cost is defined as:

$$AFC = \frac{TFC}{Q} \qquad (3.4)$$

where the notation used is defined in equation (3.1). As can be seen by equation (3.4), the measures of production costs used in calculating AFC includes only fixed costs. AFC measures the fixed production costs per unit of output.

Table 3.6a shows the AFC by size of operation. As the size or scale of the operations increase, the AFC decreases until a 10,000-tap operation. After this, AFC costs start to increase. This is due to additional capital equipment requirements. According to Huyler (2000), capital equipment requirements for operations between 500 and 10,000 taps are very similar and additional capital

Table 3.6a Total and average annual sap collection costs by size of operation –
Maple Syrup case study (1998 USD[§])

Number of taps	Gallons of syrup	TVC ($)	AVC ($/tap)	TFC ($)	AFC ($/tap)	TC ($)	ATC ($/tap)
500	127.5	520	1.04	1,800	3.60	2,320	4.64
1,000	255	890	0.89	2,270	2.27	3,160	3.16
2,000	510	1,640	0.82	3,220	1.61	4,860	2.43
3,000	765	2,400	0.80	4,170	1.39	6,570	2.19
4,000	1,020	3,120	0.78	5,120	1.28	8,240	2.06
5,000	1,275	3,900	0.78	5,850	1.17	9,750	1.95
6,000	1,530	4,620	0.77	7,020	1.17	11,640	1.94
7,000	1,785	5,390	0.77	7,910	1.13	13,300	1.90
8,000	2,040	6,080	0.76	8,880	1.11	14,960	1.87
9,000	2,295	6,840	0.76	9,810	1.09	16,650	1.85
10,000	2,550	7,600	0.76	10,800	1.08	18,400	1.84
11,000	2,805	8,800	0.80	13,750	1.25	22,550	2.05
12,000	3,060	9,600	0.80	14,880	1.24	24,480	2.04

[§]1998 USD denotes 1998 US dollars.

(Used with permission from the USDA Forest Service)

Table 3.6b Average annual sap collection costs for a 12,000-tap operation –
Maple Syrup case study (1998 USD[§])

Number of taps	Gallons of syrup	AVC		AFC		ATC ($/tap)
		Labor ($/tap)	Material ($/tap)	Equipment (annual) ($/tap)	Tax ($/tap)	
500	127.5	0.52	0.52	27.36	2.40	30.80
1,000	255	0.52	0.37	13.68	1.20	15.77
2,000	510	0.52	0.30	6.84	0.60	8.26
3,000	765	0.52	0.28	4.56	0.40	5.76
4,000	1,020	0.52	0.26	3.42	0.30	4.50
5,000	1,275	0.52	0.26	2.74	0.24	3.76
6,000	1,530	0.52	0.25	2.28	0.20	3.25
7,000	1,785	0.52	0.25	1.95	0.17	2.90
8,000	2,040	0.52	0.24	1.71	0.15	2.62
9,000	2,295	0.52	0.24	1.52	0.13	2.41
10,000	2,550	0.52	0.24	1.37	0.12	2.25
11,000	2,805	0.52	0.28	1.24	0.11	2.15
12,000	3,060	0.52	0.28	1.14	0.10	2.04

[§]Cost data based on information from Huyler (2000); 1998 USD denotes 1998 US dollars.

equipment (e.g. power tree tapper) is required for operations greater than 10,000 taps. In other words, the physical limit of a power tree tapper is to drill approximately 10,000 tap holes per season. While *FC* for the power tree tapper would be the same regardless if the size of the operation ranged between 500 and 10,000 taps, the AFC would decrease as the size of the operation increased.

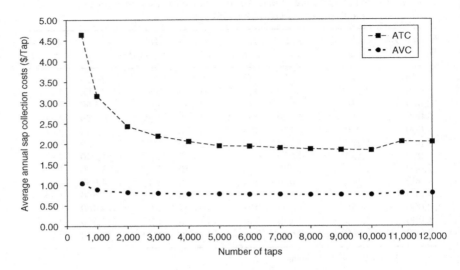

Figure 3.7a Average annual sap collection costs by size of operation – Maple Syrup Operation case study (1998 USD[§]).

[§]1998 USD denotes 1998 US dollars.

(Used with permission from the USDA Forest Service)

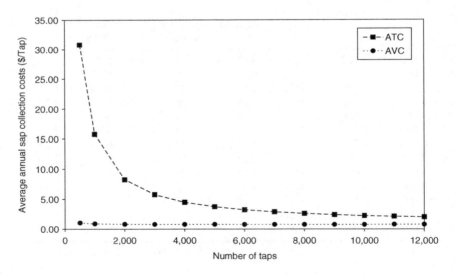

Figure 3.7b Average annual sap collection costs for a 12,000-tap operation – Maple Syrup Operation case study (1998 USD[§]).

[§]Cost data based on information from Huyler (2000); 1998 USD denotes 1998 US dollars.

Table 3.6b shows the AFC for a 12,000-tap operation. These data are illustrated in Figure 3.8.

Figure 3.8 illustrates a typical graph of AFC for a firm. As can be seen, AFC decreases quickly, then slowly as output increases. Given equation (3.4), it makes sense that AFC decreases: the numerator is constant (TFC) and the denominator (Q) increases. But what is not obvious is why Figure 3.8 has its shape. The simple answer is that the TFCs are spread out over a larger amount of output and the change in AFC with respect to output is not constant.[11]

Average variable cost (AVC)

Average variable cost is defined as:

$$AVC = \frac{TVC}{Q}$$
$$= \frac{\sum_j w_j \cdot x_j}{Q}$$

(3.5)

where the notation used is defined in equation (3.1). As can be seen by equation (3.5), the measures of production costs used in calculating AVC includes only variable costs. AVC measures the variable production costs per unit of output.[12]

Table 3.6a and Figure 3.7a shows that AVC for all operations decreases until a 10,000-tap operation. They are also relatively constant for 4,000- to 10,000-tap

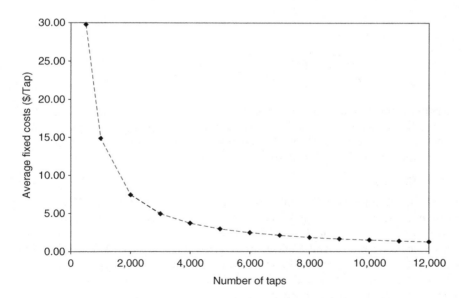

Figure 3.8 Average fixed costs for a 12,000-tap operation (1998 USD[§]).

[§]Cost data based on information from Huyler (2000); 1998 USD denotes 1998 US dollars.

operations. After this, AVCs increase; again, this is due to additional capital equipment requirements. However, the AVCs are still relatively constant. The AVC curve's shape is typical of an industry with relatively fixed capital equipment requirements and a relatively linear relationship between inputs (e.g. labor per tap – 4.74 minutes per tap (Huyler 2000)) and output.

Table 3.6b and Figure 3.7b show the AVC for a 12,000-tap operation. Due to the nature of the data available from Huyler (2000), the AVCs for a 12,000-tap operation are identical to that for all operations.[13] Regardless, the graph of AVC for any individual firm will be U-shaped. There are two reasons for this. First is the Law of Diminishing Returns as defined in Chapter 2 and illustrated in Figures 3.5(a) through 3.5(c). Second is mathematical: AVC is a ratio of TVC to output, equation (3.5). AVC can also be defined as the slope of a ray from the origin that intersects the curve depicted in Figure 3.5(c). Using a straightedge with one end anchored at the origin, you can show that the slope of this ray will decrease then start to increase giving the U-shape. Due to the data provided for the Maple Syrup Operation case study, only the left-hand side of the U is shown due to the physical limitations of the fixed capital requirements (i.e. the power tree tapper).

Average total cost (ATC)

Average total cost is defined as:

$$
\begin{aligned}
ATC &= \frac{TC}{Q} \\
&= \frac{TVC}{Q} + \frac{TFC}{Q} \\
&= \frac{\sum_j w_j \cdot x_j}{Q} + \frac{TFC}{Q} = AVC + AFC
\end{aligned}
\tag{3.6}
$$

where the notation used is defined in equation (3.1). As can be seen by equation (3.6), the measures of production costs used in calculating ATC include both variable and fixed costs. ATC are the total production costs per unit of output. Equation (3.6) also illustrates that

$$AFC = ATC - AVC$$

This relationship is best illustrated by examining Figures 3.7b and 3.8. The vertical distance between the ATC and AVC curve for a given level of output defines AFC. As you move from left to right on the output or horizontal axis of Figure 3.7b, the distance between ATC and AVC – or the number that represents AFC – is large but decreases rapidly. Examining Figure 3.8 illustrates the same concept. As you move from left to right on the output or horizontal axis of Figure 3.8, AFCs are large but decrease rapidly – as does the distance between ATC and AVC.

Tables 3.6a and 3.6b gives the ATC associated with the Maple Syrup Operation case study. Table 3.6a gives the ATC by size of operation. This table can be used by a maple syrup operation to compare their total production costs per unit to an industry-wide total production costs per unit. For example, you manage a 6,000-tap operation, the industry-wide ATC is $1.94 per tap, if your ATCs are less than these you are more production cost effective than the average 6,000-tap operation.

Table 3.6a and Figure 3.7a show that ATC decrease as the size of the operation increases. The decreasing ATC illustrates economies of scale (or increasing returns to scale), as the size of the operation increase the total production cost per unit of output decreases.[14] This occurs until an 11,000-tap operation. At this point the ATC starts to increase. The reason is that additional capital equipment (e.g. power tree tapper) is required. CFBMC (2000) describes a similar result.

Table 3.6b illustrates the ATC for a 12,000-tap operation. As with Table 3.6a, the ATCs are decreasing as output increases. A closer examination of Tables 3.3b and 3.6b and equation (3.4) provides a clue as to why this occurs. Table 3.3b shows that TVC are increasing and TFC are constant as output increases. Table 3.6b shows that while TVCs are increasing, AVCs are decreasing. This is due to the fact that while VCs are increasing they are not increasing as fast as output. Table 3.6b also shows that AFCs are decreasing as output increases.

Additional thoughts on ATC, AVC, and AFC

If you have a 12,000-tap operation, the smallest total cost per unit output, or ATC, is to produce all 12,000 taps. The same could be said if you have a 6,000-tap operation; the smallest total cost per unit output is to produce all 6,000 taps. This is due to the nature of the production system – recall the answers to the questions "What are the inputs?", "What are the outputs?", and "How do you describe the production process?"

It is critical not to confuse output level that minimizes the cost per unit output with the output level that maximizes profits. Figure 3.1 illustrates the Architectural Plan for Profit. Given the economic information we have developed so far, we only have the foundation – the Production System – and one pillar – Costs. Profits cannot be held up on only one pillar. There is missing economic information; namely, Value and Price. While the output level that maximizes profits is both technically efficient and production cost efficient, the various different output levels that are solutions to the least cost or cost effective models – which are both technically efficient and production cost efficient – are not necessarily what maximizes profit (Silberberg and Suen 2001). In Chapter 5, I will discuss this issue in greater detail.

The question a manager is often asked is how much does it cost to produce a given output. Png and Lehman (2007) and Silberberg and Suen (2001) define this as the ATC. For example, for a 12,000-tap operation, the cost per tap of producing between 0 and 12,000 taps is $2.04 per tap (Table 3.6b). However, if a 12,000-tap operation is only producing between 0 and 6,000 taps, the cost per tap is $3.25 per tap (Table 3.6b). To compare this to a 6,000-tap operation, the cost per tap of producing between 0 and 6,000 taps is $1.94 per tap (Table 3.6a).

Marginal cost (MC)

I will use the Maple Syrup Operation case study to illustrate the concept of MC. Table 3.7 shows the MCs for sap collection.

MC is defined as the change in total cost per unit change in output:

$$MC = \frac{\Delta TC}{\Delta Q(x_1, x_2, \ldots, x_j)} = \frac{\Delta TVC}{\Delta Q(\bullet)} = \frac{\Delta\left(\sum_j w_j \cdot x_j\right)}{\Delta Q(\bullet)} \tag{3.7}$$

where

ΔTC denotes the change in total cost from an incremental change in output. For example, the change in total cost for an incremental change in output from 5,000 to 6,000 taps is $\Delta TC = 19{,}500 - 18{,}780 = \$720$ (Table 3.7).

$\Delta Q(x_1, x_2, \ldots, x_j) = \Delta Q(\bullet)$ denotes the incremental change in output. For example, the incremental change in output from 5,000 taps to 6,000 taps is 1,000 taps.

$\Delta TVC = \sum_j w_j \cdot x_j$ denotes the change in total variable costs from an incremental change in output. For example, the change in total variable cost for an incremental change in output from 5,000 to 6,000 taps is $\Delta TVC = 4{,}620 - 3{,}900 = \$720$ (Table 3.7).

There are four concepts concerning MC that must be highlighted. First, as can be seen by equation (3.7), the production costs used in calculating MC includes only variable production costs. This is illustrated in equation (3.8)

$$\frac{\Delta TC}{\Delta Q(\bullet)} = \frac{19{,}500 - 18{,}780}{6{,}000 - 5{,}000} = \frac{720}{1000} = 0.72 \text{ \$/tap}$$

$$\frac{\Delta TVC}{\Delta Q(\bullet)} = \frac{4{,}620 - 3{,}900}{6{,}000 - 5{,}000} = \frac{720}{1000} = 0.72 \text{ \$/tap} \tag{3.8}$$

According to equations (3.7) and (3.8), TFCs are irrelevant to calculating MC and as such $\Delta TC = \Delta TVC$. The simple explanation for this is that fixed costs do not change with the level of production.[15] Second, MC measures an incremental change. In the case of the Maple Syrup Operation case study, the changes in output are given in 1,000-tap increments (Table 3.7). As the size of incremental change is decreased, the measure of MC becomes more accurate.[16] Third, MC is interpreted as the incremental variable production cost per unit of output. Thus, for the production of each additional tap between 5,000 and 6,000 taps, the additional cost is 0.72 \$/tap. Finally, the units on MC are dollars per output (\$/Q) or in the case of the Maple Syrup Operation case study the units on MC are \$/tap. From Chapter 1 and examination of equation (3.1), the units on price, P, must also be \$/Q or revenue per unit output. Thus if P is greater than MC, sales revenue from selling the incremental production will cover the additional variable cost of its production.

Table 3.7 Marginal sap collection costs for a 12,000-tap operation (1998 USD[§])

Number of taps	Gallons of syrup	TVC ($)	TC ($)	Marginal cost ($/tap)
500	127.5	520	15,400	–
1,000	255	890	15,770	0.74
2,000	510	1,640	16,520	0.75
3,000	765	2,400	17,280	0.76
4,000	1,020	3,120	18,000	0.72
5,000	1,275	3,900	18,780	0.78
6,000	1,530	4,620	19,500	0.72
7,000	1,785	5,390	20,270	0.77
8,000	2,040	6,080	20,960	0.69
9,000	2,295	6,840	21,720	0.76
10,000	2,550	7,600	22,480	0.76
11,000	2,805	8,800	23,680	–
12,000	3,060	9,600	24,480	–

[§]Cost data based on information from Huyler (2000); 1998 USD denotes 1998 US dollars.

Average and marginal cost

As a result of examining the graph of AVC and MC, three observations should be highlighted.[17] Figure 3.9 is the graph of the AVC and MC of sap collection for a 12,000-tap operation.

First, due to the units associated with AVC (and ATC) and MC, they can be put on the same graph. Output (Q) is on the horizontal axis – in this case "number of taps" – and dollars per unit of output ($/Q) is on the vertical axis – in this case $/tap. Both average and marginal cost describe production costs per unit of output. Viewing Figure 3.9 in the larger context of profit given in equation (3.1), revenue is defined as $P*Q$ where price (P) is what consumers pay per unit of output. The units on price are dollars per unit of output ($/Q). From Chapter 1, price is a per unit measure of value and from equation (3.1) it is half of revenue. Thus, using the graph of average and marginal cost you will be able to compare per unit production costs to the per unit value that consumers place on the output you produce or provide. This will facilitate following the Architectural Plan for Profit.

Second, shapes of the AVC and MC curves in Figure 3.9 and those of AP and MP (Figures 2.12 and 2.13) are similar. While Appendix 3 and notes 12 and 16 provide the mathematical explanation, an alternative logic is as follows: the definition of average cost is some measure of production cost divided by the amount of output produced, and the definition of marginal cost is the incremental variable production cost per unit of output. Both definitions require measuring production costs and outputs. Production costs require measuring how much input is used in producing or providing outputs, e.g. $TVC = \sum_j w_j x_j$. This requires knowledge of the production system and accurate answers to the questions "What are the inputs?", "What are the outputs?", "How do you describe the production process?". Consequently, every

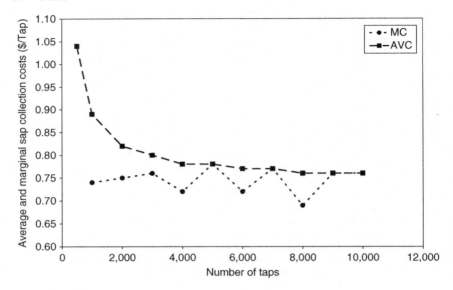

Figure 3.9 Average variable and marginal sap collection costs for a 12,000-tap operation (1998 USD§).

§Cost data based on information from Huyler (2000); 1998 USD denotes 1998 US dollars.

point on an AVC and MC curve has a corresponding point on an AP and MP curve. The AP and MP curves must follow the Law of Diminishing Returns and the AVC and MC must also follow the Law of Diminishing Returns. The reader is encouraged to review Chapter 2 to fully understand these relationships.

Finally, drawing on the relationships between averages and margins from Chapter 2 and Appendix 3: (1) when the margin is greater (less) than the average, the average is increasing (decreasing); (2) when the margin is equal to the average, the average is either a minimum or maximum. Thus for relationship between AVC (and ATC): (1) when $MC < AVC$, AVC is decreasing (see Table 3.7 and Figure 3.9); (2) when $MC > AVC$, AVC is increasing; and (3) when $MC = AVC$, AVC is at a minimum.[18] A similar result was shown with MP and AP (Figures 2.12 and 2.13).

Long-run versus short-run

The universally accepted distinction between the short-run and the long-run is that the short-run is a time period in which one or more inputs are fixed during the production process. In the long-run there are no fixed inputs. Thus, in the short-run $TC = TVC + TFC$ while in the long-run $TC = TVC$. There is no set time period distinguishing the short-from the long-run. This depends on the particular firm or industry and the production process.

In the case of the Mobile Micromill and ISO Beams case studies, labor and natural resources (e.g. logs delivered to the mills) would be variable in the short-run. For example, if consumers are willing to purchase more of their output than is being produced currently, both could potentially increase the number of shifts (labor) and the number of logs delivered to the mills to satisfy the additional demand. However, physical limits of the capital equipment and skill level of labor provide an upper limit on production and satisfying any further increase in demand for their outputs beyond this limit would require purchasing and installing additional capital equipment and training personnel. These changes cannot be accomplished in as short a time (i.e. short-run) as just increasing the number of shifts or logs delivered to the mills. In the case of the Maple Syrup Operation case study, increasing the operation to greater than 10,000 taps requires purchasing an additional power tree tapper. The purchase of this additional power tree tapper and training labor can probably happen more quickly than in either of the Mobile Micromill or ISO Beams case studies. Thus, the time period defining the difference between the short-run and long-run would be different between these case studies.

Sunk and hidden costs versus salvage value

Sunk costs

Opportunity cost is defined as the value of the next best alternative forgone. What if the cost has already been paid? For example, you probably rent an apartment during the school year and pay your rent by the first of every month. Until you write the check (and it is cashed), you have alternative uses of the money. Once the check is cashed, what are your alternative uses of the money? What are the opportunity costs? There are absolutely no choices or decisions that you can make concerning the money. The money is sunk. Thus, sunk costs measure money that has already been spent. Sunk costs are monies that are forever lost after they have been paid. Sunk costs are *utterly without relevance* for forward-looking decisions that ask: (1) What will the project *cost* from now to completion? (2) What will the project be *worth* from now to completion?

In the above definition and description of sunk costs, the phrase "utterly without relevance" was used. I bring your attention to this phrase for two reasons. First, answering the questions "What are the inputs?" from Chapter 2 requires identifying those inputs that a manager controls directly; namely, those inputs that are *relevant* to production decisions and following the Architectural Plan for Profit. Second, those inputs the manager has direct control over are those used in defining the concept of cost (total, variable, fixed, average, marginal, explicit, and implicit) discussed in this chapter. Basically, including sunk costs in any future decision will not provide you with the means for seeking ways to improve or increase your profits.

Hidden costs

Hidden costs are just as problematic as sunk costs. According to Maital (1994: 21) some of your highest business costs "are for things that you have already bought and paid for." Again, think of the definition of an opportunity cost – the value of the next best alternative forgone. The concept of an opportunity cost is answering the "What if" question. If you own capital, e.g. a piece of equipment or building: What are you giving up to keep capital in its current use? For example, you own a sawmill that cuts between 10 MMBF and 15 MMBF per year. The sawmill is located on 25 acres of land. The log yard uses about five to seven acres. You have a 15,000 ft² building which houses the sawmill, a 15,000 ft² building which you use to store the lumber by grade and species, and an old house used as an office for the mill. You own all of this debt-free. The land is zoned for commercial use, has easy access to a major highway, and is on the edge of a medium-sized town. Can you identify the hidden costs in this example?[19]

While sunk costs may or may not show up on one of your cost account ledgers, hidden costs never do. This means that you as a manager must be keenly aware of *all* your opportunity costs.

Salvage value

While sunk costs never enter into forward-looking decisions, what about the salvage values? The salvage value or residual value of an asset is the estimated value that the asset will sell for at the end of its useful life. You just bought a new car for $20,000. However, a month later you realize that your job instead requires a 4×4 pickup that costs $27,000. The check you wrote for $20,000 has been cashed. Is the money spent on the new car a sunk cost in the decision as to whether to buy the pickup? The trade-in value or salvage value of the car is $17,000. Thus, the sunk cost of the car is $3,000. A sunk cost of $20,000 assumes the salvage value is $0.00. The salvage value is important economic information in the decision to buy the pickup. Given this discussion, if you could sub-lease your apartment that you rent during the school year, what would be your sunk costs?

Production systems – revisited

It is important to reiterate the direct connection between the cost concepts (total, variable, fixed, average, and marginal) discussed in this chapter and the production system discussed in Chapter 2. Knowledge of the production system is embodied in the answers to the three questions: "What are the inputs?", "What are the outputs?", and "How do you describe the production process?". In this chapter, we focus our attention on the answers to the first and last questions.

The answer to the first question identifies those inputs that the manager controls directly and those inputs that are relevant to production decisions. The definition of total cost given in equation (3.1) includes variable and fixed costs. The variable and fixed costs used in the profit equation and the variable costs used in the least cost/cost effective equations come directly from the variable and fixed inputs that are described in the answer to the first question. Lack of due diligence on your part in identifying these inputs makes it very hard to follow the Architectural Plan for Profit.

The answer to the last question determines how many inputs (variable and fixed) are used to produce various levels of output. Specifically, this answer requires describing input–output combinations where it is not possible to increase output without increasing inputs, or technical efficiency input–output combinations. If input–output combinations are technically efficient, then they will be production cost efficient. Production cost efficient input levels are used in defining total variable costs and are used in solving the least cost/cost efficient decision models described in equations (3.2) and (3.3). Finally, production cost efficient input–output combinations are necessary, but not sufficient, for maximizing profits.

The answers to both the questions lead to a final observation. Every point on the TC, TVC, ATC, AVC, and MC curves *must have* corresponding points on TP, AP, and MP curves. Without the TP, AP, and MP curves the economic information required to define the TC, TVC, ATC, AVC, and MC curves correctly is missing. Calculating costs based on incorrect economic information will not allow you, as the manager, to make decision to determine ways to improve or increase your profits.

Costs given non-continuous production systems

Not every production process can be described using small incremental changes in inputs and outputs. The Micromill and ISO Beams case studies are examples of these types of production systems. In these cases, the technical efficient input–output combinations are described by discrete points. Figure 2.15 in Chapter 2

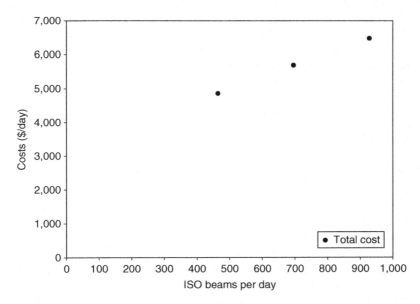

Figure 3.10 Estimated total cost for the ISO Beam case study.[§]

[§]These total cost estimates are approximations only.

shows the ISO beam mill's daily total product (Patterson *et al.* 2002). The estimated total cost associated with these three production levels is given in Figure 3.10.

Due to the nature of the production system, AP, MP, AC, and MC have no economic meaning. For example, increasing labor in the ISO Beams case study by one person per day would have no change on output. The change in output is either 0 or 232 ISO beams per day. The limiting factors are the boilers and kiln(s) and the pre-dryer. Thus, no amount of increases in the variable input labor is going to have any impact on the amount of output produced. Adding an additional person will decrease AP (i.e. the numerator is constant while the denominator increases) but the decrease is due to increased technical inefficiency – not the Law of Diminishing Returns. Adding an additional person will increase AVC (the numerator increases while the denominator is constant), but again this increase is due to technical inefficiency – not the Law of Diminishing Returns. The calculation of MP requires that incremental changes in input and output are feasible. In this case, no such increments are feasible (see Figure 2.15). While you could use equation (2.11) to calculate changes in output relative to changes in input, the result would have no economic meaning. The same argument can be made for MC. While it is possible to calculate changes in cost relative to changes in output using equation (3.7), the result would have no economic meaning. Figure 3.10 shows that incremental changes in cost relative to output are not feasible. This is similar to the discussion presented by Zudak (1970).

In the Mobile Micromill case study, Becker *et al.* (2004) describe the optimal crew size as four people (i.e. one skilled mill operator/supervisor, one log sorter and loader, and two laborers on the green chain). Again, adding an additional person to the crew will have no effect on output. Thus, arguments similar to those above can be given to justify the concepts of AP, MP, AVC, and MC.

In contrast, increasing the amount of labor employed per season in the Maple Syrup Operation case study would allow a proportional increase in the number of taps (i.e. 4.75 minutes/tap, Huyler 2000). For example, an increase of 1,000 (500) taps per season would require an additional 79 (39.5) person-hours per season. Thus, given the nature of this production system, increasing output by either 1,000 or 500 taps per season could be described as a marginal change.

Supply chain management

Supply chain management is defined as the systemic, strategic coordination of the traditional business functions and the tactics across these business functions within a particular company and across businesses within the supply chain, for the purposes of improving the long-term performance of the individual companies and the supply chain as a whole (Mentzer *et al.* 2001: 18). Whether a business uses raw resource inputs and produces a final consumer good (e.g. maple syrup) or is part of a value-added process involving a number of businesses interacting to produce a final good or service, the profits of all businesses involved could be improved if costs all along this chain are minimized and customer value and satisfaction are maximized (Mentzer *et al.* 2001). Supply chain management is concerned with improving both efficiency (i.e. cost reduction (Penfield 2007)) and effectiveness (i.e. customer service) in a strategic context (i.e. creating customer value and satisfaction through integrated supply chain management) to obtain competitive advantage that ultimately brings profitability (Mentzer *et al.* 2001). While the logistics and strategic planning of managing a supply chain are beyond the scope of this book (for example, see Bettinger *et al.* 2009 for a discussion of this topic), I will focus on the efficiency aspect of supply chain management.

The concept of a supply chain describes an input – transformation – output process depicted in Figure 3.11.

Examining Figure 3.11 reveals three points of interest. First, is the flow of input to output as input to output as input to output … until a final consumer good is produced or provided. While the end result is the final consumer good (output), each intermediate value-added process produces an output which the next value-added process (who are consumers) uses as an input. Along this chain, each consumer's value and satisfaction are maximized and costs are minimized. Second, in order for this to occur, information, services, and finances, etc. should flow in both directions. This requires coordinated strategic and logistic planning across all firms involved in the supply chain. Third, Figure 3.11 can be compared and contrasted with Figure 2.4. The flow chart description of a production system is analogous to the supply chain illustrated in Figure 3.11. There are inputs, transformation process (i.e. production or value-adding processes), and outputs. The three questions to systematically analyze a production process from Chapter 2 in a very simple case provide a starting point to examine the more complex chain.

The first pillar of the Architectural Plan for Profit examined has been that of Cost for the simple reason that managing costs are paramount for generating the most profits for any given level of output. This is also an explicit goal of supply chain management and can be illustrated by using a profit margin given in equation (3.9)[20]

$$\text{Profit Margin } (\Pi_M) = \frac{\Pi}{TR} = \frac{TR - TC}{TR} \tag{3.9}$$

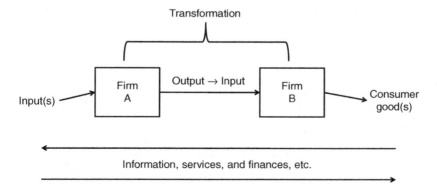

Figure 3.11 Supply chain.

where the terms have been defined in equation (3.1). A profit margin describes the percent of every dollar of total revenue generating profits for the entrepreneur; for example, a profit margin of 5 percent defines $0.05 of every $1.00 total revenue is profit. A higher profit margin implies a firm has a greater ability to control costs than a lower profit margin. Equation (3.10) uses equation (3.9) to illustrate the impacts of controlling costs on total revenue at a given level of output.

$$\Delta TR = TR_{new} - TR$$
$$= \frac{TR - TC + \Delta TC}{\Pi_M} - \frac{TR - TC}{\Pi_M} \tag{3.10}$$
$$= \frac{\Delta TC}{\Pi_M}$$

where ΔTC denotes cost savings and ΔTR denotes the cost savings equivalent in terms of total revenue. Table 3.8 illustrates the impact of cost savings given various different profit margins.

Table 3.8 shows that a cost savings at lower profit margins have a greater equivalence in terms of total revenue than at higher profit margins. Table 3.8 also illustrates the importance of controlling costs in that the potential benefits are magnified many times. For example, the time and effort involved in a $1.00 cost savings is equivalent to a revenue increase of $10.00 given a profit margin of 10 percent. Would the same amount of time and effort have increased revenue by $10.00? It should be noted that decreased costs and increasing revenues are not mutually exclusive decisions and are part of building the Architectural Plan for Profit.

How to use economic information – *costs* – to make better business decisions?

Simply put, if you do not manage costs you will lose control of profits. The Pillar of Cost (see Figure 3.1) is built on the foundation of the Production System. The economic information contained in this foundation is also included in the Pillar of Cost. The economic information contained in the production system is determined by answering the following three questions: (1) What are the inputs? (2) What are

Table 3.8 The impact of cost savings

Cost savings	Profit margin			
	0.05	0.10	0.15	0.20
	Revenue equivalent			
$1.00	$20.00	$10.00	$6.67	$5.00
$10.00	$200.00	$100.00	$66.67	$50.00
$100.00	$2,000.00	$1,000.00	$666.67	$500.00
$1,000.00	$20,000.00	$10,000.00	$6,666.67	$5,000.00

the outputs? (3) How do you describe the production process? The answer to these three questions should be structured to provide the manager with technically efficient input(s)–output(s) combinations. Answering the question "What are the inputs?" will identify those inputs that are relevant to the production process, that you the manager control directly, and that are relevant to following the Architectural Plan for Profit. These relevant inputs include both variable and fixed factors of the production system.

What is the economic information contained in the Pillar of Cost? A cursory look at the definition of total cost shows it comprises three components: wages, variable inputs, and fixed inputs (see equation (3.1)), which can be combined to determine total variable and fixed costs. However, a critical examination of the pillar is more rewarding. First, cost must always be viewed in terms of opportunity cost – the value of the next best alternative forgone. In other words, for what purpose are you using your variable and fixed inputs? Can they be employed to perform different parts of the production process? What are you giving up to keep the variable and fixed inputs in their current use? Are your variable and fixed inputs providing the most value they can? Second, you are a consumer of inputs just as there are consumers of the outputs you produce or provide. As a consumer of goods produced by other entrepreneurs you will only buy the product if you are better off with it than without it. The same logic must hold for purchasing inputs. Are the benefits of your variable and fixed inputs providing greater value than their costs? Third, cost includes the price or wage you pay for your inputs. This implies that you must purchase your inputs from a market. Your knowledge of your input markets must be equal to your knowledge of your output markets. Basically, you want to pay no more than you have to for your inputs. Finally, if profits are to be as large as possible, revenues must be as large as possible and costs must be as small as possible for any given input(s)–output(s) combination. Thus, you are searching for input–output combinations that are production cost efficient. The economic information from the Pillar of Cost must allow you to do this.

4 Revenue

According to the Architectural Plan for Profit, revenue does not even appear as one of the pillars.

However, comparing Figures 3.1 and 4.1 shows that the pillars associated with revenue are Value and Price implying that revenue is a function of both these pillars. In Chapter 1, value was defined as the expressed relative importance or worth of an object to an individual in a given context (Brown 1984) and the degree to which buyers think those goods and services make them better off than if they did without (Maital 1994). Both of these definitions state that individuals have preferences for different outputs and these preferences will be observable given choices individuals make in the market. Based on these definitions, the Pillar of Value contains two elements. The first is the expressed choices that individuals make, or in different terms, if consumers choose to buy the outputs that you produce or provide, this is your revenue. The focus of this chapter will be on determining how to measure the revenue concept of value so you can manage it for the purpose of improving profit. The second is modeling the preferences of your consumers so that you produce or provide goods that will make them better off than if they did without. This concept of value will be discussed in Chapter 6.

Also in Chapter 1, price was defined as a per-unit measure of assigned value and a measure of relative scarcity. In the case of output price, it is economic information that both consumers and producers use in their decisions to purchase and produce an output, respectively.[1] The market price is determined when both consumers and producers reveal their choices. Thus far, the market has been characterized by workable competition – no one buyer or group of buyers and no one seller or group of sellers can influence price; that is, output price is constant. In other words, the market sets the price and producers have almost no ability to manipulate the price of their output. In this chapter, I will continue with this market characterization and assume output price is constant. In Chapter 6, I will discuss how market price is determined, and in 7 I will discuss how market structures other than workable competition impact price and pricing strategies used by producers.

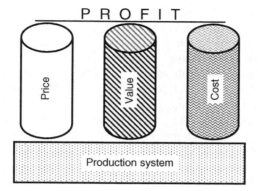

Figure 4.1 The Architectural Plan for Profit.

Why are revenues important?

Based on the definition of the profit model (equation (4.1)):

$$Max \ \Pi = TR - TC$$
$$= P \cdot Q(x_1, x_2,..., x_j) - TVC - TFC \qquad (4.1)$$
$$= P \cdot Q(\bullet) - \sum_j w_j \cdot x_j - TFC$$

where

$Max \ \Pi$ denotes maximization of profit (Π);

TR denotes Total Revenue, $TR = P \cdot Q(x_1, x_2,..., x_j) = P \cdot Q(\bullet)$;

   P denotes the market price of the good or service,

   $Q(x_1, x_2,..., x_j) = Q(\bullet)$ denotes the system of producing or providing a good or service, Q, using inputs that the manager has direct control over, $x_1, x_2,..., x_j$,

TC denotes Total Cost, $TC = TVC + TFC$;

   TVC denotes Total Variable Costs, $TVC = \sum_j w_j \cdot x_j$,

   $\sum_j$ denotes the summation operator,

   w_j denotes the jth wage or price paid for jth input x_j,

   x_j denotes the jth input; for example, labor,

TFC denotes Total Fixed Costs.

The most obvious place to start a discussion on revenues would seem to be with total revenue. It may seem odd then to start this discussion by reviewing the definition of cost, specifically the definition of opportunity cost – the value of the next best alternative forgone. The basic idea of an opportunity cost is what you are giving up, in terms of explicit and implicit resources, as a result of your current choices. There are opportunity costs or value to be measured and managed in terms of costs (Chapter 3). This chapter will examine the opportunity costs or value to be measured and managed in terms of revenues.

Examining total revenue of equation (4.1) reveals the following three observations. First, TR is half of profit. Second, to improve profits, TRs must be as large as possible while cost must be as small as possible for various input(s)–output(s) combinations. That is, generating revenue and minimizing cost must happen simultaneously. How you create value is equally as important as cost-cutting because cost-cutting is pointless if the good or service that results is not valued by consumers (Maital 1994). Finally, $TR = P*Q$ is a measure of assigned value. If a consumer purchases your product, then you know: (1) the benefit they receive from the product is greater than their cost; and (2) the consumer's cost is your TR. Therefore, your TR depends on your knowledge of consumers' willingness to part with their hard-earned cash for your output. In this chapter, I will focus economic information contained in TR (the combination of Pillars of Price and Value) so you can use it to improve profits.

However, before proceeding, I have some thought questions:

1   In the previous chapters when I introduced the profit model, I also introduced the least cost/cost effective model. Why have I not included it here?[2]
2   Compare and contrast economic descriptors used to measure technical efficiency from Chapter 2 with those used to measure production cost efficiency in Chapter 3.
3   Based on your answer to thought question 2, predict what economic descriptors will be used to measure value with respect to total revenue.

Case studies

The case studies used in this chapter will include the Inside-Out (ISO) Beams (Patterson *et al.* 2002 and Patterson and Xie 1998), Mobile Micromill (Becker *et al.* 2004), Maple Syrup Operation (Huyler 2000), Select Red Oak (Hahn and Hansen 1991), Loblolly Pine Plantation (Amateis and Burkhart 1985 and Burkhart *et al.* 1985), and Great Lakes Charter Boat Fishing (Lichtkoppler and Kuehn 2003). The case studies can be found on the Routledge website for this book. Their summaries will not be repeated here as they have been used in earlier chapters.

Economic descriptors of total revenue

There are six main descriptors of revenue, three describing revenue with respect to output: total revenue, average revenue, marginal revenue; and three describing revenue with respect to inputs used in the production process: total revenue product, average revenue product, and marginal revenue product. I will address each in turn.

Total revenue (TR)

Total revenue is defined by equation (4.2):

$$TR = P \cdot Q(x_1, x_2, \ldots, x_j) = P \cdot Q(\bullet) \qquad (4.2)$$

where the notation used is defined in equation (4.1). Given the assumptions about the market (i.e. workable competition), P is constant and TR is linear with respect to output. I will use the case studies to illustrate this point starting with the Great Lakes Charter Boat Fishing case study. Table 4.1 shows the TR for the average charter boat fishing business. Figure 4.2 illustrates the TR for full day charter fishing trips for lake trout and salmon.

As can be seen by Figure 4.2, TR is in fact linear with respect to output. In other words, if 14 full day charter fishing trips for lake trout and salmon are provided, the TR would be $5,698 (= 407 · 14) per season. Given the information presented in Table 4.1, what would be the graph of TR for full day charter fishing trips for steelhead or smallmouth bass look like? What about for half day charter fishing trips? They would all have the same linear shape as shown in Figure 4.2. Of course the scales on the axes of the graphs would be different as the magnitudes of the numbers would change. What would be the behavior of a charter boat captain given this data on TR? Would they try to book (i.e. sell) more lake trout and salmon full day excursions or half day yellow perch excursions? Finally, would the TR of the other case studies exhibit similar results?

In the ISO Beams case study, a weighted price of $7.76 per ISO beam was used to calculate TR (Patterson *et al.* 2002). Figure 4.3 shows the graph of TR.

TR is linear with respect to the amount of output sold. However, the value of an ISO beam would probably vary in a manner similar to solid 4 × 4 beams.

Table 4.1 Total revenue for New York's Great Lakes Charter Boat Fishing industry (2002 USD[§])

Fish species	Average no. trips/business	Average charge/trip	Total revenue
Lake trout and salmon			
Full day	27.6	$407	11,233
Half day	6.9	$306	2,111
Steelhead			
Full day	9.4	$401	3,769
Half day	1.9	$299	568
Smallmouth bass			
Full day	7.6	$342	2,599
Half day	1.7	$251	427
Walleye			
Full day	5.1	$380	1,938
Half day	0.6	$273	164
Yellow perch			
Full day	0.3	$364	109
Half day	<0.1	$254	25

[§]2002 USD denote 2002 US dollars.

(Date used with permission from Lichtkoppler and Kuehn (2003) and Ohio Sea Grant Extension)

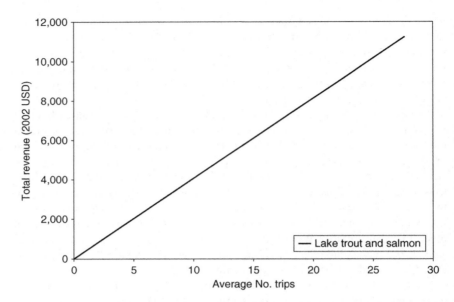

Figure 4.2 Total revenue from a full day charter boat fishing for lake trout and salmon on New York's Great Lakes (2002 USD§).

§2002 USD denote 2002 US dollars.

(Date used with permission from Lichtkoppler and Kuehn (2003) and Ohio Sea Grant Extension)

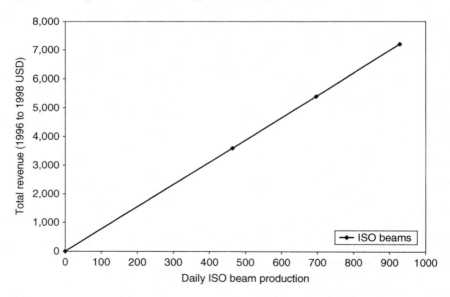

Figure 4.3 Total revenue for the ISO Beam case study (1996–1998 USD§).

§1996 to 1998 denotes 1996 to 1998 US dollars.

(Based on data from Patterson *et al.* (2002))

Patterson *et al.* (2002) showed that the market price of solid 4 × 4 beams vary by season and length. In general, however, longer solid 4 × 4 beams had a higher market price than short beams. The same could be postulated about ISO beams of different length. Thus, TR would behave in a similar manner as in the Great Lakes Charter Boat Fishing case study for different fish species and length of excursion. If the market price data showed a variation of ISO beams by season and length, what would be the ISO beam mill owner's behavior?

In the Mobile Micromill case study (Becker *et al.* 2004), output price by lumber size and grade is given in Table 2.3. However, TR is still linear with respect to the amount of output sold even given the information presented in Table 2.3. Based on the information given in Table 2.3, what do you think the mill owner's behavior will be in terms of producing the various possible outputs described in Table 2.3, given the percentage of lumber size and grade recovery given various sizes of ponderosa pine log (see Table 2.2)? Is this mill owner's behavior similar to that of the charter boat captain or the ISO beam mill owner?

In the Maple Syrup Operation case study, the output was described as the number of taps by CFBMC (2000) and Huyler (2000). However, consumers do not buy "taps"; they buy syrup or other maple sugar products. Thus, in terms of describing TR, number of taps is insufficient in characterizing the output. As with the previous case studies, output has more than one attribute: maple syrup is graded primarily on color. The grade that receives the highest consumer price is Grade A light amber, followed in succession by Grade A medium amber, and Grade A dark amber (Marckers *et al.* 2006). Variations in grades of maple syrup are due to method of production (e.g. type of evaporator), year of production, and season in which the sap was collected and syrup produced (Markers *et al.* 2006). Nonetheless, TR is still linear with respect to the amount of maple syrup sold by grade.

Before I examine the next two case studies, I believe it is important to revisit the three fundamental questions that are needed to systematically examine any production system: (1) What are the inputs? (2) What are the outputs?, and (3) How do I describe the production process? It is critical to think about the how the economic information from the answers will be used. In other words, have you provided sufficient detail to help build the Pillars of Cost, Value, and Price? Focusing on the answer to the question "What are the outputs?" as has been illustrated above, not enough detail has always been provided to define TR adequately. You may have to revisit the production process description to account for quality and product classification indicators that impact TR directly.

In the Select Red Oak case study, output is defined as board foot volume International ¼ per tree (Hahn and Hansen 1991). As described in Chapter 2, hardwood trees, such as red oak, are characterized by an index of quality. A tree designated as a Grade 1 denotes the highest quality, Grade 2 is next in quality followed by Grade 3 and so on. Diameter at breast height is one – of many – attribute used to distinguish between tree Grades.[3] Other things being equal, the larger the diameter the better the tree grade. Individuals who purchase red oak place a higher value on Grade 1 trees, then Grade 2 and so on. The difference between the stumpage price of a Grade 1 tree versus a Grade 2 tree versus a Grade 3 tree can be substantial (Stier 2003).[4] The description of the production process (see equations (2.4) and (2.5)) allows volume to be described in terms of diameter at breast height (DBH) (see Table 2.5). Using hardwood tree grading criteria from the US Department of Agriculture Forest Service *Forest Inventory and Analysis Handbook* (2006) and stumpage price information from New York State Department of Environmental Conservation (2008), Table 4.2 gives the breakdown of stumpage price by tree grade.

Figure 4.4 illustrates the TR for Select Red Oak per tree by tree grade using the information in Tables 4.2 and 2.5.

Table 4.2 Stumpage price for Select Red Oak by tree grade

Tree grade	DBH range[§]	Stumpage price[‡] ($/bd.ft Int ¼)
Grade 3	11 inches ≤ DBH ≤ 13 inches	0.215
Grade 2	13 inches ≤ DBH ≤ 16 inches	0.30
Grade 1	16 inches ≤ DBH	0.425

[§]DBH denotes diameter at breast height. DBH ranges are based on hardwood tree grading information found in United States Department of Agriculture, Forest Service (2006).
[‡]The stumpage prices are taken from Bureau of Land Resources (2008). The New York State Department of Environmental Conservation collects and distributes stumpage price data in terms of low, average, and high price range, but not by tree grade. The median price in each range for the Hudson/Mohawk Regions is used for stumpage price by grade; namely, the median price for the low price range is Grade 3, the median price for the average price range is Grade 2, the median price for the high price range is Grade 1. Stumpage price is defined as 2008 US dollars per board feet International ¼.

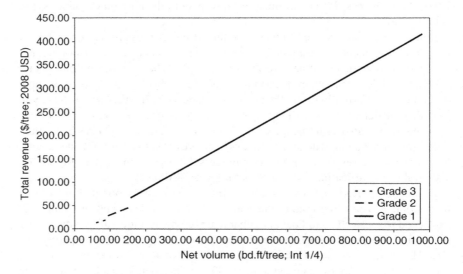

Figure 4.4 Total revenue for Select Red Oak per tree by tree grade.[§]
[§]2008 USD denotes 2008 US Dollars. Given $S = 66$.

As can be seen in Figure 4.4 the TR is linear with respect to tree grade.

In the Loblolly Pine Plantation case study, output is defined total cubic-foot yield, outside bark, per acre (Amateis and Burkhart 1985 and Burkhart *et al.* 1985). As described in Chapter 2, small diameter loblolly pines are used for pulpwood, as the trees increase in diameter and height, they can be used as Chip-n-Saw, sawtimber, plywood logs, or power poles. The lowest valued output is pulpwood and the highest valued output is power poles. In terms of TR based on the different potential product classes, output needs to be described using the attributes of yield (cubic feet per acre), height (feet), and diameter at breast height (DBH). Table 4.3 provides stumpage price by product class.

The production system as described (see equations (2.6), (2.7), and (2.8)) includes yield and height but not DBH. Additional information is required in terms of relating yield to DBH so yield by product class can be determined. Sharma and Oderwald (2001) and Tasissa *et al.* (1997) provide the required information. Table 4.4 defines Loblolly Pine Plantation yield by plantation age, height, and DBH and TR associated with each yield, plantation age, height, and DBH combination.

Given the information in Tables 4.3 and 4.4, a graph of TR by product class is shown in Figure 4.5. Examining Tables 4.3 and 4.4 and Figure 4.5 reveals two observations. First, no TR can be determined for yields for plantation ages younger than 12 years old, even though the yields are positive. The reason is that the trees are not large enough in terms of DBH to satisfy the minimum requirements for any product class (see Table 4.3). Second, TR is linear with respect to product class.

Average revenue (AR)

Average revenue is defined by equation (4.3):

$$AR = \frac{TR}{Q(x_1, x_2, ..., x_j)} = \frac{P \cdot Q(\bullet)}{Q(\bullet)} = P \cdot \left[\frac{Q(\bullet)}{Q(\bullet)} \right] = P \qquad (4.3)$$

Table 4.3 Stumpage price for loblolly pine by product class

Product class	DBH range[§]	Stumpage price[‡] ($/cu.ft)
Pulpwood	4.6 inches ≤ DBH ≤ 9 inches	0.21
Chip-n-Saw	9 inches ≤ DBH ≤ 12 inches	0.50
Sawtimber	12 inches ≤ DBH	1.46

[§]DBH denotes diameter at breast height. Product class ranges are based on work by Dr E. David Dickens from the Warnell School of Forest Resources at the University of Georgia, Athens (http://www.warnell.uga.edu/Members/dickens accessed on 9 September 2008).

[‡]Stumpage prices are for the 1st quarter of 2008 published by Timber Mart-South (http://www.tmart-south.com/tmart/prices.html accessed on 9 September 2008). Timber Mart-South provides stumpage prices in US dollars per ton. The following conversions by product class were used to change the units from dollars per ton to dollars per cubic feet ($/cu.ft): pulpwood 0.028889 ton/cu.ft; Chip-n-Saw 0.028889 ton/cu.ft; sawtimber 0.043236 ton/cu.ft.

Table 4.4 Loblolly Pine Plantation yield and revenue§

Plantation age	Height (feet)	DBH (inch)	Yield (cu.ft/acre)	TR ($/acre)	AR ($/cu.ft)	MR ($/cu.ft)
6	16	2.5	421	–	–	–
9	25	4.0	1,063	–	–	–
12	34	5.0	1,877	402.87	0.21	–
15	43	5.7	2,757	591.77	0.21	0.21
18	50	6.3	3,629	778.94	0.21	0.21
21	57	6.9	4,448	954.81	0.21	0.21
24	64	7.5	5,195	1,115.03	0.21	0.21
27	70	8.2	5,866	1,259.03	0.21	0.21
30	76	8.9	6,470	1,388.71	0.21	0.21
33	82	9.7	7,022	3,489.26	0.50	–
36	87	10.6	7,540	3,746.63	0.50	0.50
39	92	11.7	8,041	3,995.28	0.50	0.50
42	97	13.0	8,539	12,455.78	1.46	–
45	102	14.5	9,047	13,196.80	1.46	1.46
47	105	15.7	9,396	13,706.07	1.46	1.46

§DBH denotes diameter at breast height. Yield is measured in terms of cubic feet per acre (cu.ft/acre) for $S = 66$; $A_1 = 25$; $N_p = 1250$. TR denotes total revenue. AR denotes average revenue. MR denotes marginal revenue.

where the notation used is defined in equation (4.1). AR is defined as the revenue per unit of output. The units on AR are the same as the units on price, P, that is dollars per unit output, or $/Q.[5] Given the assumptions about the market (i.e. workable competition), P is constant and AR is linear with respect to output. In fact, AR is a horizontal line and is not a function of the amount of output you sell.

At first glance it may seem odd that AR is constant or does not change as more or less output is produced and sold. To illustrate this I will use the information from Tables 4.3 and 4.4 for the Loblolly Pine Plantation case study. Table 4.4 and equations (4.4a) and (4.4b) define the AR of selling 2,757 or 5,866 cubic feet per acre of pulpwood, respectively:

$$AR = \frac{591.77}{2757} = 0.21 \ \$/cu.ft \tag{4.4a}$$

$$AR = \frac{1259.03}{5866} = 0.21 \ \$/cu.ft \tag{4.4b}$$

Thus, the AR for selling pulpwood is 0.21 $/cu.ft. Or for every cubic foot of pulpwood I can sell, I will receive $0.21 in revenue. Similar calculations and interpretations of AR can be made for Chip-n-Saw and sawtimber. Figure 4.6 shows the AR for each product class.

Figure 4.6 shows that AR is in fact linear with respect to output for each product class. In addition, AR is constant for each product class. Examining Tables 4.3 and 4.4 and Figure 4.6 show that no AR can be determined for yields for plantation ages younger than 12 years old, even though the yields are positive. The reason is

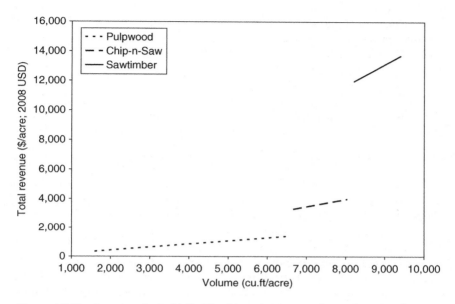

Figure 4.5 Total revenue for Loblolly Pine Plantation per acre by product class.[§]
[§]2008 USD denotes 2008 US dollars. Given $S = 66$, $A_I = 25$, and $N_p = 1210$.

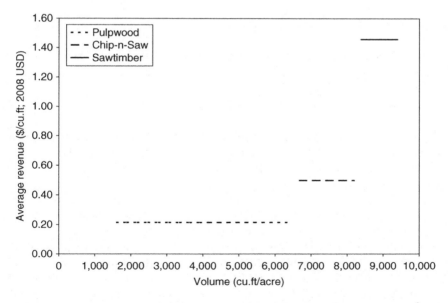

Figure 4.6 Average revenue for Loblolly Pine Plantation per acre by product class.[§]
[§]2008 USD denotes 2008 US dollars. Given $S = 66$, $A_I = 25$, and $N_p = 1210$.

that the trees are not large enough in terms of DBH to satisfy the minimum requirements for any product class (see Table 4.3).

Based on the above discussion, I will leave it to you to determine what the tables and graphs of AR look like for rest of the case studies used in this chapter.

Marginal revenue (MR)

Marginal revenue is defined as the change in TR from the sale of an additional unit of output:

$$MR = \frac{\Delta TR}{\Delta Q(x_1, x_2, ..., x_j)} = \frac{\Delta[P \cdot Q(\bullet)]}{\Delta Q(\bullet)} = P\left[\frac{\Delta Q(\bullet)}{\Delta Q(\bullet)}\right] = P \tag{4.5}$$

where the notation used is defined in equation (4.1). The units on MR are the same as the units on price, P, that is dollars per unit output, or $/Q.[6] Given the assumptions about the market (i.e. workable competition), P is constant and does not change if the amount of output sold changes. MR is linear with respect to output. In fact MR is a horizontal line and is not a function of the amount of output you sell.

Again it may seem odd that MR is constant. As before, I will use the Loblolly Pine Plantation case study to illustrate MR based on the information from Tables 4.3 and 4.4. Equation (4.6a) defines MR for a Loblolly Pine Plantation between the plantation ages of 21 and 24 characterized by 4,448 and 5,195 cubic feet of pulpwood respectively:

$$MR = \frac{\Delta TR}{\Delta Q(\bullet)} = \frac{1115.03 - 954.81}{5195 - 4448} = 0.21 \ \$/cu.ft \tag{4.6a}$$

In fact based on equation (4.5), the MR for pulpwood will be constant. Equation (4.6b) shows the MR for all the pulpwood in a loblolly pine plantation is constant:

$$MR_{Pulpwood} = \frac{\Delta TR}{\Delta Q(\bullet)} = \frac{1388.71 - 402.87}{6470 - 1877} = 0.21 \ \$/cu.ft \tag{4.6b}$$

Equation (4.6b) states for every additional cubic foot of pulpwood sold, you will receive $0.21 in additional revenue. The same set of calculations can be made for the Chip-n-Saw and sawtimber product classes. The MRs for each product class are show in Figure 4.7.

Examining Tables 4.3 and 4.4 and Figure 4.7 reveals three observations. First, no MR can be determined for yields for plantation ages younger than 12 years old, even though the yields are positive. The reason is that the trees are not large enough in terms of DBH to satisfy the minimum requirements for any product class (see Table 4.3). Second, a MR cannot be calculated for changes in yield between the plantation ages of 9 and 12, 30 and 33, and 39 and 42. The calculation of MR requires that the TR curve be continuous. Figure 4.5 shows that the TR curve is not continuous when the trees grow from one product class to another. For example, between the plantation ages of 30 and 33 the trees change product class from pulpwood to Chip-n-Saw (see Table 4.3 and DBH column of Table 4.4).

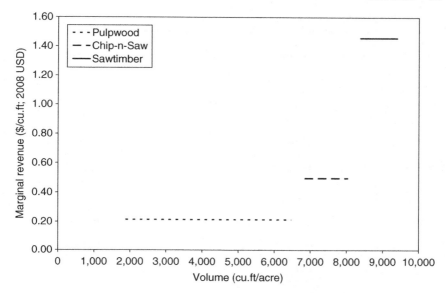

Figure 4.7 Marginal revenue for Loblolly Pine Plantation per acre by product class.§

§2008 USD denotes 2008 US dollars. Given $S = 66$, $A_I = 25$, and $N_p = 1210$.

At this point the TR curve has a break in it (see Figure 4.5) and this is reflected by the fact that MR cannot be calculated at this point (see Table 4.4) and the graph of MR also has a break at this point (see Figure 4.5). Finally, $MR = AR$ and the units on MR and AR are identical as $MR = AR = P$ (see equations (4.3) and (4.5), Table 4.4, and Figures 4.6 and 4.7).

Based on the above discussion, I will leave it to you to determine what the tables and graphs of MR look like for rest of the case studies used in this chapter.

The next three descriptors of revenue are defined with respect to inputs used in the production process. For the purposes of the following discussion, I will use the Select Red Oak case study and DBH as the input.

Total revenue product (TRP)

Total revenue product is given in equation (4.7):

$$TRP = P \cdot TP = P \cdot Q(\bullet) \tag{4.7}$$

where the notation used is defined in equation (4.1) with TP denoting Total Product. Given the assumptions about the market (i.e. workable competition), P is constant and does not change if the amount of output sold changes. Equations (4.7) and (4.2) appear to be identical. However, I want to draw your attention to the TP term.[7] Table 4.5a gives the TRP for Select Red Oak based on the stumpage price and grade information contained in Table 4.2 and given a site index of 66.

Table 4.5a Select Red Oak total revenue product[§]

DBH (inches)	Volume(net) (bd.ft/tree)	Total revenue product ($/tree)		
		Grade 3	Grade 2	Grade 1
10	47.1	–	–	–
11	60.7	13.04	–	–
12	76.2	16.39	–	–
13	93.9	20.18	28.16	–
14	113.6	–	34.08	–
15	135.4	–	40.63	–
16	159.3	–	47.78	67.69
17	185.1	–	–	78.67
18	212.8	–	–	90.46
19	242.4	–	–	103.01
20	273.6	–	–	116.27
25	449.0	–	–	190.83
30	637.7	–	–	271.02
35	810.4	–	–	344.43
40	941.0	–	–	399.95
45	1,013.7	–	–	430.83
50	1,025.8	–	–	435.99
51	1,021.6	–	–	434.19

[§]DBH denotes diameter at breast height. Volume(net) is measured in terms of net board feet (International ¼) per tree, (bd.ft/tree) for $S = 66$. Grade and stumpage price are defined in Table 4.2.

As the focus is on an input, in this circumstance DBH, TRP is graphed with respect to input.

Comparing Figures 4.8 and 4.4 reveals two interesting observations. First, while TR is linear with respect to output, TRP is *not* linear with respect to input. Second, the shape of the TRP curve is similar to the net yield TP curve given in Figure 2.5. The TRP curve (as well as the TP curve) represents a technical efficiency frontier and must also satisfy the Law of Diminishing Returns.

While TR relates output to the revenue it generates, TRP relates the productivity of a variable input (used to produce the output) to the revenue it generates. TRP is used to relate variable input productivity to revenue. Table 4.5a illustrates this relationship between DBH and the revenue it generates (i.e. TRP); for example, a 15-inch select red oak generates a revenue of $40.63 per tree. As with depictions of the TP, TRP for a given variable input assumes that all other inputs are held constant. In the case of the Select Red Oak case study, the data given in Table 4.5a is calculated using a site index of 66. At a site index of 99, the TRP of a 15-inch select red oak would be $45.30 per tree (see Tables 2.5 and 4.2). The difference of $4.67 per tree would be due to the increased site productivity. Finally, DBH generates no revenue until the tree reaches 11 inches (Table 4.2).[8] This is due to the input limitations of sawmill production system in terms of processing the tree into lumber.

While TRP can be calculated for either DBH or site index, the interpretation of TRP is tricky for two reasons. First, in Chapter 2, the inputs that are relevant for

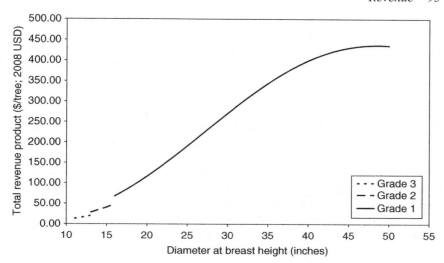

Figure 4.8 Total revenue product for Select Red Oak by grade.§

§2008 USD denotes 2008 US dollars. Given $S = 66$.

following the Architectural Plan for Profit are those you have direct control over. However, you do not have direct control over either DBH or site index. You can manipulate them indirectly by thinning or fertilizing. DBH is a proxy used to measure the productivity of natural system in terms of the desired output – volume – and your manipulation of that system to produce the desired output.[9] Second, while TP is continuous based on the description of the production system (see equations (2.4) and (2.5) in Chapter 2), TRP shown in Figure 4.8 is not continuous due to different grades (see Table 4.2). Thus, while the productivity of a variable input changes due to the Law of Diminishing Returns, its contribution to revenue can also change due to different prices reflecting different grades of quality.

The labor input in the Charter Boat Fishing, Mobile Micromill, and ISO Beams case studies is under the direct control of the manager; however, not enough data are provided in the descriptions of the production system to calculate TRP in a manner similar to that given above. Nonetheless, the interpretation of an input's productivity in terms of revenue is just as important.

Average revenue product (ARP)

Average revenue product is given by equation (4.8):

$$ARP = AR \cdot AP = \frac{P \cdot Q(\bullet)}{Q(\bullet)} \cdot \frac{Q(\bullet)}{x_j} = P \cdot AP \qquad (4.8)$$

where the notation used is defined in equation (4.1) with AP denoting Average Product. Given the assumptions about the market (i.e. workable competition), P is constant and does not change if the amount of output sold changes. As with TRP, the focus is on measuring the productivity of a variable input in term the revenue generation. In this case the measure of productivity is AP. Table 4.5b gives the ARP for Select Red Oak based on the stumpage price and grade information contained in Table 4.2 and given a site index of 66.

Figure 4.9 is the graph of ARP with respect to DBH. How does the graph of ARP compare with a graph of AR for the Select Red Oak case study? If you had drawn the graph of AR as recommended, you would have discovered that it was similar to the graph of AR for the Loblolly Pine case study (Figure 4.5). While AR is a set of horizontal lines with respect to output, ARP is *not* with respect to input. The shape of the ARP curve is similar to the net yield AP curve given in Figure 2.10. The ARP curve (as well as the AP curve) represents a technical efficiency frontier and must also satisfy the Law of Diminishing Returns.

While AR relates output to the revenue per unit of output it generates, ARP relates the average productivity of a variable input to the revenue it generates. The interpretation of ARP is tied directly to AP. According to Table 4.5b, the ARP for a 12-inch DBH select red oak is 1.37 \$/tree/inch ($= 0.215$ \$/bd.ft $\cdot$ 6.35 bd.ft/tree/inch). The AP is 6.35 bd.ft/tree/inch; this does not imply that the productivity (measured as net volume) of each inch of DBH up to 12 inches is the same (see

Table 4.5b Select Red Oak average revenue product[§]

DBH (inches)	AP(net) (bd.ft/tree/inch)	Average revenue product (\$/tree/inch)		
		Grade 3	Grade 2	Grade 1
10	4.71	–	–	–
11	5.51	1.19	–	–
12	6.35	1.37	–	–
13	7.22	1.55	2.17	–
14	8.11	–	2.43	–
15	9.03	–	2.71	–
16	9.95	–	2.99	4.23
17	10.89	–	–	4.63
18	11.82	–	–	5.03
19	12.76	–	–	5.42
20	13.68	–	–	5.81
25	17.96	–	–	7.63
30	21.26	–	–	9.03
35	23.15	–	–	9.84
40	23.53	–	–	10.00
45	22.53	–	–	9.57
50	20.52	–	–	8.72
51	20.03	–	–	8.51

[§]DBH denotes diameter at breast height. AP(net) denotes average product measured in terms of net board feet (International ¼) per tree per inch (bd.ft/tree/inch) for $S = 66$. Grade and stumpage price are defined in Table 4.2.

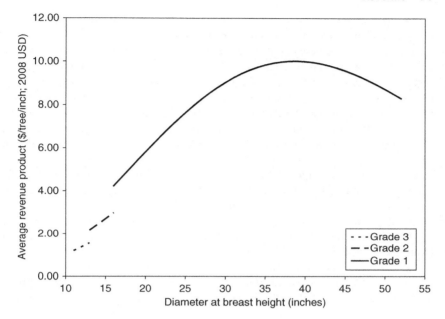

Figure 4.9 Average revenue product for Select Red Oak by grade.[§]

[§]2008 USD denotes 2008 US dollars. Given $S = 66$.

Table 4.5b).[10] Thus ARP does not imply the revenue generated by each inch of DBH up to 12 inches is the same (see Table 4.5b).

Comparing the ARP of a 12-inch to a 14-inch DBH select red oak tree is instructive. The ARP of a 14-inch DBH select red oak tree according to Table 4.5b is 2.43 $/tree/inch. The two additional inches of DBH increase the ARP 1.77 time. This is due to the price change associated with the higher grades for red oak (Table 4.2) and at 14 inches of DBH AP is still increasing (Table 2.8 and Figure 2.10). As above, this does not imply that the productivity (measured as net volume) or the revenue generated by every inch of DBH is the same. However, what this implies is that the same inches of DBH on a 14-inch red oak tree on average generate more revenue than those same inches of DBH on a 12-inch red oak tree. How would this discussion of the ARP that an input generates translate to the use of the same unit of labor used to provide a half versus full day of lake trout or salmon fishing (see Table 4.1)? The use of the same unit of labor in the production of longer versus shorter ISO beams? And the productivity of DBH in the Loblolly Pine Plantation?

Marginal revenue product (MRP)

Marginal revenue product is given by equation (4.9):

$$MRP = MR \cdot MP = \frac{\Delta[P \cdot Q(\bullet)]}{\Delta Q(\bullet)} \cdot \frac{\Delta Q(\bullet)}{\Delta x_j} = P \cdot MP \qquad (4.9)$$

where the notation used is defined in equation (4.1) with MP denoting Marginal Product. Given the assumptions about the market (i.e. workable competition), P is constant and does not change if the amount of output sold changes. As with TRP and ARP, the focus is on measuring the productivity of a variable input in term of revenue generation. In this case the measure of productivity is MP. Table 4.5c gives the MRP for Select Red Oak based on the stumpage price and grade information contained in Table 4.2 and given a site index of 66.

Figure 4.10 is the graph of MRP with respect to DBH. How does the graph of MRP compare with a graph of MR for the Select Red Oak case study? If you had drawn the graph of MR as recommended, you would have discovered that it was similar to the graph of MR for the Loblolly Pine Plantation case study (Figure 4.7). While MR is a set of horizontal lines with respect to output, MRP is *not* with respect to input. The shape of the MRP curve is similar to the net yield MP curve given in Figure 2.10. The MRP curve (as well as the MP curve) represents a technical efficiency frontier and must also satisfy the Law of Diminishing Returns.

Table 4.5c shows that a 1-inch increment of DBH growth on an 18-inch DBH red oak generates more revenue than a 1-inch increment of DBH growth on either a 14-inch or 12-inch DBH red oak. The value of an incremental inch is greater at

Table 4.5c Select Red Oak marginal revenue product[§]

DBH (inches)	MP(net) (bd.ft/tree/inch)	Marginal revenue product ($/tree/inch)		
		Grade 3	Grade 2	Grade 1
10	11.56	–	–	–
11	13.54	2.91	–	–
12	15.58	3.35	–	–
13	17.65	3.79	–	–
14	19.73	–	5.92	–
15	21.81	–	6.54	–
16	23.85	–	7.16	–
17	25.83	–	–	10.98
18	27.73	–	–	11.79
19	29.53	–	–	12.55
20	31.21	–	–	13.26
25	37.04	–	–	15.74
30	37.41	–	–	15.90
35	31.90	–	–	13.56
40	21.81	–	–	9.27
45	9.58	–	–	4.07
50	−2.12	–	–	−0.90
51	−4.22	–	–	−1.79

[§]DBH denotes diameter at breast height. MP(net) denotes Marginal Product measured in terms of net board feet (International ¼) per tree per inch (bd.ft/tree/inch) for $S = 66$. Grade and stumpage price are defined in Table 4.2.

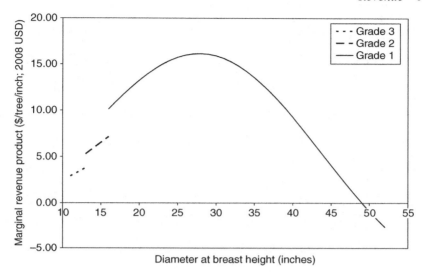

Figure 4.10 Marginal revenue product for Select Red Oak by grade.[§]

[§]2008 USD denotes 2008 US dollars. Given $S = 66$.

higher grades. As with ARP this is due to the price change associated with the higher grades for red oak (Table 4.2) and MP is still increasing at 18 inches of DBH (Table 2.8 and Figure 2.10). MRP is non-continuous even though MP is continuous. This is because of the price differences due to grade (Table 4.2). Thus, no MRP can be calculated for a 13-inch DBH Grade 2 or 16-inch DBH Grade 1 red oak tree.

Reviewing the relationship between AP and MP, shows that whenever *MP > AP*, AP is increasing; whenever *MP < AP*, AP is decreasing; and when *MP = AP*, AP is at its maximum value (Chapter 2, Table 2.8, and Figure 2.10). The same relationship holds for ARP and MRP. Tables 4.5b and 4.5c and Figure 4.11 show that whenever *MRP > ARP*, ARP is increasing; whenever *MRP < ARP*, ARP is decreasing; and when *MRP = ARP*, ARP is at its maximum value.

The interpretation of MRP is different than ARP. While ARP determines the average revenue generated by the productivity of input for a range of inputs, for example, from 0 to 12-inches DBH or 0 to 14-inches DBH for a red oak tree, MRP is for an incremental change. For example, a 1-inch incremental change in DBH from a 12- to a 13-inch red oak tree generates an additional 3.79 $/tree, while a 1-inch incremental change in DBH from a 19- to 20-inch red oak generates an additional 13.26 $/tree.

Why does apparently the same 1-inch increment generate more revenue on going from a 19- to 20-inch DBH red oak than a 12- to 13-inch red oak? The answer to this question requires your understanding of the underlying fundamental characteristics for systematically examining the production system; in short, by reviewing the three fundamental questions posed in Chapter 2 concerning the

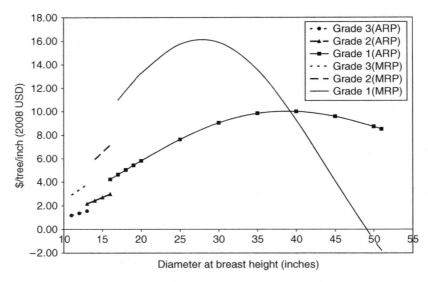

Figure 4.11 Average and marginal revenue product for Select Red Oak by grade.[§]
[§]2008 USD denotes 2008 US dollars. Given $S = 66$.

production system. The functional form used in equation (2.4) is based on a Weibull model (Hahn and Hansen 1991); however, in simple geometric terms the volume of a cylinder or cone (e.g. a tree) is a function of diameter (DBH) and height (site index, S). Figure 4.12 illustrates how the same 1-increment, for a given height, will generate more volume (and revenue) on the 19- to 20-inch DBH red oak than on the 12- to 13-inch DBH red oak.

We know that as a tree grows in DBH they also grow in height which will further increase the volume (and revenue) for the same 1-inch increment on the larger red oak. Can you do a similar type of analysis for labor in the ISO Beams, Mobile Micromill, and Great Lakes Charter Boat Fishing case studies? Or one of the other variable inputs under the direct control of you the manager? Is there enough information provided in the case studies to answer the question? If not, what additional information would you need and how would you collect it?

Finally, think about how you might use this piece of economic information. If you know the revenue generated by using an additional increment of an input that you control directly, what is the most you would be willing to pay to use that input?

How to use economic information – *revenues* – to make better business decisions?

The concept of TR in the profit model is tied directly to the degree to which buyers think those goods and services make them better off than if they did without (Maital 1994) and market price. If consumers of the output(s) that you produce or provide do not value it, how much revenue are you going to generate by selling it?

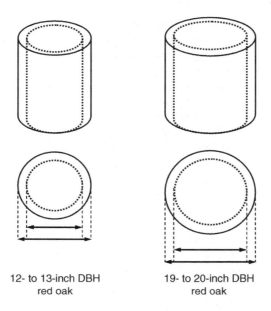

12- to 13-inch DBH   19- to 20-inch DBH
     red oak   red oak

Figure 4.12 One-inch increment to diameter at
breast height for red oak.

As Maital (1994) points out, if this is the case what are the benefits of cost-cutting. Thus, how your business creates value, or new innovative outputs that create additional value is an ongoing activity of managers (Maital 1994).

What is the economic information contained in TR? Based on equation (4.2), TR is a function of price, P, and the amount of output produce or provided by the production system, $Q(\bullet)$. The economic information then seems to be price and the production system. A closer look at these two components is warranted starting with price. The economic information summarized in price is: (1) a measure of relative scarcity and a per unit measure of value; (2) the intersection of supply and demand for the output; (3) the market structure; and (4) the assertions concerning human decision behavior given in Chapter 1. In short, price is economic information that consumers observe and use to make choices about whether or not to buy your output given their budget. Your ability to generate TR depends on the information you can obtain about your consumers' preferences and the market in which those preferences are expressed. The economic information summarized in the production system is technical efficiency and the Law of Diminishing Returns. This information is obtained from your answers to the three fundamental questions for systematically examining the production system given in Chapter 2.

Total revenue is most commonly defined with respect to output. The general idea is simply how much revenue can I create by selling the output that I produce or provide. Given this idea, there are three descriptors of revenue: total revenue, average revenue, and marginal revenue. If you have limited ability to set the market price of your output, then $AR = MR = P$. In other words, the additional revenue you can generate by selling a single unit of output is the same as the per unit revenue generated by selling a given quantity of output. The usefulness of these revenue descriptors is the ability to compare them to the cost of producing the output to help you search for ways to increase profits.

Total revenue can also be defined with respect to inputs. These descriptors are: total revenue product, average revenue product, and marginal revenue product. These descriptors of revenue relay on the precision that you answer the three fundamental questions to systematically examine the production system. They will allow you to relate the productivity of an input to the revenue it generates. The usefulness of these revenue descriptors is to compare what you pay for the inputs with the revenue that they generate to help you search for ways to increase profits.

Finally, I began this chapter by talking about opportunity cost which may have seemed a little odd given the focus is supposed to be on revenue generated by selling your output. Odd as it may seem, I am going to end this chapter by talking about opportunity cost – the value of the next best alternative forgone. As a manager you must continually ask the following questions:

- What potential value am I forgoing based on my current production system, markets, and perceived consumer preferences?
- What potential value might I forgo based on changes in my production system, changes in the markets, and changes in consumer preferences?
- What economic information do I need to collect to help me answer these questions?

The economic information from the Pillars of Price and Value will help you answer these questions.

5 Profit

In Chapter 1, I discussed some assertions about individual decision behavior, one of which is that humans are maximizers. We seek to maximize our net benefits or profits. This is done by weighing the benefits and costs of a choice. This is illustrated in the Architectural Plan for Profit given in Figure 5.1.

The simple idea is to make the revenues (i.e. the Pillars of Price and Value given the outputs produced or provided by the production system) as large as possible relative to the costs (i.e. the Pillar of Cost given the relevant inputs used by the production system). This is illustrated by Lichtkoppler and Kuehn (2003):

> Results of the 2002 Great Lakes charter captain surveys suggest that to continue profitability, charter captains should aggressively market their industry, increase revenues, and reduce expenses.
>
> (Lichtkoppler and Kuehn 2003: 4)

A similar argument is made by the authors of the Mobile Micromill (Becker *et al.* 2004) and the Maple Syrup Operation (Graham *et al.* 2006; Huyler 2000; CFBMC 2000) case studies. The purpose of these publications is to provide potential entrepreneurs with revenue and cost information concerning profits and how to increase profitability. This is consistent with the behavioral assertion described in Chapter 1.

In Chapters 2 through 4, the Architectural Plan for Profit, as represented by the profit and least cost/cost effective models, was taken apart and individual components were examined critically with the economic information highlighted for each component provided to a manager. In this chapter, I will put the parts back together to illustrate how the resulting economic information can be used by managers to help make choices to increase profits. I will start with a brief review of the profit and least cost/cost effective models.

Profit model

The profit model is given by equation (5.1).

$$
\begin{aligned}
Max\ \Pi &= TR - TC \\
&= P \cdot Q(x_1, x_2, ..., x_j) - TVC - TFC \\
&= P \cdot Q(\bullet) - \sum_j w_j \cdot x_j - TFC
\end{aligned}
\tag{5.1}
$$

where

    Max Π denotes maximization of profit (Π);

    TR denotes total revenue, $TR = P \cdot Q(x_1, x_2, \ldots, x_j) = P \cdot Q(\bullet)$

        P denotes the market price of the good or service,

        $Q(x_1, x_2, \ldots, x_j) = Q(\bullet)$ denotes the system of producing or providing a good or service, *Q*, using inputs that the manager has direct control over, $x_1, x_2, \ldots, x_j$,

    TC denotes total cost, $TC = TVC + TFC$;

    TVC denotes total variable costs, $TVC = \sum_j w_j \cdot x_j$,

        $\sum_j$ denotes the summation operator,

        w_j denotes the *j*th wage or price paid for *j*th input x_j,

        x_j denotes the *j*th input; for example, labor,

    TFC denotes total fixed costs.

Equation (5.1) is the economic model of decision behavior that allows the explicit weighing of benefits relative to costs.

Least cost/cost effective models

The decision behavior illustrated by the least cost model, equation (5.2), is to minimize the variable costs of producing or providing a given level of output.

$$Min\ TVC = \sum_j w_j \cdot x_j$$

s.t.   (5.2)

$$Q(x_1, x_2, \ldots, x_j) = Q^0$$

where

    Min TVC denotes the objective of minimizing total variable costs, $TVC = \sum_j w_j \cdot x_j$,

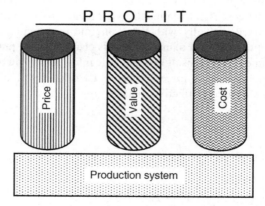

Figure 5.1 The Architectural Plan for Profit.

Σ_j denotes the summation operator,

w_j denotes the jth wage or price paid for jth input x_j,

x_j denotes the jth input,

s.t. denotes "subject to" and what follows is a constraint placed on the objective;

$Q(x_1, x_2,\ldots, x_j) = Q^0$ denotes the system of producing or providing a good or service, Q, using inputs the manager has direct control over, $x_1, x_2,\ldots, x_j$, at a given quantity or quality, Q^0.

The decision behavior illustrated by the cost effective model, equation (5.3), is to produce or provide the most output for a given budget.

$$Max\ Q(x_1, x_2,\ldots, x_j)$$

s.t.

$$\sum_j w_j \cdot x_j = C^0 \tag{5.3}$$

where

$Max\ Q(x_1, x_2,\ldots, x_j)$ denotes the system of providing or producing the maximizing quantity or quality of a good or service, Q, using inputs that the manager has direct control over, $x_1, x_2,\ldots, x_j$;

s.t. denotes subject to;

$\sum_j w_j \cdot x_j = C^0$ denotes that the TVCs of using inputs $x_1, x_2,\ldots, x_j$ to produce or provide the good or service cannot exceed a given budget, C^0;

Σ_j denotes the summation operator,

w_j denotes the jth wage or price paid for jth input x_j,

x_j denotes the jth input.

Optimizing equation (5.1) requires managing and minimizing costs. Appendix 4 shows that the optimal input–output combination resulting from equation (5.1) must also satisfy equations (5.2) and (5.3) and must be production cost efficient. Thus, I will focus only on the profit model.

Case studies

The case studies used in this chapter will include the Inside-Out (ISO) Beams (Patterson *et al.* 2002 and Patterson and Xie 1998), Mobile Micromill (Becker *et al.* 2004), Maple Syrup Operation (Huyler 2000 and CFBMC 2000), and Great Lakes Charter Boat Fishing (Lichtkoppler and Kuehn 2003). The case studies can be found on the Routledge website for this book. Their summaries will not be repeated here because they were used in earlier chapters.

Profit maximization

How do you know if a business has maximized profits – producing or providing the optimal amount of outputs using the optimal level of inputs – using equation (5.1)? The answer, while seeming paradoxical, is that you do not know if they have *maximized* profits (Silberberg and Suen 2001). Then what is the purpose of the profit model? While on the outset it would seem that the greatest utility of equation (5.1) is defining the absolute level of outputs and inputs that maximize profits this, however, is not the case. The greatest utility of equation (5.1) is developing a rule from it that can be used to compare the profitability of one production level relative to another systematically. Thus the profit model provides a rule to examine choices that are consistent with the assertion that your goal is to maximize profits.

The rule comes from marginal analysis. In Chapter 1, the concept of marginal analysis was introduced. I will summarize that discussion briefly in the next section. However, I would recommend that you review that discussion in Chapter 1.

Marginal analysis

As a manager you will always search for ways to increase your profits, or as stated in the last section, to compare the profitability of one production level relative to another. Marginal analysis is a systematic examination of one choice relative to another given observable market conditions. Given the assertion of profit maximization, marginal analysis can determine if profits increase or decrease as production is increased (decreased) by an incremental amount. The rule states that you should increase the amount of output you provide if its incremental or marginal revenue is greater than its incremental or marginal cost. I will develop and examine the profit maximization searching rule with respect to output.[1]

Output approach

Total revenue

Total revenue is the first of two variables of profit as given in equation (5.1). In Chapter 4, I discussed the economic information in TR. Graphically, TR is illustrated by Figure 5.2.

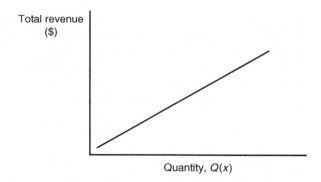

Figure 5.2 Total revenue.

Comparing Figure 5.2 with the TR for the Great Lakes Charter Boat Fishing (Figure 4.2) and ISO Beams (Figure 4.3) case studies shows the same linear relationship between revenue and output. Comparing Figure 5.2 with the TR for select red oak by tree grade (Figure 4.4) and loblolly pine by product class (Figure 4.5) shows that the relationship between revenue and output for any given tree grade or product class is linear. What this means is that selling double the output will double the revenue.[2]

Total cost

The second variable of profit as given by equation (5.1) is total costs (TC). In Chapter 3, I discussed the economic information in TC. Graphically, TC is illustrated by Figure 5.3.

The shape of the TC curve is derived directly from the shape of the production system which is determined by the Law of Diminishing Returns. The TC curve depicts the technically efficient and production cost efficient fixed and variable costs of producing any given level of output.[3] For example, Tables 3.3a and 3.3b illustrate the TCs of annual sap collection (Huyler 2000).

Profit

As defined by equation (5.1), profits are the residual after all opportunity costs (both explicit and implicit, see Chapter 1) have been taken into account. Graphically, profit can be depicted by overlaying the graph of TC on the graph of TR (Figure 5.4a).

Figure 5.4b illustrates that profit for any given level of output is the vertical distance between TR and TC. Moving from left to right on the output axis: (1) *TR < TC* and $\Pi < 0$ at low production levels the fixed and variable production costs outweigh the revenue received if they are sold; (2) *TR = TC* and $\Pi = 0$ at some output level profits switch from negative to positive; (3) *TR > TC* and $\Pi > 0$, the following

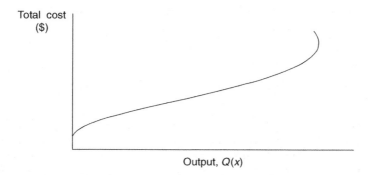

Figure 5.3 Total cost.

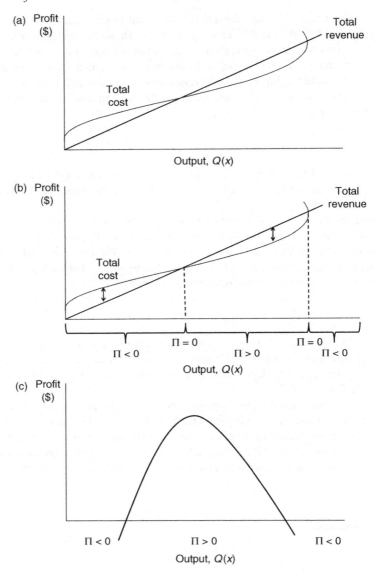

Figure 5.4 (*a*) Total revenue and total cost; (*b*) Profit; (*c*) Profit.

output levels are of the most interest to entrepreneurs; (4) *TR = TC* and Π = 0 at some output level profits switch from positive to negative because the production system follows the Law of Diminishing Returns (i.e. the fixed factors of production constrain the ability to produce output in such a manner that using more variable input causes the amount of output produced per unit input (average product) to decline); (5) $TR < TC$ and $\Pi < 0$ if the entrepreneur continues to produce output the total costs will be greater than the revenue generated by their sales.

Analyzing the previous paragraph and Figure 5.4b shows that profits follow a reasonable path based on the economic information in TC and TR. This is illustrated in Figure 5.4c. Based on Figure 5.4c the obvious output level where profits are the greatest is at the top of the curve. However, the intrinsic value of developing Figures 5.4a, 5.4b, and 5.4c is not to identify the top of the curve in Figure 5.4c but the profit searching rule that can be derived from its development.

Profit searching rule

Individuals are motivated by profits and seek ways to improve their profits (Kant 2003). If I produce or provide more (or less) output, what will happen to my profits? My managerial goal is for this output change to increase profits. Thus, I am looking for output changes that will lead to positive changes in profit. How can I use this reasoning to develop a rule for searching if output changes will lead to positive profit changes?

This idea can be summarized by using algebra and Figure 5.5a.

Let Q_1 denote the current output levels and $\Pi_1 = TR_1 - TC_1$ denotes the profit given Q_1 and Q_2 denote alternative output levels produced or provided and $\Pi_2 = TR_2 - TC_2$ denotes the profit given Q_2. How my profit changes if I change output can be represented algebraically by equation (5.4):

$$
\begin{aligned}
\frac{\Delta\Pi_{2,1}}{\Delta Q_{2,1}} &= \frac{\Pi_2 - \Pi_1}{Q_2 - Q_1} \\
&= \frac{TR_2 - TC_2 - [TR_1 - TC_1]}{Q_2 - Q_1} \\
&= \frac{TR_2 - TR_1 - TC_2 + TC_1}{Q_2 - Q_1} \\
&= \frac{TR_2 - TR_1}{Q_2 - Q_1} - \frac{TC_2 - TC_1}{Q_2 - Q_1} \\
&= \frac{\Delta TR_{2,1}}{\Delta Q_{2,1}} - \frac{\Delta TC_{2,1}}{\Delta Q_{2,1}}
\end{aligned}
\tag{5.4}
$$

What the last term in equation (5.4) shows is that I want to compare the change in TR resulting from the change in output to the change in TC resulting from the change in output. Examining the last term in equation (5.4) a little closer reveals two interesting relationships. First, $\Delta TR_{2,1}/\Delta Q_{2,1}$ is exactly the same as equation (4.5) and defines marginal revenue (MR) or the change in TR $(TR_2 - TR_1)$ from the sale of an additional unit of output $(Q_2 - Q_1)$. In addition, equation (4.5) shows that, given the market assumption of workable competition, MR equals output price (i.e. $MR = P$). Second, $\Delta TC_{2,1}/\Delta Q_{2,1}$ is exactly the same as equation (3.7) and defines marginal cost (MC) or the change in production costs $(TC_2 - TC_1)$ from producing an additional unit of output $(Q_2 - Q_1)$. The last term in equation (5.4)

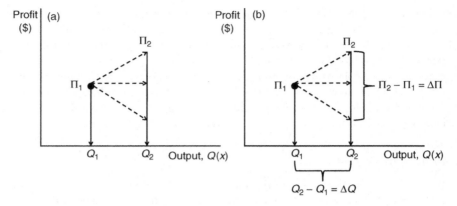

Figure 5.5 (*a, b*) Profit searching.

summarizes the profit searching rule and can be rewritten, in general terms, as a comparison between MR and MC given in equation (5.5):

$$\frac{\Delta\Pi}{\Delta Q} = MR - MC \qquad (5.5)$$

According to the searching rule, for profit to increase the added revenue from the sale of the additional output (MR) must more than cover the additional variable production costs (MC). The left-hand side of equation (5.5) is illustrated by Figure 5.5b.

There are three possible paths illustrated in Figure 5.5b. If $MR > MC$ then $\Delta\Pi = \Pi_2 - \Pi_1 > 0$ and $\Delta\Pi/\Delta Q > 0$. This path shows an increase in profits. If $MR = MC$ then $\Delta\Pi = \Pi_2 - \Pi_1 = 0$ and $\Delta\Pi/\Delta Q = 0$. This path shows no increase in profits. If $MR < MC$ then $\Delta\Pi = \Pi_2 - \Pi_1 < 0$ and $\Delta\Pi/\Delta Q < 0$. This path shows a decrease in profits.

The profit searching rule is illustrated by using the Maple Syrup Operation case study (CFBMC 2000). Table 5.1a defines the annual costs to produce maple syrup for a 2,550 gallon operation.

Table 5.1b defines the marginal production costs. I am currently producing 765 gallons of maple syrup per year. If I increase production to 1,530 gallons of maple syrup per year, this will add $11.26 per gallon of maple syrup per year to my costs (i.e. marginal costs). If I sell a gallon of maple syrup for more than $11.26 per gallon (i.e. marginal revenue) then my annual profits will increase. If I sell a gallon of maple syrup for less then $11.26 per gallon than my annual profits will decrease.[4]

Profit searching rule – revisited

The profit searching rule in equation (5.5) can be used by entrepreneurs to increase their profits. Focusing on the MC term of the profit searching rule briefly, marginal cost has been defined as the change in total cost per unit change in output (Chapter 3)

Table 5.1a Annual production costs for 2,550-gallon maple syrup operation[‡]

Item	127.5 (gal/yr)	255 (gal/yr)	765 (gal/yr)	1,530 (gal/yr)	2,550 (gal/yr)
Total Variable Cost					
Labor	$1,601.79	$2,347.50	$4,641.17	$8,259.34	$13,083.57
Supplies	$1,208.08	$2,271.45	$5,068.54	$9,204.94	$14,812.56
Other	$471.12	$689.85	$1,261.92	$2,120.02	$3,667.98
Total	$3,280.99	$5,308.81	$10,971.62	$19,584.30	$31,564.11
Total Fixed Costs	$33,223.23	$33,223.23	$33,223.23	$33,223.23	$33,223.23
Total Cost	$36,504.22	$38,532.04	$44,194.85	$52,807.53	$64,787.34

[‡]Source of data CFBMC (2000). Costs are in 1999 US dollars using an exchange rate of 1.49 Canadian dollar per US Dollar. (gal/yr) denotes gallons of maple syrup per year. These costs do not include the cost of establishing a sugarbush.

Table 5.1b Marginal costs for 2,550-gallon maple syrup operation[‡]

Annual production (gal/yr)	Total variable cost (USD/yr)	Marginal cost (USD/gal)
127.5	$36,504.22	
		$15.90
255	$38,532.04	
		$11.10
765	$44,194.85	
		$11.26
1,530	$52,807.53	
		$11.74
2,550	$64,787.34	

[‡]Source of data CFBMC (2000). Costs are in 1999 US dollars (USD). (gal/yr) denotes gallons of maple syrup per year. (USD/gal) denotes 1999 US dollars per gallon of maple syrup.

or, as above as the change in variable production costs associated with changing the output level. Although both definitions are accurate, the difference highlights the fact that TFCs are not included in determining MC. This is illustrated in equations (3.7) and (3.8) from Chapter 3. As is shown in equation (5.1), TFCs are an integral component in determining profit. Thus, while the profit searching rule can be used to determine if changing output will improve or increase profits, equation (5.1) must be used to determine if the new output level will result in profits being positive. This requires including TFCs.

Profit maximization given a non-continuous production process, costs, and revenues

As has been discussed previously (e.g. Chapter 2), not all production processes can be described in a continuous manner. Using the Maple Syrup Operation case study as an example, Table 5.2a lists the total variable, total fixed, and total costs for three maple syrup operations.

Table 5.2a shows that if maple syrup entrepreneurs want to increase the size of their operation from 1,530 to 2,550 or 5,100 gallons of maple syrup per year, not only would their variable production costs increase but the fixed costs would also increase.[5] Table 5.2b illustrates the average and incremental costs of increasing the operation size.[6] In contrast, Table 5.1a lists the costs for a 2,550-gallon maple syrup per year operation. Such an operation has the equipment and infrastructure to produce 127.5, 255, 765, 1,530, or 2,550 gallons of maple syrup per year. Thus, the fixed costs are the same no matter how much is produced. NOTE: I have changed the terminology from marginal cost to incremental cost when discussing

Table 5.2a Total annual production costs of three different maple syrup operations[‡]

Cost item	1,530 (gal/yr)	2,550 (gal/yr)	5,100 (gal/yr)
Variable Costs			
Labor	$8,259.34	$13,083.57	$25,144.14
Supplies	$9,204.94	$14,812.56	$28,748.85
Other	$2,120.02	$3,667.98	$6,528.32
Fixed Costs	$10,688.95	$15,048.79	$24,726.19
Total Cost	$30,273.25	$46,612.90	$85,147.50

[‡]Source of cost data CFBMC (2000). Costs are in 1999 US dollars using an exchange rate of 1.49 Canadian dollar per US dollar. (gal/yr) denotes gallons of maple syrup per year. These costs do not include the cost of establishing a larger sugarbush or the costs purchasing the additional equipment.

Table 5.2b Average and incremental annual production costs of three different maple syrup operations[‡]

Annual production (gal/yr)	Total cost (USD/yr)	Average cost (USD/gal)	Incremental cost (USD/gal)
1,530	$30,273.25	$19.79	
			$16.02
2,550	$46,612.90	$18.28	
			$15.11
5,100	$85,147.50	$16.70	

$$\frac{\Delta TC}{\Delta Q} = \frac{\$85,147.50 - \$46,612.90 \text{ (USD/yr)}}{5,100 - 2,550 \text{ (gal/yr)}} = \$15.11 \text{ (USD/gal)}$$

[‡]Source of cost data CFBMC (2000). Annual production costs are in 1999 US dollars (USD/yr) using an exchange rate of 1.49 Canadian dollar per US dollar. (gal/yr) denotes gallons of maple syrup per year. (USD/gal) denotes 1999 US dollars per gallon of maple syrup. These costs do not include the cost of establishing a larger sugarbush or the costs purchasing the additional equipment.

the difference between Tables 5.1a and 5.1b and Tables 5.2a and 5.2b. This illustrates the nature of the change required to adjust the size of an operation versus increasing/decreasing output from an operation of a given size. For example, increasing the size of the operation would require purchasing additional capital equipment (e.g. taps, vacuum lines, etc., see Table 3.1), and land (e.g. obtaining a larger sugarbush), etc. The annual costs of these purchases (e.g. payments to the bank for the loan to increase the size of your operation, which will be discussed in Chapter 8) must also be included. As the costs given in Tables 5.2a and 5.2b do not include these costs, they underestimate the total costs.

As was discussed in Chapter 4, differences in quality or other output attributes may cause differences in output price. In the case of maple syrup, it is graded based primarily on color with Grade A light amber receiving the highest price (Marckers *et al.* 2006). A similar concept is illustrated in the Loblolly Pine Plantation case study. Small diameter loblolly pines are used for pulpwood. As the trees increase in diameter and height, they can be used as Chip-n-Saw, sawtimber, plywood logs, or power poles. The lowest valued output is pulpwood and the highest valued output is power poles as illustrated by Figure 4.5. Figure 4.7 illustrates the MR associated with the different loblolly pine product classes. However, unlike the incremental costs in Table 5.2b, Table 4.4 illustrates the fact that no MR with any economic interpretation can be calculated between product classes. The same can be stated for any output that has differences in quality or other attributes that may cause differences in output price. In these cases, each profit position or investment must be estimated and compared. The choice would be that position giving you the greatest profit. In Chapter 8 I will discuss common tools that can be used to choose among different investments.

Breakeven analysis

Breakeven analysis provides different benchmarks in profit searching by an entrepreneur. Basically, its objective is to show how many units of output must be produced and sold to cover some measure of costs. I will examine three common breakeven points associated with average total costs, average variable cost, and marginal cost.

Average total cost

This breakeven point determines how many units of output must be produced and sold to cover variable and fixed costs. This is illustrated in equation (5.6a):

$$ATC = \frac{TC}{Q(\bullet)} = P \tag{5.6a}$$

There are two observations drawn from equation (5.6a). First, the units on output price, P, are dollars per unit output and the units on average total cost (ATC) are also dollars per unit output (see Chapter 3). Second, the condition expressed in

114   *Profit*

equation (5.6a) is equivalent to finding the point where total cost equals total revenue as shown in equation (5.6b).

$$ATC = \frac{TC}{Q(\bullet)} = P$$

$$\left[\frac{TC}{Q(\bullet)}\right] \cdot Q(\bullet) = P \cdot Q(\bullet) \tag{5.6b}$$

$$TC = TR$$

Table 5.3 gives the annual average fixed, variable, and total costs for a maple syrup operation that produces 5,100 gallons of maple syrup annually.

In the *New Hampshire Forest Market Report, 1998–1999* published by the University of New Hampshire Cooperative Extension, the retail price was $33.10 per gallon of maple syrup. Using linear interpolation given the information in Table 5.3 and illustrated in Figure 5.6, a maple syrup entrepreneur must produce and sell approximately 1,351 gallons of maple syrup annually at a price of $33.10 per gallon to cover the fixed and variable production costs.

As described in Chapter 3 and listed in Table 3.5, the cost data for the Great Lakes Charter Boat Fishing case study do not distinguish explicitly between fixed and variable operating expenses. The annual operating expenses for New York's charter boat fishing captains are $11,093 without boat payments. These expenses are total operating costs averaged over the responding charter boat fishing captains answering a survey about operating expenses. Table 4.1 lists the price per full-day or half-day trip by various fish species. Based on this information, New York's charter boat fishing captains' must sell 27 full-day or 36 half-day

Table 5.3 Annual average and marginal costs for 5,100-gallon maple syrup operation[‡]

Annual production (gal/yr)	Average fixed costs (USD/gal)	Average variable cost (USD/gal)	Average total cost (USD/gal)	Marginal cost (USD/gal)
127.5	$193.93	$25.73	$219.66	
				$15.90
255	$96.97	$20.82	$117.78	
				$11.10
765	$32.32	$14.34	$46.66	
				$11.26
1,530	$16.16	$12.80	$28.96	
				$11.74
2,550	$9.70	$12.38	$22.07	
				$11.32
5,100	$4.85	$11.85	$16.70	

[‡]Source of cost data CFBMC (2000). Costs are in 1999 US dollars using an exchange rate of 1.49 Canadian dollar per US Dollar. (gal/yr) denotes gallons of maple syrup produced per year. (USD/gal) denotes US dollars per gallons of maple syrup produced annually.

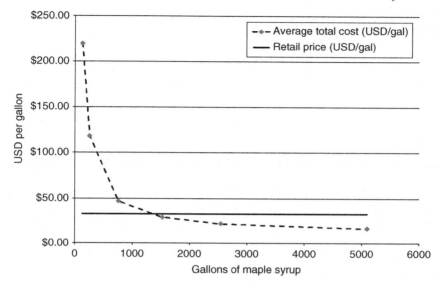

Figure 5.6 Average annual costs for 5,100-gallon maple syrup operation.[‡]

[‡]Source of cost data CFBMC (2000). Costs are in 1999 US dollars using an exchange rate of 1.49 Canadian dollar per US dollar. (USD/gal) denotes US dollars per gallons of maple syrup produced annually. Retail Price is 33.10 (USD/gal) based on *New Hampshire Forest Market Report, 1998–1999* published by the University of New Hampshire Cooperative Extension (http://extension.unh.edu/resources/resource/267/NH_Forest_Market_Report,_1998-1999 accessed on 6 October 2009).

fishing trips for lake trout and salmon to cover their annual operating costs given in Table 3.5. It is left to the reader to calculate the number of full- and half-day fishing trips for the rest of the listed species that charter boat fishing captains must sell to breakeven and cover operating expenses.

The ISO Beams and Mobile Micromill case studies also provide breakeven analyses. However, the analyses discussed include payments to buy capital equipment. I will discuss these breakeven analyses in Chapter 8.

Average variable cost

This breakeven point determines how many units of output must be produced and sold to cover variable costs. This is illustrated in equation (5.7a):

$$AVC = \frac{TVC}{Q(\cdot)} = P \tag{5.7a}$$

The two observations that can be drawn from equation (5.7a) are similar to those from equation (5.6a). First, the units on output price, P, are dollars per unit output and the units on average variable cost (AVC) are also dollars per unit output (see Chapter 3). Second, the condition expressed in equation (5.7a) is equivalent to finding the point where total variable cost first equals total revenue as shown in equation (5.7b):

$$AVC = \frac{TVC}{Q(\cdot)} = P$$

$$\left[\frac{TVC}{Q(\cdot)}\right] \cdot Q(\cdot) = P \cdot Q(\cdot) \tag{5.7b}$$

$$TVC = TR$$

Examining Table 5.3 shows that the average variable production costs are less than the retail price of \$33.10 per gallon of maple syrup. This is illustrated in Figure 5.7.

As no cost information is given below producing and selling 127.5 gallons of maple syrup per year, this maple syrup entrepreneur must annually produce and sell about 127.5 gallons of maple syrup at a retail price of \$33.10 per gallon to cover the variable production costs.

Marginal cost

The final breakeven point compares MC with output price as illustrated in equation (5.8):

$$MC < P \tag{5.8}$$

There are three observations drawn from equation (5.8). First, as with the previous two breakeven points, the units on output price, P, are dollars per unit output and the units on marginal costs (MC) are also dollars per unit output (see Chapter 3). Second, the condition $MC \leq P$ bounds profit between its minimum and maximum points.[7] Third, the reader will note the similarity of the breakeven point given in equation (5.8) with the profit searching rule given in equation (5.5) and illustrated in Figure 5.5. Combining the second and third observations show that if $MC < P$ then the additional production costs of increasing output by one unit (MC) will be less than the price that same unit could be sold for (P). As a result, profits are increasing.

Table 5.3 also gives the marginal costs for a maple syrup operation that produces 5,100 gallons of maple syrup annually. The marginal costs and a retail price of \$33.10 per gallon are illustrated in Figure 5.8.

Based on these cost data, each additional gallon of maple can be sold for greater than its production costs.

To produce or not to produce?

The question of whether to produce or not seems fairly simple. If profits are positive, then produce. However, if profits are negative, then should you stop producing? The answer to this question may not be obvious. The key to thinking about this question is to remember that costs comprise variable and fixed production

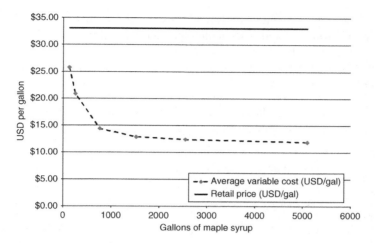

Figure 5.7 Average annual variable costs for 5,100-gallon maple syrup operation.[‡]

[‡]Source of cost data CFBMC (2000). Costs are in 1999 US dollars using an exchange rate of 1.49 Canadian dollar per US dollar. (USD/gal) denotes US dollars per gallons of maple syrup produced annually. Retail Price is 33.10 (USD/gal) based on *New Hampshire Forest Market Report, 1998–1999* published by the University of New Hampshire Cooperative Extension (http://extension.unh.edu/resources/resource/267/NH_Forest_Market_Report,_1998-1999 accessed on 6 October 2009).

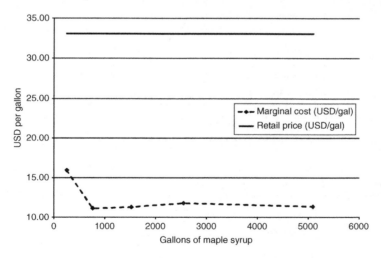

Figure 5.8 Marginal costs for 5,100-gallon maple syrup operation.[‡]

[‡]Source of cost data CFBMC (2000). Costs are in 1999 US dollars using an exchange rate of 1.49 Canadian dollar per US dollar. (USD/gal) denotes US dollars per gallons of maple syrup produced annually. Retail Price is 33.10 (USD/gal) based on *New Hampshire Forest Market Report, 1998–1999* published by the University of New Hampshire Cooperative Extension (http://extension.unh.edu/resources/resource/267/NH_Forest_Market_Report,_1998-1999 accessed on 6 October 2009).

costs. If you stop producing then variable costs would go to zero, but you would still have to cover the fixed costs in the short-run. Thus, if profits are negative, do you lose less money by shutting down or producing?

Table 5.4 describes the three conditions relevant to analyzing the question of whether produce or not.

Condition I is obvious as revenue will cover variable and fixed production costs. Condition II states the loss associated with producing is less then with not producing, as shown in Table 5.5a and 5.5b.

Table 5.4 To produce or not to produce?[‡]

Condition	Profit	Decision
I. P ≥ Minimum ATC	$\Pi > 0$	Produce
II. Minimum AVC < P < Minimum ATC	$\Pi < 0$	Produce; loss associated with producing is less than those associated with not producing
III. P ≤ Minimum AVC	$\Pi < 0$	Shutdown; loss associated with producing is greater than just paying fixed costs

[‡]P denotes output price. AVC denotes average variable costs. ATC denotes average total cost. Π denotes profit.

Table 5.5a To produce or not to produce?: Condition II[‡]

Condition II	
AVC < P	*P < ATC*
TVC < P·Q P·Q – TVC > 0	P·Q < TVC + TFC P·Q – TVC – TFC < 0

[‡]P denotes output price. AVC denotes average variable costs. ATC denotes average total cost. TFC denotes total fixed cost. Q denotes the output level when $MC = MR = P$. Mathematically, ATC and AVC are at their lowest value when $MC = ATC$ and $MC = AVC$, respectively.

Table 5.5b To produce or not to produce?: Condition II[‡]

Condition II	
Shutdown	*Produce*
$\Pi = -TFC$	P·Q – TVC – TFC P·Q – TVC = K > 0 (Table 5.5a) |K – TFC| < TFC

[‡]P denotes output price. AVC denotes average variable costs. ATC denotes average total cost. TVC denotes total variable costs. TFC denotes total fixed cost. Q denotes the output level when $MC = MR = P$. Π denotes profit. $|K - TFC|$ denotes the absolute value of the difference.

The left-hand column of Table 5.5a shows that if output price is greater than average variable costs, then total revenue is greater than the total variable costs of producing the output. The right-hand column shows that the total revenue does not cover the combined variable and fixed production costs; thus, profits are negative. The left-hand column in Table 5.5b shows that the costs of shutting down production are equal to the total fixed costs. The right-hand side of Table 5.5b shows that the loss of producing is less than total fixed costs. The entrepreneur should continue producing in the short-run.

Condition III, given in Table 5.4, states that the loss associated with producing is greater than with not producing. The entrepreneur's decision would be to shutdown. This is illustrated in Tables 5.6a and 5.6b.

The left-hand column of Table 5.6a shows that if output price is less than average variable costs, then total revenue is less than the total variable costs of producing the output. The right-hand column shows that the total revenue does not cover the combined variable and fixed production costs, and as a result, the entrepreneur loses money. As with Table 5.5b, the left-hand column in Table 5.6b shows that the costs of shutting down production are equal to the total fixed costs. The right-hand side of Table 5.6b shows that the loss of producing is greater than total fixed costs. The entrepreneur should stop producing.

Table 5.6a To produce or not to produce?: Condition III[‡]

Condition III	
$P < AVC$ (Table 5.5a)	$P < AVC$
$P \cdot Q < TVC$	$P \cdot Q < TVC + TFC$
$P \cdot Q - TVC < 0$	$P \cdot Q - TVC - TFC < 0$

[‡]P denotes output price. AVC denotes average variable costs. ATC denotes average total cost. TFC denotes total fixed cost. Q denotes the output level when $MC = MR = P$. Mathematically, ATC and AVC are at their lowest value when $MC = ATC$ and $MC = AVC$, respectively.

Table 5.6b To produce or not to produce?: Condition III[‡]

Condition II	
Shutdown	Produce
$\Pi = - TFC$	$P \cdot Q - TVC - TFC$
	$P \cdot Q - TVC = K < 0$ (Table 5.5a)
	$\lvert K - TFC \rvert > TFC$

[‡]P denotes output price. AVC denotes average variable costs. ATC denotes average total cost. TVC denotes total variable costs. TFC denotes total fixed cost. Q denotes the output level when $MC = MR = P$. Π denotes profit. $\lvert K - TFC \rvert$ denotes the absolute value of the difference.

How to use economic information – *profits* – to make better business decisions?

I have described the entrepreneur's objective as maximizing profit. The description of profit is given in equation (5.1) and illustrated in the Architectural Plan for Profit (Figure 5.1). As can be seen by Figure 5.1, Profits rests on the economic information within the Pillars of Price, Value, and Cost. Finally, these pillars rest on the foundation that is the Production System. Chapters 2 through 4 examined each component of the profit model individually and its relationship to the Architectural Plan for Profit. This chapter now looks at the profit model and the Architectural Plan for Profit as a whole.

The economic information contained in profits is a union of the economic information contained in Chapters 2, 3, and 4, and I recommend that the reader reviews the final section of each of these listed chapters.

The utility of the profit model is the profit searching rule derived from the assertion of profit maximization. This rule states that the profits of an activity increase if the marginal or incremental revenues are greater than the marginal or incremental opportunity costs. This simple rule provides entrepreneurs with a method to systematically analyze one production level with an alternative. The marginal or incremental costs are derived from the economic information in the Pillar of Cost and the Production System. The marginal or incremental revenues are derived from the economic information in the Pillars of Price and Value given the outputs from the Production System.

I have called the weighing of marginal or incremental revenues against marginal or incremental costs a "searching" rule for a reason. Profit maximizing is a dynamic process, not a static one. Input and output markets change resulting in fluctuations in the prices you receive for your outputs and pay for your inputs. In addition, while there may be fixed components of a production system in the short-run, you – as a manager – are continually searching for more technically efficient and production cost efficient input–output combinations. Finally, you are continually searching for ways to create additional value to the outputs that you produce or provide and new outputs that are congruent with your current business.

6 Supply and demand

In the previous five chapters, the focus was primarily on the manager or entrepreneur and the assertion that they maximize profits. This assertion was used to develop the Architectural Plan for Profit given in Figure 6.1 and the economic model of profit maximization given in equation (6.1).

$$
\begin{aligned}
Max\ \Pi\ &= TR - TC \\
&= P \cdot Q(x_1, x_2, ..., x_j) - TVC - TFC \\
&= P \cdot Q(\bullet) - \sum_j w_j \cdot x_j - TFC
\end{aligned}
\tag{6.1}
$$

where the terms have been defined at the beginning of Chapters 2, 3, 4, and 5. The Architectural Plan for Profit identifies key pieces of economic information that the entrepreneur should obtain in their search for output levels that will increase their profits. The Architectural Plan for Profit (Figure 6.1) can be thought of as a graphical version of the mathematical equation for profit given in equation (6.1).

In this and the next chapter, I want to step back and take a broad look at the interactions among buyers and sellers. Following the same logic as I have used before, I want start at the ending point and take that apart to reveal the economic information it contains. Figure 6.2 shows the familiar economic model of supply and demand or the interactions among buyers (demand) and sellers (supply).[1]

Figure 6.2 describes a market where an individual can choose to provide or produce a particular good or service and another individual can choose to purchase that particular good or service. Based on the assertion described in Chapter 1, each individual is a maximizer: sellers maximize their profits (Π) and buyers maximize their net benefits (NB). This is described by the paired expressions in equation (6.2):

$$
\begin{aligned}
Max\ \Pi &= [P \cdot Q(\bullet)] - TC \\
Max\ NB &= B - C = B - [P \cdot Q(\bullet)]
\end{aligned}
\tag{6.2}
$$

where B denotes the buyer's benefit from purchasing the good or service and C denotes the explicit portion of their opportunity cost. The paired expressions in equation (6.2) show that the seller's total revenue, $[P \cdot Q(\bullet)]$ (Chapter 4) is the same as the buyer's explicit cost, $[P \cdot Q(\bullet)]$. Comparing Figure 6.2 to equation (6.2)

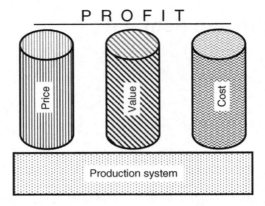

Figure 6.1 The Architectural Plan for Profit.

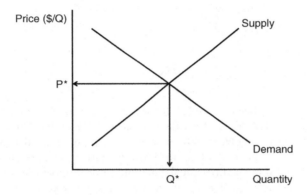

Figure 6.2 The market – supply and demand.

reveals that the price which sellers and buyers agree upon, P* from Figure 6.2, equals the market price, P, from equation (6.2). In addition, the amount of the output exchanged, Q* from Figure 6.2, equals the level of output, $Q(\cdot)$ (Chapter 2) from equation (6.2).[2]

The previous paragraph establishes an initial connection between the supply and demand curves and the assertion concerning individual decision behavior. I will start with discussing the economic information contained in the supply curve and show how it reflects what was discussed in Chapters 2 through 5. Next, I will examine the demand curve to see if any parallels can be drawn between it and what we know about the supply curve.

Case studies

The case studies used in this chapter will include the Inside-Out (ISO) Beams (Patterson *et al.* 2002 and Patterson and Xie 1998), Mobile Micromill (Becker *et al.* 2004), Maple Syrup Operation (Huyler 2000 and CFBMC 2000), Great Lakes Charter Boat Fishing (Lichtkoppler and Kuehn 2003), Select Red Oak (Hahn and Hansen 1991), and Loblolly Pine Plantation (Amateis and Burkhart 1985 and Burkhart *et al.* 1985). The case studies can be found on the Routledge website for this book. Their summaries will not be repeated here as they have been used in earlier chapters.

Supply

A supply curve is traditionally drawn with a positive slope implying that there is a direct relationship between output price and the quantity of the output produced or provided; in other words, as the market price goes up, as a seller you would buy additional inputs to increase the quantity of output you produce or provide. This is illustrated in Figure 6.2 and seems like a reasonable response. In fact, this behavior is codified in the Law of Supply:

> *Ceteris paribus* (other things being equal), as the price of a good or service increases (decreases) the supplier will increase (decrease) the quantity produced or provided.

The Law of Supply leads to defining the supply curve depicted in Figure 6.2:

> The willingness and ability of entrepreneurs to make available a given quantity at a given place and time.

However, what is behind the Law of Supply and the definition of the supply curve that makes the relationship depicted in Figure 6.2 reasonable? Why is supply depicted as a positive relationship between market price and quantity produced or provided? How is the concept of supply and its graphic representation related to the Architectural Plan for Profit? I will answer these questions in the following sections.

Pillar of Price and the Production System foundation

An initial comparison of Figures 6.1 and 6.2 shows both have a price element; in Figure 6.1 it is the Pillar of Price and in Figure 6.2 it is the vertical axis of Price and the price agreed upon by buyers and sellers, P^*. In Figures 6.1 and 6.2, price represents a measure of relative scarcity and a per unit measure of value or a marginal concept (see Chapter 1). In addition, Figures 6.1 and 6.2 both have an output element; in Figure 6.1 it is the foundation described by the Production System (Chapter 2) and in Figure 6.2 it is the horizontal axis of Quantity and the output level of exchange agreed upon by buyers and sellers, Q^*. For argument's sake, let us assume that you produce an output like maple syrup. To produce or make available a gallon of maple syrup you would have to have knowledge of its production system. Thus, you would have systematically answered the following three questions discussed in Chapter 2: (1) What are the input(s)? (2) What are the output(s)? and (3) How do you describe the production process? Based on this line of reasoning there would seem to be a relationship between the supply curve given in Figure 6.2 and a production system. This relationship is shown in Figure 6.3.

Examining Figure 6.3 reveals three observations. First, the horizontal axis from the supply graph is the same as the vertical axis from the production system graph; for example, from 127.5 to 2,550 gallons of maple syrup per year. Second, the output level of exchange agreed upon by buyers and sellers, Q^*, from the supply curve can be tied directly to the input level, x^*, required to produce Q^* as given by the production system; for example, Huyler (2000) and CFBMC (2000) describe the various input levels for various annual levels of maple sap and maple syrup respectively. A similar relationship could be developed between the supply of full- or half-day charter boat fishing trips and the input levels used to provide them. Finally, the economic information contained in the production system (Chapter 2) and the foundation of the Architectural Plan for Profits should also be in the supply curve.

Pillar of Cost

Building the Pillar of Cost requires first identifying all the relevant inputs that depict the underlying fundamental characteristics of the production system and that you, as manager, have direct control over (Chapter 2). Or answering the question "What are the input(s)?", given the price or wage you must pay for each input, you can then develop descriptors of cost: total cost; total variable and total fixed cost; average total, variable, and fixed cost; and marginal cost (Chapter 3). Building the Pillar of Cost also allows the entrepreneur to develop a supply curve for the business.

The definition of supply is the "willingness and ability to make available" a given level of output. From Chapter 5, the profit searching rule describes

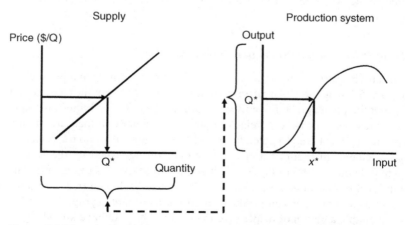

Figure 6.3 The relationship between supply and the production system.

comparing marginal revenue against marginal cost. A profit maximizing entrepreneur will increase the level of output produced or provided if marginal revenue is greater than marginal cost. Thus, I would argue that your "willingness and ability" as a seller of an output is described by the marginal cost curve. Furthermore, I would contend that the minimum price that you would be willing to accept would be defined by the minimum of average variable cost. This contention is based on the "To produce or not to produce" discussion and Tables 5.4, 5.6a, and 5.6b from Chapter 5. Thus, the supply curve is defined as the marginal cost curve above the minimum of average variable cost. To illustrate this, I will use a hypothetical softwood dimension lumber sawmill.[3] While it is hypothetical, it is based on a sawmill in New York State that produces between 5 and 10 million board feet of softwood dimension lumber annually or between 416.6 and 833.3 thousand board feet (MBF) of softwood dimension lumber per month. Table 6.1 describes the production system that uses two inputs: labor and delivered logs to produce the output lumber.

Table 6.2 shows the total variable, average variable, and marginal costs associated with producing dimension lumber.[4]

Figure 6.4 is a graph of the average variable and marginal cost curves. This graph shows the shutdown point or the minimum point on the average variable cost curve as described in Tables 5.4, 5.6a, and 5.6b from Chapter 5.

Table 6.2 also shows the supply schedule derived from the average variable and marginal cost information. Figures 6.5a and 6.5b illustrate the supply curve for this softwood dimension lumber sawmill.[5]

Examining Table 6.2 and Figure 6.5b shows that this softwood dimension lumber sawmill is *not* willing and able to provide less than 450.10 MBF of lumber per month given the current production technology (Table 6.1) and the costs of labor and delivered logs (Table 6.2). If the market price is less than $340.36 per MBF of lumber, then it is cheaper for the sawmill to shutdown than produce any amount of lumber (Tables 5.4, 5.6a, and 5.6b).[6] Technically efficient and production cost efficient output levels are defined by the marginal cost curve (Chapter 3). Thus, this sawmill's supply curve is defined as the marginal cost above the minimum average variable cost curve.

Following the logic used to develop a supply curve for the hypothetical softwood dimension lumber sawmill, I can develop a supply curve based on the Maple Syrup Operation case study (CFBMC 2000). Annual maple syrup variable production cost information is given in Table 6.3.

If I extrapolate the data to a maple syrup operation that produces greater than 5,100 gallons of syrup annually, I can develop average variable and marginal cost curves given in Figure 6.6.[7]

Figure 6.6 shows that the minimum average cost or shutdown price occurs at approximately $12.00 per gallon for maple syrup. Figures 6.7a and 6.7b show the supply curve for this maple syrup operation.

Figure 6.7b shows that the owner(s) of this maple syrup operation are *not* willing and able to provide less than 5,100 gallons of maple syrup annually given current production technology and variable and fixed costs.

Table 6.1 Production system of a softwood dimension lumber sawmill

Labor[§] (persons)	Logs[‡] (cu.ft/month)	Lumber[†] (MBF/month)
7.0	2,089,221.80	261.15
7.5	2,292,922.94	286.62
8.0	2,494,166.09	311.77
8.5	2,691,917.28	336.49
9.0	2,885,275.85	360.66
9.5	3,073,463.04	384.18
10.0	3,255,811.46	406.98
10.5	3,431,755.27	428.97
11.0	3,600,821.09	450.10
11.5	3,762,619.64	470.33
12.0	3,916,837.84	489.60
12.5	4,063,231.71	507.90
13.0	4,201,619.60	525.20
13.5	4,331,876.10	541.48
14.0	4,453,926.27	556.74
14.5	4,567,740.46	570.97
15.0	4,673,329.43	584.17
15.5	4,770,739.88	596.34
16.0	4,860,050.36	607.51
16.5	4,941,367.51	617.67
17.0	5,014,822.58	626.85
17.5	5,080,568.26	635.07
18.0	5,138,775.76	642.35
18.5	5,189,632.16	648.70
19.0	5,233,337.99	654.17
19.5	5,270,105.00	658.76
20.0	5,300,154.14	662.52
20.5	5,323,713.72	665.46
21.0	5,341,017.78	667.63
21.5	5,352,304.55	669.04
22.0	5,357,815.11	669.73

[§]The number of persons employed during a typical ten-hour shift.
[‡]Denotes the cubic foot (cu.ft) volume of logs delivered to the sawmill per month.
[†]The volume, measured as 1000 board feet (MBF), of dimension lumber produced by the sawmill per month. This is calculated based on a lumber recovery factor of 8 board feet per log cubic foot (Spelter and Alderman 2005). A board foot is defined as board with the dimension of 1 × 1 × 12 there are 12 board feet per 1 cu.ft.

This discussion shows that there is a positive relationship between the output price and your willingness and ability as a producer or provider of a given good or service. Simply put, the supply curve has a positive slope. What is also important and not explicitly visible is that the supply curve (e.g. Figures 6.5a and 6.5b, and 6.7a and 6.7b) is based on the production system as described in the previous section. There is a production system for producing a physical good such as maple syrup or dimension lumber (e.g. a 2 × 4 × 8), and there are also production systems for producing a non-physical good such as information provided by a consultant or a contract produced by a lawyer. Thus, the concept of supply and a supply curve is not dependent on

Table 6.2 Variable and marginal costs and supply of a softwood dimension lumber sawmill

Lumber (MBF/month)	TVC§ (USD/month)	AVC‡ (USD/MBF)	MC† (USD/MBF)	Supply (USD/MBF)
261.15	91,975.50	352.19	315.74	–
286.62	100,030.66	349.01	316.35	–
311.77	108,018.80	346.47	317.55	–
336.49	115,911.70	344.47	319.31	–
360.66	123,684.81	342.94	321.60	–
384.18	131,316.89	341.81	324.45	–
406.98	138,789.75	341.03	327.85	–
428.97	146,087.94	340.56	331.84	–
450.10	153,198.56	340.36	336.47	340.36
470.33	160,110.99	340.42	341.78	341.78
489.60	166,816.70	340.72	347.86	347.86
507.90	173,309.03	341.22	354.79	354.79
525.20	179,583.03	341.93	362.69	362.69
541.48	185,635.27	342.83	371.71	371.71
556.74	191,463.72	343.90	382.04	382.04
570.97	197,067.57	345.15	393.89	393.89
584.17	202,447.10	346.56	407.58	407.58
596.34	207,603.60	348.13	423.49	423.49
607.51	212,539.20	349.86	442.11	442.11
617.67	217,256.81	351.74	464.12	464.12
626.85	221,760.01	353.77	490.44	490.44
635.07	226,052.98	355.95	522.37	522.37
642.35	230,140.36	358.28	561.77	561.77
648.70	234,027.27	360.76	611.43	611.43
654.17	237,719.18	363.39	675.77	675.77
658.76	241,221.86	366.17	762.14	762.14
662.52	244,541.34	369.11	883.75	883.75
665.46	247,683.83	372.20	1,067.08	1,067.08
667.63	250,655.74	375.44	1,373.97	1,373.97
669.04	253,463.54	378.85	1,990.16	1,990.16
669.73	256,113.82	382.42	3,847.57	3,847.57

§Total variable cost (TVC) based on an average monthly salary for labor of 5,000 US dollars (USD) per month, a stumpage price of 0.02 USD per cubic foot (cu.ft) (http://www.dec.ny.gov/docs/lands_forests_pdf/spr2008winter.pdf accessed on 11 November 2009) for Adirondack eastern white pine (*Pinus strobus*), and an average transportation cost of 0.01 USD/cu.ft.
‡Average variable cost (AVC) is measured as USD per 1000 board feet (MBF).
†Marginal cost (MC) is measured as USD per 1000 board feet (MBF)

producing a physical good. The supply curve is an economic model that represents the opportunity cost of producing or providing a good or service.

Profit and the Pillar of Value

Based on the previous two sections, it appears that the supply curve is related to production costs directly. How is the supply curve related to profit

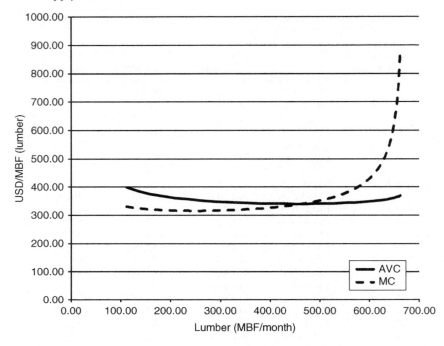

Figure 6.4 Average variable and marginal cost of a softwood dimension lumber
      sawmill.[§]

[§]AVC denotes average variable cost and MC denotes marginal cost.

and the Pillar of Value? I will start with the definition of profit given in
equation (6.3):

$$\Pi = TR - TC$$
$$= P \cdot Q(\bullet) - TVC - TFC \tag{6.3}$$

where the terms have been defined previously. Figure 6.8 isolates the supply
curve from Figure 6.2.

Examining Figure 6.8 also shows the concept of total revenue, $P^* \cdot Q^*$, can be
derived from the supply curve. Total revenue is a value concept. Maital (1994: 6)
describes value as the "degree to which buyers think those goods and services
make them better off, than if they did without." As described in Chapter 1, if a
person purchases a good, we know the minimum assigned value they place on
that good, other things being equal. In addition, as was described in the interpretation of equation (6.2), your total revenue, a measure of value to you as an
entrepreneur, is dependent directly on the value your consumer places on your
good or service.

Is total cost also a measure of value? I could make the algebraic argument that
if the units on total revenue represent value then the units on total cost would
also represent value. When you are purchasing an input (e.g. labor) as part of

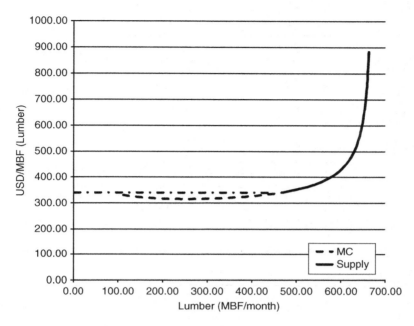

Figure 6.5a Marginal cost and supply curves of a softwood dimension lumber
sawmill.[§]

[§]MC denotes marginal cost.

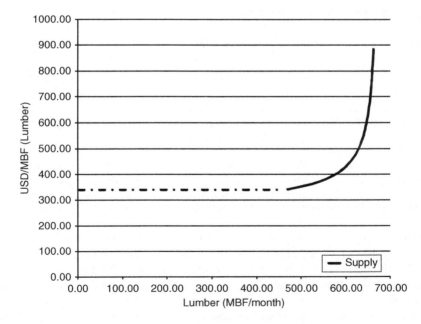

Figure 6.5b Supply curve of a softwood dimension lumber sawmill.

Table 6.3 Total annual variable maple syrup production costs[‡]

Maple syrup (gal/yr)	Total annual variable production costs (USD)
127.5	$3,280.99
255	$5,308.81
765	$10,971.62
1,530	$19,584.30
2,550	$31,564.11
5,100	$60,421.31

[‡]Source of cost data CFBMC (2000). Costs are in 1999 US dollars (USD) using an exchange rate of 1.49 Canadian dollar per US dollar. (gal/yr) denotes gallons of maple syrup per year. These costs do not include the cost of establishing a larger sugarbush or the costs purchasing the additional equipment.

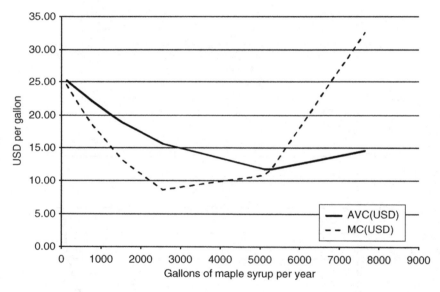

Figure 6.6 Average variable and marginal cost curves for maple syrup production.[‡]

[‡]Source of cost data CFBMC (2000). Costs are in 1999 US dollars (USD) per gallon using an exchange rate of 1.49 Canadian dollar per US dollar. AVC(USD) denotes average variable costs measured in 1999 US dollars. MC(USD) denotes marginal cost measured in 1999 US dollars.

producing or providing a good or service, Maital's (1994: 6) description of value is still applicable. As a buyer, you will only purchase labor if it makes you better off (additional revenue from sales of the additional output produced by the labor relative to its additional cost) than if you did without. In fact, total cost is the explicit component of opportunity cost or the value of the next best alternative forgone. Therefore, whether you are discussing total revenue or total cost, both concepts are important in building the Pillar of Value.

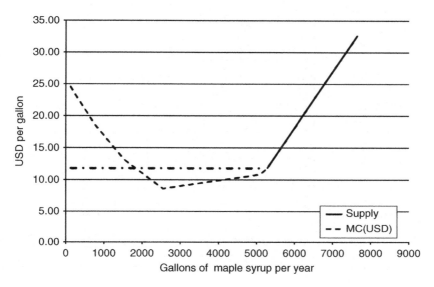

Figure 6.7a Marginal cost and supply curves for maple syrup production.[‡]

[‡]Source of cost data CFBMC (2000). Costs are in 1999 US dollars (USD) per gallon using an exchange rate of 1.49 Canadian dollar per US dollar. MC(USD) denotes marginal cost measured in 1999 US dollars.

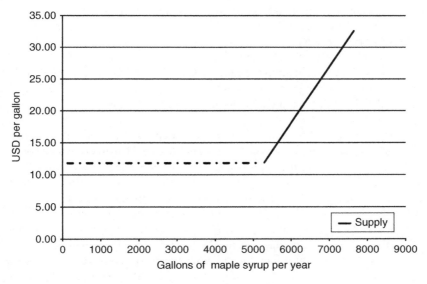

Figure 6.7b Supply curve for maple syrup production.[‡]

[‡]Source of cost data CFBMC (2000). Costs are in 1999 US dollars (USD) per gallon using an exchange rate of 1.49 Canadian dollar per US dollar.

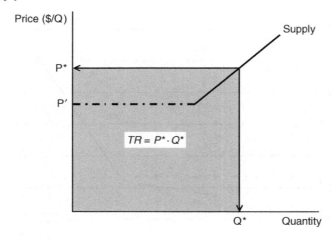

Figure 6.8 The supply curve.

Returning to the profit equation given in equation (6.3), we have established that total revenue can be obtained from the supply curve as illustrated in Figure 6.8. Can the same be said about total cost? The simple answer is no; however, while total costs cannot be determined from the supply curve, total variable cost can. From Figures 6.5a and 6.5b, and 6.7a and 6.7b I have shown that the supply curve is defined as the marginal cost curve above the minimum average variable cost. Mathematically, the area under the supply curve defines total variable costs.[8] This is illustrated in Figure 6.9.

Figures 6.8 and 6.9 define TR and TVC respectively. Subtracting TVC from TR gives a measure of net benefit to the entrepreneur. This is illustrated in Figure 6.10.

This is obviously not profit as given by equations (6.1) or (6.3) and missing is total fixed costs (TFC). Figure 6.10 defines producer surplus:

> Producer surplus (PS) is the benefit to the entrepreneur from producing or providing a good or service for sale at a price defined by the market net of the opportunity costs of producing or providing it (e.g. the variable production costs).

As shown by Figure 6.10, PS is defined as the area below the market price line and above the supply curve. PS is the incentive for entrepreneurs to produce or provide outputs. As described above, PS differs from profit in that it does not include TFC. Is the profit searching rule, developed in Chapter 5, of weighing incremental or marginal revenue against incremental or marginal cost consistent with the definition of PS? Determining the point in Figure 6.2 that maximizes PS given P* will lead to the optimal output level of Q* (Figure 6.10). As the supply curve is defined as marginal cost, the optimal output level is given by equation (6.4a)

$$P = Supply = MC \qquad (6.4a)$$

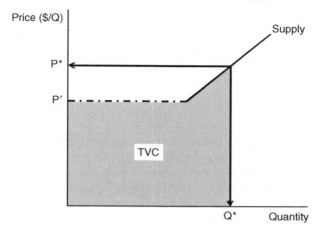

Figure 6.9 Total variable cost.[‡]

[‡]TVC denotes total variable cost. P′ denotes the minimum average variable cost. P* and Q* denote the market equilibrium price and quantity respectively.

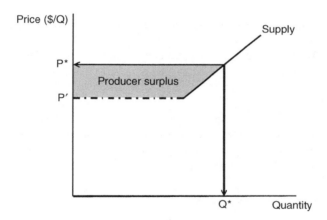

Figure 6.10 Producer surplus.

Assuming workable competition implies that output price, P, is equal to incremental or marginal revenue, MR (Chapter 4). Equation (6.4b) is modified to reflect this:

$$MR = P = Supply = MC \tag{6.4b}$$

Examining the far left-hand and the far right-hand terms give $MR = MC$ which is the profit searching rule given in Chapter 5. Thus, using the profit searching rule to maximizing profits (producer surplus) implies maximizing producer surplus (profits).

As an entrepreneur, your willingness and ability to produce or provide a given amount of output is the result of the economic information you have collected based on the Architectural Plan for Profit. This is summarized in your supply curve. The amount and type of detailed economic information you collect will dictate the detail you can use in generating a supply curve for your business.

The supply curve

As described above, supply, the supply curve, or the supply schedule is an economic model describing the *opportunity cost* of producing or providing a good or service. As such it embodies the economic information contained within the Architectural Plan for Profit. The supply curve in Figure 6.2 is a representation of this economic model and can reflect responses an entrepreneur makes given changes in economic information. The change and response relationship modeled by the supply curve is then Change: changes in the economic information contained within the Architectural Plan for Profit → Response: change in the quantity of the output produced or provided. The importance of modeling is in analyzing the change and response relationship; namely, to recognize and react to changes in economic information by modifying the quantity of output produced or provided.

The modeling of the change and response relationship is simplified by using the following implicit relationship between quantity supplied and a condensed set of variables reflecting the relevant economic information. This is given in Table 6.4 and equation (6.5):

$$Q_S = f(P; ProdSys, P_I, E_P) \qquad (6.5)$$

Equation (6.5) states that the quantity of any given good or service that an entrepreneur is willing and able to provide (Q_S) is a function of (f(•)) the price of the good (P or own-price), the production system used to produce or provide the good or service (*ProdSys*), price of the inputs (P_I), and expected future prices for inputs

Table 6.4 Factors that effect supply (Q_S)

$$Q_S = f(P_Q; ProdSys, P_I, E_p, ...)$$

Factor	Description	Supply effect
P_Q	Own-price; market price of the output	Movement along the supply curve. As own-price increases (decreases) quantity supplied increases (decreases).
P_I	Price of inputs	Price of input increases (decreases), variable production costs increase (decrease), supply curve shifts left (right) and quantity supplied decreases (increases).
ProdSys	Technical efficiency of the production system	Increasing technical efficiency within the production system shifts the supply curve to the right and quantity supplied increases.

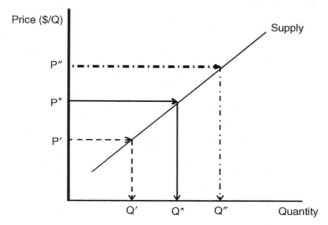

Figure 6.11 Modeling the effect of changes in output price.[‡]

$Q_S = f(P; ProdSys, P_I, E_P)$ all other variables held constant.
[‡]Change: Output or Own-Price → Response: changes in quantity supplied.

and the output (E_P). I will analyze the effect of changing each variable on the quantity supplied and then reflect this in the supply curve model.

The supply curve given in Figure 6.2 is graphed with the price of the good or service (P) and the quantity supplied on the vertical and horizontal axes, respectively. Given the nature of the graph, the production system used, price of inputs, and expected future price changes are held constant. Thus, I will start with changes in the price of the good or service and the entrepreneur's response. If an entrepreneur observes the price of the output is increasing, then profit searching behavior would lead to increasing the quantity of output produced or provided. This change and response relationship is modeled by the supply curve given in Figure 6.11.

The entrepreneur, after reviewing the production process and talking with any employees and managers, determines that changes could be made to the production process to make it more technically efficient (Chapter 2) and production cost efficient (Chapter 3). Thus, the profit searching entrepreneur will change the production process so that more output could be produced using the same amount of inputs or producing the same amount of output using fewer inputs reducing the per unit production costs. This change and response relationship is modeled by the supply curve given in Figure 6.12.

In Figure 6.12, increasing technically efficiency and production cost efficiency is modeled by a shift in the supply curve from Supply to Supply" (all other variables held constant).

If the price of one or more inputs used in the production process increases, then this will cause the total variable production costs to increase (all other variables held constant). A profit searching entrepreneur faced with increasing production costs will look for ways of reducing the use of the now higher priced inputs. Simply using fewer of these inputs will reduce production costs, but it will also

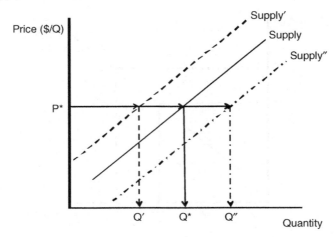

Figure 6.12 Modeling the effect of changes in the production system.[‡]

$Q_S = f(P; \mathbf{ProdSys}, P_I, E_P)$ all other variables held constant
[‡]Change: Production System → Response: changes in quantity supplied.

result in a decrease in the quantity of output produced or provided. This change and response relationship is modeled by the supply curve given in Figure 6.13.

In Figure 6.13, the entrepreneur's response to increased input price from P_I to P_I'' will result in producing less output. This is modeled by shifting the supply curve to the left. One way an entrepreneur may attempt to deal with changing input prices is to search for substitutes for the inputs used currently. Finding lower priced substitutes (i.e. changing input prices from P_I to P_I') will decrease the production costs and allow the profit searching entrepreneur to increase the amount of output produced or provided.

If the entrepreneur expects the future price of the output to increase, they may decrease the quantity of output they are willing to sell today so that they would have them available for sale at the future date and price (all other variables held constant). This change and response relationship is modeled by the supply curve given in Figure 6.14.

In Figure 6.14, a profit searching entrepreneur's reaction to an expected increase in the output price (E_P to E_P'') will be to make less output available for sale today. This is modeled by shifting the supply curve to the left. It is left to the reader to determine how the amount of output supplied would change if the entrepreneur expected future input prices to increase or decrease, then model these change and response relationships using the supply curve.

Demand

Re-examining Figure 6.2 shows that half of the market is described by supply and the other half is described by demand. Market equilibrium is defined by the

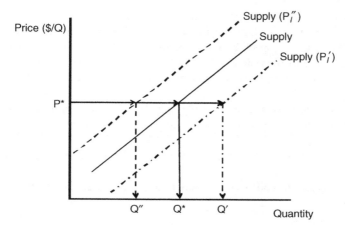

Figure 6.13 Modeling the effect of changes in the price of inputs.[‡]

$Q_S = f(P; ProdSys, \boldsymbol{P_I}, E_P)$ all other variables held constant.
[‡]Change: Price of Inputs → Response: changes in quantity supplied.

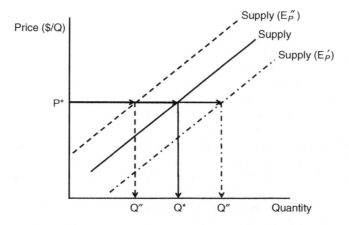

Figure 6.14 Modeling the effect of changes in the expected prices.[‡]

$Q_S = f(P; ProdSys, P_I, \boldsymbol{E_P})$ all other variables held constant.
[‡]Change: Expected Price → Response: changes in quantity supplied.

intersection of supply and demand. Thus, equal attention must be given to demand. Reviewing equation (6.2) illustrates that a seller's profit is dependent upon the total revenue generated by consumers purchasing the outputs they produce or provide. Total revenue is only generated if consumers "think those goods and services make them better off, than if they did without" (Maital 1994: 6). Examining the Architectural Plan for Profit (Figure 6.1) illustrates the importance of demand. In addition to profit, the Pillar of Value is built based on your knowledge of the consumers that purchase your outputs.

A demand curve is drawn traditionally with a negative slope implying that there is an inverse relationship between output price and the quantity of the output purchased by consumers; in other words, as the market price goes down, consumer would buy more of the output. This is illustrated in Figure 6.2 and seems like a reasonable response. In fact, this behavior is codified in the Law of Demand:

> *Ceteris paribus* (other things being equal), consumers purchase more (less) of a good during a given time interval the lower (higher) its relative price.

The Law of Demand leads to defining the demand curve depicted in Figure 6.2:

> Willingness and ability of consumers to pay for a given quantity at a given place and time.

However, what is behind the Law of Demand and the definition of the demand curve that makes the relationship depicted in Figure 6.2 reasonable? Why is demand depicted as an inverse relationship between market price and quantity produced or provided?[9]

Opportunity cost

Cost is not a concept that is often combined with demand. Cost is most often associated with the costs of producing or providing an output (e.g. Chapter 3). However, cost, or more appropriately opportunity cost, can be used to explain why the demand curve is depicted with a negative slope implying that there is an inverse relationship between output price and the quantity of the output purchased. An entrepreneur must remember that consumers are giving up purchasing other commodities if they buy yours. As the relative per unit value or price consumers pay increases, the more consumers must give up or forgo in order to obtain your good or service. In this situation, consumers will choose to purchase less of your good or service, other things being equal. Conversely, if the relative per unit value or price decreases, consumers will choose to purchase more of your good or service, other things being equal. Consumers will weigh the benefits that your good or service can provide relative to the opportunity cost of obtaining it. Thus, the demand curve is depicted as an inverse relationship between market price and quantity and is a model of a consumer's opportunity cost.

Pillar of Value

Based on Chapters 2 and 3, you should know all the economic information about producing or providing your outputs. Now, think about this, you sell a product (e.g. charter boat fishing trips, maple syrup) and you don't know people's willingness and ability to pay (i.e. demand) for your product, so how long will you stay in business? The Pillar of Value not only represents the value you receive from using various different inputs that you have direct control over, it also represents economic information you must obtain about your consumers or customers. The statement by Maital bears repeating again; consumers will only buy your products if they "think those goods and services make them better off, than if they did without" (Maital 1994: 6). Thus, the importance of having this economic information seems obvious. What may be less obvious is where to obtain such information.

I will use an example taken from the forestry profession to illustrate obtaining the economic information on consumers to developing the Pillar of Value. The premise of the example is a consultant who works with private landowners to manage their forestlands sustainably. Butler (2008) gives the following statistics concerning the amount of privately owned forestland.[10] In the United States, family forest landowners own 35 percent and other private landowners own 21 percent of forestland, giving a total of 56 percent of forestland owned by private landowners. Regionally, in the Northern and Southern United States private forestland ownership is 75 percent and 86 percent of forestland, respectively. In the Rocky Mountain and Pacific Coast Regions, private forestland ownership is 25 percent and 33 percent, respectively. Land holdings of less than 100 acres are 33 percent of the total private forestland area. However, 94 percent of all private forest landowners hold lands of less than 100 acres and 61 percent hold only one acre. Private forest landowners have multiple reasons for owning forestland (Butler and Leatherberry 2004; Belin *et al.* 2005; Hagan *et al.* 2005; Butler 2008 and 2010). The primary reasons are for aesthetics, privacy, recreation, and protection of nature. Near the bottom of reasons for forestland ownership is the generation of income (e.g. sale of timber). While these owners are not necessarily against selling timber, it is not the reason they own forestland and any sale of timber must be consistent with the primary reasons for owning forestland. Finally, they are older (55 years old plus), better educated, and have higher incomes than the general population (Butler 2008). Thus, the sustainable forest management services that you are trying to sell them should reflect their ownership preferences, relative size of the land holdings, and inheritance or family legacy concerns these owners will probably have in the near future. This point bears repeating: it is *their* preference set you are managing, not yours. These are not your forests that you are trying to manage sustainably; it is *their* woods.[11]

While very succinct, this example illustrates how you would now be better prepared to sell these landowners sustainable forest management plans that they would be willing and able to buy than if you did not have this economic information. All the sources used to develop the economic information in this example are publicly available. As an entrepreneur, you should actively search out this type of information on your customers.

Finally, a value concept similar to producer surplus can be developed by combining the discussion from the previous sections with the Pillar of Value and Max $NB = B - C$ from equation (6.2). Figure 6.15a illustrates a consumer's incremental choice for purchasing a good; for example, gallons of maple syrup.

At price P' the consumer would only purchase the Q' gallons of maple syrup. The price P' represents the consumer's willingness and ability to pay for Q' gallons of maple syrup; but, they only have to pay the market equilibrium price P^*.

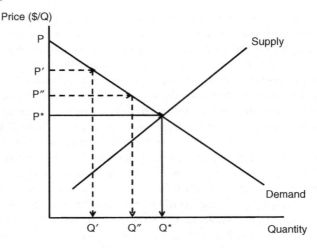

Figure 6.15a Deriving consumer surplus.[‡]

[‡]Price P denotes the price at which zero quantity is purchased by the consumer.

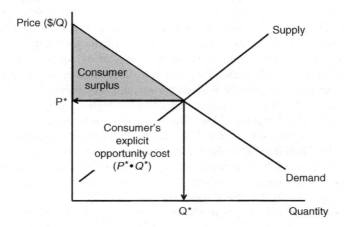

Figure 6.15b Consumer surplus.

Thus, there is a surplus or net benefit of $[P^* \cdot Q' + \frac{1}{2} \cdot Q' \cdot (P - P')] - P^* \cdot Q' = \frac{1}{2} \cdot Q' \cdot (P - P')$ that the consumer captures from being able to purchase at the market equilibrium price of P^*. At price P'' the consumer would only purchase the Q'' gallons of maple syrup. The price P'' represents the consumer's willingness and ability to pay for Q'' gallons of maple syrup; but, they only have to pay the market equilibrium price P^*. Thus, there is a surplus or net benefit of $[P^* \cdot Q'' + \frac{1}{2} \cdot Q'' \cdot (P - P'')] - P^* \cdot Q'' = \frac{1}{2} \cdot Q'' \cdot (P - P'')$ that the consumer captures from being able to purchase at the market equilibrium price of P^*. Finally, at price P^* the consumer would purchase the Q^* gallons of maple syrup. The price P^*

represents the consumer's willingness and ability to pay for Q* gallons of maple syrup. Thus, there is a surplus or net benefit of $[P^* \cdot Q^* + \frac{1}{2} \cdot Q'' \cdot (P - P^*)] - P^* \cdot Q^* = \frac{1}{2} \cdot Q'' \cdot (P - P^*)$ that the consumer captures from being able to purchase at the market equilibrium price of P*.[12] The surplus or net benefit to the consumer is illustrated in Figure 6.15b.

This leads to the definition of consumer surplus:

> Consumer surplus (CS) is the net benefit to the consumer from being able to buy a commodity at the market equilibrium price.

As shown by Figure 6.15b, CS is the area above the market price line and below the demand curve. What Figure 6.15b also shows is that a consumer's surplus is maximized at the market equilibrium price, P*, and quantity, Q*. The reader should notice the parallels between the discussion of consumer and producer surplus.

Demand curve

As described above, demand, the demand curve, or the demand schedule is an economic model describing the *opportunity cost* of consumers for the goods or services that you produce or provide. The demand curve in Figure 6.2 is a representation of this economic model and can reflect reactions consumers make given changes in economic information they obtain from the markets and their preferences (Chapter 1). Based on your knowledge of the market and your customers' preferences, you are building the Pillar of Value contained within the Architectural Plan for Profit. As with the supply curve, the demand curve is used to model change and response relationships. For example, External Change: your prototypical customer's annual income changes → consumer's reaction: change in the quantity of the output they are willing and able to buy. How do you as an entrepreneur respond to this change? The importance of modeling is in analyzing the change and response relationship; namely, to recognize external changes affecting your consumers, their reaction to these external changes, and your response to this new economic information. In other words, how do you "shift" your supply in response to these external changes? Your response might be as simple as modifying the quantity of output produced or provided, or it might require modifying the attributes or characteristics of your output, or creating new products and services in addition to the amount produced or provided of existing products.

Consumers view purchasing your products *not* as charitable giving, but in response to the perceived value they place on acquiring whatever it is that you produce or provide. Thus, as entrepreneurs, you will need to recognize external changes that effect consumers' economic and social environment, their reaction to these external changes, and how you might respond to these changes. Modeling this change and response relationship is developed by identifying the relevant economic information consumers obtain from the markets and social and demographic factors that can be used to complete the description of their preferences.

The economic variables and social factors are summarized in equation (6.6) and Table 6.5.

$$Q_D = f(P; P_S, P_C, Y, E_p, E_Y, SocDem) \tag{6.6}$$

Table 6.5 and equation (6.6) states that the quantity of any given good or service that customers are willing and able to buy (Q_D) is a function of the price of the good (P or own-price); price of related goods (i.e. substitutes (P_S) and complements (P_C)); expected prices of the good (E_P); income and expected income (Y and E_Y, respectively); and social and demographic factors (*SocDem*).

The traditional model of the demand curve is graphed with the price of the good or service (P) and the quantity demanded on the vertical and horizontal axes of Figure 6.2, respectively. Given the nature of this graph all the other variables given in equation (6.6) and Table 6.5 are held constant. To analyze change and response relationships, I will start with the economic variables and in particular changes in the price of the good or service (own-price) and consumers reaction. If consumers observe the price of the output is increasing (decreasing), then maximizing net benefit searching behavior would lead to decreasing (increasing) the quantity of output purchased. This is modeled by the demand curve given in Figure 6.16.

In response to the external change in output price and consumers reaction, the profit seeking entrepreneur would modify the amount of output they produce or provide – that is "shift" their supply – to match consumers' willingness and ability to pay.

The change and response relationships associated with the other economic variables are summarized in Table 6.6 and illustrated in Figure 6.17.

Table 6.5 Demand's economic variables and social factors[‡]

Prices and income	
Variable	*Notation*
Own-price	P
Price of related goods	P_S – price of substitutes
	P_C – price of complements
Income	Y
Expected own-price	E_P
Expected income	E_Y
Elasticity (price and income sensitivity)	

Social and demographic factors	
Demographics	
Exploit openings	
Habit and loyalty	
Trends and exuberance	

[‡] I am indebted to Maital (1994) for the idea of presenting demand in this fashion.

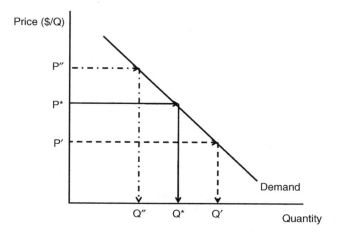

Figure 6.16 Modeling the effect of changes in output price.[‡]

[‡]External Change in market price of the output or Own-Price →
Consumers' React by modifying their amount purchased → Entrepreneurial
Response by modifying the amount of output produced or provided to
match consumer's willingness and ability to purchase. This change and
response relationship is for what are termed "normal" goods. There is a
unique class of goods called "Giffen" goods where as the price of the
good increases, the amount purchased by the consumer also increases.
While the discussion of this is beyond the scope of this book, I would
encourage the interested reader to research this topic.

Table 6.6 Change and response relationships for the economic variables of demand

Economic variable	External change	Consumers' reaction $(Q_D)^{\ddagger}$	Entrepreneur's response $(Q_S)^{\ddagger}$
Price of substitutes (P_S)	Increase	Increase	Increase
	Decrease	Decrease	Decrease
Price of complements (P_C)	Increase	Decrease	Decrease
	Decrease	Increase	Increase
Income $(Y)^{\S}$	Increase	Increase	Increase
	Decrease	Decrease	Decrease
Expected future prices (E_P)	Increase	Increase	Increase
	Decrease	Decrease	Decrease
Expected future income (E_Y)	Increase	Increase	Increase
	Decrease	Decrease	Decrease

[‡]Q_D denotes quantity demanded by consumer and Q_S denotes quantity produced or provided by the
entrepreneur.
[§]The consumers' reaction given changes in income is for what are termed "normal" goods. There
is a unique class of goods called an "inferior" good in which as income increases the amount of the
good consumers are willing and able to purchase decreases. The classic example is using the bus
for transportation. As consumers' incomes increase the amount they use the bus decreases, all else
held constant. While the discussion of this is beyond the scope of this book, I would encourage the
interested reader to research this topic.

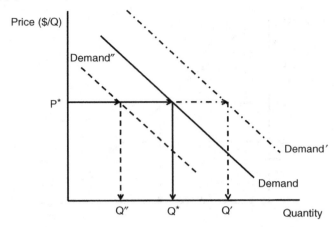

Figure 6.17 Modeling the effect of changes in economic
variables given in Table 6.6.‡

‡External Change in prices or income → Consumers' React by modifying
their amount purchased → Entrepreneurial Response by modifying the
amount of output produced or provided to match consumer's willingness
and ability to purchase.

I will discuss each very briefly. Substitutes are goods and services produced or
provided by your competitors that consumers could use in place of your product.
For example, Vermont, New Hampshire, or Canadian maple syrup or imitation
maple syrup could be a substitute for New York maple syrup. If the price of these
substitutes increases (decreases) relative to the price of New York maple syrup,
consumers will be likely to increase (decrease) their purchases of New York
maple syrup, all else held constant. Would you consider ISO beams and solid
wood beams as substitutes? Complements are goods and services that are used
concurrently with your product. For example, hotels and restaurants are often
complementary goods to the purchase of a charter boat fishing trip. If the price of
these complements increases (decreases) relative to the price of a charter boat
fishing trip, then consumers will be likely to decrease (increase) their purchases
of charter boat fishing trips, all else held constant. Consumers' income must be
allocated among various items. For example, if consumers' incomes decrease,
they may switch from buying New York maple syrup to purchasing imitation
maple syrup, all else held constant. If there is an expected increase in the price of
New York maple syrup next year, then consumers will be more likely to purchase
more (stock up on) New York maple syrup this year, all else held constant. Finally,
if consumers' income is expected to increase next year, then consumers will be
more likely to purchase more (stock up on) New York maple syrup this year, all
else held constant.

While the actual change in the quantity demanded by consumers in Figures 6.16
and 6.17 cannot be estimated based on the above information, what the entrepreneur

is observing critically are the reactions consumers have made to these external changes. Namely, the entrepreneur is concerned how their total revenue will be affected given consumers' reactions: any movement along or shift in the demand curve (Figures 6.16 and 6.17) will affect the market equilibrium (Figure 6.2) and buyers' explicit cost, $[P \cdot Q(\bullet)]$, transferred to the entrepreneur as their total revenue, $[P \cdot Q(\bullet)]$, as described by equation (6.2). Knowing this economic information will allow entrepreneurs to fine tune their responses to consumers' reactions.

The sensitivity of consumers' reactions to external changes in prices and income is called elasticity (Table 6.5). While the concept of sensitivity seems simple enough there still is a question of how it is measured. Total revenue is defined as $P \cdot Q(\bullet)$ where the units on price have been described as $/Q (e.g. in 2009 a gallon of maple syrup was selling at retail stores for almost $160 per gallon (Chapter 5)) and the units on $Q(\bullet)$ are generally described by some physical measurement (e.g. in the maple syrup example the physical measurement is gallons). However, neither price nor quantity are easily comparable even given similar products. For example, in the Select Red Oak case study, output is measured as the gross and net board foot volume measured in International ¼ log rule per individual tree and price would then be dollars per gross or net board foot volume measured in International ¼ log rule (Chapter 2; Hahn and Hansen 1991). In contrast, in the Loblolly Pine Plantation case study output is measured as total cubic-foot volume, outside bark, per acre of planted loblolly pine and price would then be dollars per total cubic-foot volume, outside bark (Chapter 2; Amateis and Burkhart 1985 and Burkhart *et al.* 1985). Thus, to account for the differences in price and output units, elasticity is defined as:

Elasticity – the percentage change in either price or income relative to the percentage change in quantity, holding all else constant.

The various different types of demand elasticities and how to estimate them are given in Table 6.7.

Elasticity describes a price or income and quantity effect on total revenue. As illustrated by Table 6.7, elasticity is a ratio of price or income effects to quantity effects. If the price or income effects are greater (less) than the quantity effects the elasticity will be greater (less) than one.

I will analyze each briefly starting with own-price elasticity. A percentage increase in the price of the good or service will cause total revenue to increase (as P increases $P \cdot Q(\bullet)$ increases). However, as there is an also negative relationship between own-price and quantity demanded, consumers' reaction to an increase in price is to reduce their purchases resulting in a decrease in total revenue (as Q decreases $P \cdot Q(\bullet)$ decreases). If the price effects on total revenue are small relative to the quantity effects on total revenue, then own-price elasticity is less than one and the demand curve is defined as elastic. In this case, a percentage increase in the price of a good would cause the total revenue to decrease. The opposite would occur with a percentage decrease in price. If the price effects on total revenue are large relative to the quantity effects on total revenue, then own-price elasticity is

Table 6.7 Demand elasticities

Name	Definition[†]	Arc-elasticity[§]	Point-elasticity
Own-price elasticity (E_P)[‡]	$\dfrac{\%\Delta P}{\%\Delta Q_D}$	$\dfrac{(Q_D'' - Q_D')}{(P'' - P')} \cdot \dfrac{\dfrac{(P'' + P')}{2}}{\dfrac{(Q_D'' + Q_D')}{2}}$	$\dfrac{\partial Q_D(\bullet)}{\partial P} \cdot \dfrac{P}{Q_D}$
Cross-price elasticity (Substitute, E_{P_S})[††]	$\dfrac{\%\Delta P_S}{\%\Delta Q_D}$	$\dfrac{(Q_D'' - Q_D')}{(P_S'' - P_S')} \cdot \dfrac{\dfrac{(P_S'' + P_S')}{2}}{\dfrac{(Q_D'' + Q_D')}{2}}$	$\dfrac{\partial Q_D(\bullet)}{\partial P_S} \cdot \dfrac{P_S}{Q_D}$
Cross-price elasticity (Complement, E_{P_C})[††]	$\dfrac{\%\Delta P_C}{\%\Delta Q_D}$	$\dfrac{(Q_D'' - Q_D')}{(P_C'' - P_C')} \cdot \dfrac{\dfrac{(P_C'' + P_C')}{2}}{\dfrac{(Q_D'' + Q_D')}{2}}$	$\dfrac{\partial Q_D(\bullet)}{\partial P_C} \cdot \dfrac{P_C}{Q_D}$
Income (E_Y)	$\dfrac{\%\Delta Y}{\%\Delta Q_D}$	$\dfrac{(Q_D'' - Q_D')}{(Y'' - Y')} \cdot \dfrac{\dfrac{(Y'' + Y')}{2}}{\dfrac{(Q_D'' + Q_D')}{2}}$	$\dfrac{\partial Q_D(\bullet)}{\partial Y} \cdot \dfrac{Y}{Q_D}$

[†] $\%\Delta P$ and $\%\Delta Y$ denotes the percentage change in price and income respectively. $\%\Delta Q_D$ denotes the percentage change in quantity demanded.
[§] Arc-elasticities and point-elasticities are estimated assuming all else remains constant.
[‡] Own-price elasticity is technically a negative number as the demand curve has a negative slope. However, it is often given as the absolute value. If $0 < |Ep| < 1$ then the demand is inelastic. If $|Ep| > 1$ then the demand curves is elastic. If $|Ep| =$ infinity then the demand curve is perfectly elastic.
[††] If the cross-price elasticity is negative, then the goods are complements, $E_{Pc} < 0$. If the cross-price elasticity is positive, then the goods are substitutes, $E_{Ps} > 0$.

greater than one and the demand curve is defined as inelastic.[13] In this case, a percentage increase in the price of a good would cause the total revenue to increase. The opposite would occur with a percentage decrease in price.

How can the entrepreneur use the information on own-price elasticity to their advantage? Whether the demand curve for your good or service is elastic or inelastic depends on a three factors. First is the number of substitutes. If there are a large number of substitutes for your goods or services and if the price of your good or service increases, then consumers will shift to one of the many substitutes and your total revenue would decrease. The demand for a good or service with many substitutes is defined to be elastic. If there are a few substitutes for your goods or services and if the price of your good or service increases, then consumers will have limited ability to shift to a substitutes and your total revenue would increase. The demand for a good or service with few substitutes is defined to be inelastic. As an entrepreneur you should be obtaining information about your market continually; in other words, how much market power do you have to set price? If you increase your price by a small percentage and observe a large percentage reduction in the amount of output you sell and total revenue, then you produce or provide a good or service with many substitutes. In this case, you have little latitude to influence market price and the market is characterized by workable competition. I will discuss this topic further in Chapter 7. Based on this discussion, would you think the demand for New York maple syrup is elastic or inelastic? Would a good like insulin have an elastic or inelastic demand curve?

Second is the percent of consumers' income that is spent on your good or service. The larger the percent of consumers' budget spent on a good, the more sensitive they are to price changes and the more elastic the demand curve. For example, a home is one of the largest purchases consumers often make and paying for the house requires a significant amount of their annual income. Thus, as the price of a home increases (including the interest rate on any mortgage), the more consumers will modify the type of home (e.g. square footage, and urban vs suburban, etc.) they will purchase. In 2008–2009, the housing market collapsed in the United States. While the cause and effect relationships surrounding this collapse are complex, before the collapse many consumers could get home loans with little to no down payments and relatively small initial interest rates on the large amounts borrowed. Based on this information, could you make the argument that consumers behaved as if the demand for housing was inelastic when it should have been elastic? What about the demand for the pen or pencil that you are using to take notes with, is the demand for that writing implement elastic or inelastic? Why?

Third, the more time consumers have to search for alternatives, the more elastic the demand for the good or service. The internet has allowed most consumers the ability to search for alternatives relatively quickly. In addition, with increasing access to smart phones and other handheld devices, consumers can often search immediately.

Cross-price elasticity describes the percentage change in the amount of a good or service that you sell to consumers given a percentage change in either the price of a substitute or complement. Compare these two goods: sport utility vehicles (SUVs) and unleaded gas. Would you consider these two goods substitutes or complements? The US Department of Energy provides data on the historic nominal retail price of regular unleaded gas from January 1976 to November 2009 (http://www.eia.doe.gov/emeu/mer/prices.html accessed on 28 December 2009). Based on these data, there were significant percent increases in the nominal price of regular unleaded gas starting in 2000 and peaking in 2008. These percentage change in gas prices were followed by a significant percent decrease in the number of SUVs purchased as described by the *New York Times* 2 May 2008 article (http://www.nytimes.com/2008/05/02/business/02auto.html accessed on 28 December 2009) and *The Wall Street Journal* Market Data Center (http://online. wsj.com/mdc/public/page/2_3022-autosales.html accessed on 28 December 2009). Feng *et al.* (2005) showed that for consumers who own a single vehicle, a 1 percent increase in the price of gas would result in decreasing their probability of purchasing an SUV by 7 percent. If goods are substitutes, they will exhibit the opposite effect. For example, Feng *et al.* (2005) also showed that for consumers who own a single vehicle a 1 percent increase in the price of gas would result in increasing their probability of purchasing a car by 0.9 percent. Finally, Feng *et al.* (2005) showed that for consumers that own two vehicles as the price of gas increased the probability of owning a car plus an SUV decreased by 79.3 percent and the probability of owning a car plus an additional car increased by 69.5 percent. Based on the above illustration using cars and SUVs, what do you think the cross-price elasticities for maple syrup made from New York sugar maple trees would be relative to that made from Vermont, New Hampshire, or Canadian sugar maple trees or imitation maple syrup?

As consumers' incomes change they buy different goods and services. For example, as you go from a college student to a full-time job your income changes as do your purchases. Income elasticity measures the responsiveness of consumers purchasing what you produce or provide to changes in income. If the income elasticity is positive, the good is described as a normal good. For example, maple syrup would be considered a normal good in that as your income increases your purchase of maple syrup would increase. The larger the income elasticity, the greater the responsiveness to percent increases in income. If the income elasticity is negative, the good is described as an inferior good. For example, imitation maple syrup may be considered as an inferior good in that as your income increases you would purchase less imitation maple syrup.

A related elasticity not given in Table 6.7 is called advertising elasticity. The purpose of advertising is to increase the sales of the goods or services that you produce or provide by showing a targeted group of consumers the superiority of your product relative to similar products. Advertising is a production expense; one way to measure the usefulness of advertising is to track, for example, how a 1 percent increase in your advertising expenditures will affect your total revenue. Advertising is more successful if it is targeted to those consumers that would be the most interested in your product (e.g. men vs women, and old vs young, etc.) and reaching them using the appropriate media (e.g. print, radio, television, internet, etc.). Targeting consumers implies that their demand depends on social and demographic factors.

Table 6.5 lists a number of social and demographic factors that influence demand. This list is not meant to be all encompassing. Its purpose is to get you to think past focusing only on the economic variables. I have already given an example of some of the social and demographic factors that influence private forestland owners' demand for management services that you might sell. For example, the primary reason of owning forestland for these owners is esthetics and not revenue from timber sales. Thus, if you could exploit a potential opening that would generate revenue for the landowners (but not from timber sales) this would reduce their landownership costs; for example, the sale of ecosystem services such as carbon sequestration derived from forest management. In 2002, the Chicago Climate Exchange (http://www.chicagoclimatex.com/ accessed on 29 December 2009) was established to facilitate the selling of carbon offset credits. The American Carbon Registry (http://www.americancarbonregistry.org/ accessed on 15 April 2011), the Climate Action Reserve (http://www.climateactionreserve.org/ accessed on 15 April 2011), and the Intercontinental exchange (https://www.theice.

com/homepage.jhtml accessed on 15 April 2011) are examples of other markets for carbon. Forest management has the potential to generate carbon offsets that could be traded. One of the earliest forest and natural resource management companies to exploit this opening was FORECON, Inc. (http://www.foreconinc.com/ EcoMarket/ accessed on 29 December 2009). The continued potential to generate revenues from the sale of ecosystem services such as carbon offsets was also part of the United Nations Climate Change Conference held in Copenhagen Denmark in December 2009 (http://en.cop15.dk/ accessed on 29 December 2009) and discussed in a *New York Times* article published on 12 December 2009 (http://www. nytimes.com/2009/12/16/science/earth/16forest.html accessed on 29 December 2009). While the sale of ecosystem services will probably not be a panacea to reduce the costs of forestland ownership for all landowners, an entrepreneur should actively search for similar opportunities that they could exploit.

Searching for new products and product information takes time that consumers could use to do something else with (opportunity cost); thus, consumers develop habits and loyalty to various products. For example, I will buy a certain make of car because my father always bought them. An entrepreneur should actively search for opportunities to develop that loyalty with their consumers. An entrepreneur should be aware (and cautious) of trends and exuberance. While large profits can be made if you are on the upswing of a trend, coming into this type of market late could be costly. An entrepreneur should be conscious of the relevant social and demographic factors that describe their customers. The timber output from forests represents a unique situation with regard to the social and demographic factors. Producing a commercially viable tree may take from 20 years for southern yellow pines grown in the Southern United States to 100 years for hardwoods grown in the Northeastern United States. Forestland owners who derive revenue from the sale of stumpage (i.e. the value of standing timber) are in essence trying to predict consumers' preferences decades in the future (Smith 1988). In addition, Smith (1988) speculated that species that were considered as not very commercially viable today may become commercially viable in the future. For example, willows (genus *Salix*) are currently being viewed as a source of bioenergy and biofuels (Volk and Luzadis 2008).

How to use economic information – *supply and demand* – to make better business decisions?

Successful entrepreneurs are actively searching for ways to increase their profits. This includes placing equal time and energy on understanding the concept of supply as well as demand. Fundamentally, supply represents the opportunity cost of the entrepreneur. The opportunity cost could be described by the physical production costs of manufacturing an ISO beam or gallon of maple syrup. They also describe your time, labor, and capital of producing or providing a good or service. These opportunity costs are modeled by a supply curve or supply schedule and embody the economic information contained within the Architectural Plan for Profit. The economic model of supply provides an entrepreneur with a systematic way of analyzing change and response relationships; namely, to recognize changes in economic information obtained from the market and then modify the quantity of output produced or provided.

Fundamentally, demand, the demand curve, or a demand function represents the opportunity cost of consumers of the goods or services you produce or provide. The definition of demand is given as the willingness and ability to purchase the outputs that you produce or provide. Consumers must be willing – placing value on your output – and able – if a consumer is not able you have no possibility of converting a consumer's expenditure into your total revenue. As just stated, your total revenue is dependent directly on your understanding of consumers' demand. The economic model of demand provides an entrepreneur with a systematic way of analyzing change and response relationships; namely, to recognize external changes affecting your consumers, their reaction to these external changes, and your response to this new economic information. In other words, how do you "shift" your supply in response to these external changes? Your response might be as simple as modifying the quantity of output produced or provided, or it might require modifying the attributes or characteristics of your output, or creating new products and services in addition to the amount produced or provided of existing products.

A brief examination of these last two paragraphs shows that they are almost identical in terms of wording. This was done on purpose and helps illustrate the important relationship between the concepts of supply and demand. Successful entrepreneurs realize this relationship and the equal importance of supply and demand in searching for ways to increase profits.

7 Market equilibrium and structure

In Chapter 6, I was somewhat liberal with my use of the term market and market equilibrium with respect to describing Figure 6.2; namely, supply denoted a representative entrepreneur and demand denoted a representative consumer. This was done so that I could focus on the individual concepts of supply and demand as economic models that define the opportunity cost of the entrepreneur (supply) and the opportunity cost of a consumer (demand). In addition, I showed that the economic information in Figure 6.2 parallels what is in the Architectural Plan for Profit (Figure 7.1).

In this chapter, I will look more broadly at supply and demand than in the last chapter; namely, that a market for a particular good or service (e.g. maple syrup or ISO beams) comprises all the producers and consumers of that product. Thus, a market supply curve is an aggregate of all the individual entrepreneurs' supply curves and the market demand curve is an aggregate of all the individual consumers' demand curves. In this chapter, I will focus on the role of the market, the various different market structures that an entrepreneur might face, and the various pricing strategies that an entrepreneur might use. I will end with the concept of derived demand and its counterpart derived supply. The concept of derived demand is important especially for entrepreneurs whose focus is natural resource management and the goods and services that ecosystems produce.

Case studies

The case studies used in this chapter will include the Inside-Out (ISO) Beams (Patterson *et al.* 2002 and Patterson and Xie 1998), Mobile Micromill (Becker *et al.* 2004), Maple Syrup Operation (Huyler 2000 and CFBMC 2000), Great Lakes Charter Boat Fishing (Lichtkoppler and Kuehn 2003), Select Red Oak (Hahn and Hansen 1991), Loblolly Pine Plantation (Amateis and Burkhart 1985 and Burkhart *et al.* 1985), and Changing Timberland Ownership (Hagan *et al.* 2005). The case studies can be found on the Routledge website for this book. Their summaries will not be repeated here.

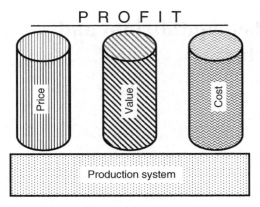

Figure 7.1 The Architectural Plan for Profit.

The market

A market supply curve is the horizontal sum of all entrepreneurs' supply curves (e.g. Chapter 6; Figures 6.5b and 6.7b). Figure 7.2a illustrates deriving a market supply curve.

A market demand curve is the horizontal sum of all consumers' demand curves. Figure 7.2b illustrates deriving a market demand curve.

The market is the focus of economic activity and provides a place where economic information concerning the relative scarcity of a good or service is available and the choices individuals make are observed. An individual can choose to provide or produce a particular good or service and another individual can choose to purchase that particular good or service. A market can be a physical place, such as a grocery store or local farmers' market, or it can be place like eBay® where individuals offer to sell Gibson Les Paul guitars, bamboo flyrods, or one red paperclip (http://oneredpaperclip.blogspot.com/2005/07/about-one-red-paperclip.html, accessed on 8 January 2010) and others can choose to purchase the item or make a counter-offer. Finally, markets are where goods and services move between individuals (supply and demand) and also across geographic regions and time; for example, the sale of maple syrup produced in the Northeast to customers in the Pacific Northwest. Markets also reduce participants' transactions costs.[1]

The role markets play can be divided into four broad components.

Market equilibrium

Market equilibrium is given by the intersection of the market supply curve (Figure 7.2a) and demand curve (Figure 7.2b).[2] In Chapter 6, I showed that the equilibrium also described the price and quantity combination that maximizes producer surplus (Figure 6.10) and consumer surplus (Figure 6.15b). Extrapolating this to the market, Figure 7.3a shows that the market producer surplus is the sum of each entrepreneurs' maximized producer surplus.[3]

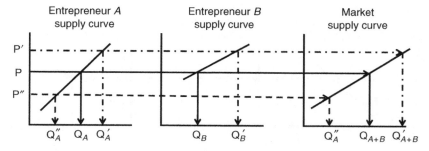

Figure 7.2a Market supply curve.

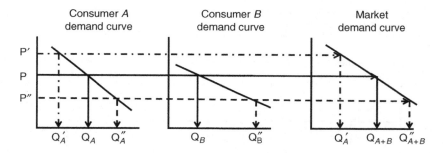

Figure 7.2b Market demand curve.

Again extrapolating this to the market, Figure 7.3b shows that the market consumer surplus is the sum of each consumers' maximized consumer surplus.

Thus, if the market is described by workable competition (Chapter 1) then market equilibrium can be found by maximizing the sum of producer and consumer surplus. This is depicted in Figure 7.3c.

Figure 7.3c describes interactions among buyers (market demand) and sellers (market supply) within a market. As I have asserted in earlier chapters and modeled by Figures 7.3a, 7.3b, and 7.3c, sellers maximizing their profits and buyers maximizing their net benefits will result in the cooperative solution of market equilibrium price, P*, and quantity, Q*. As shown in Figure 7.3a, not every entrepreneur seeking to maximize profit produces or provides the same amount of output. Some entrepreneurs have higher production costs and could drop out if the market price fell (Table 5.4 and Figures 6.5b and 6.7b). Similarly, as shown by Figure 7.3b, not all net benefit maximizing consumers purchase the same amount of output. If the market price increases some will no longer purchase the output.

What and how much to produce or provide?

Should maple syrup producers produce maple syrup? Should manufacturers of ISO beams produce ISO beams? Should charter boat fishing captains continue to provide fishing services? Should landowners continue to grow select red oak or loblolly

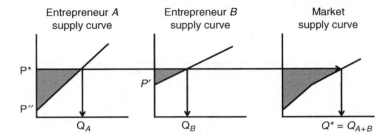

Figure 7.3a Market producer surplus.[‡]

[‡]The shaded area below the P* price line and above the supply curve defines producer surplus (for example, see Figure 6.10). The area below the supply curve bounded by Q defines total variable cost (for example, see Figure 6.9).

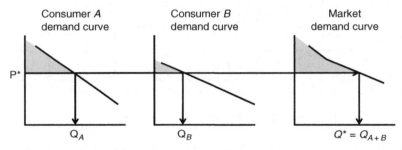

Figure 7.3b Market consumer surplus.[‡]

[‡]The shaded area below the demand curve and above the P* price line defines consumer surplus (for example, see Figure 6.15b). The area below the P* price line bounded by Q defines consumers explicit costs (for example, see Figure 6.15b).

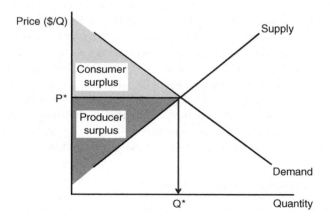

Figure 7.3c Market equilibrium.

pine? How does Figure 7.3c help us answer these similar questions? Simply, Figure 7.3c models the situation that given the equilibrium market price, P*, entrepreneurs are willing and able to supply positive amounts of the output depending on their production system and cost structure. This is shown in Figure 7.3a.

How much output should each of these entrepreneurs produce or provide? Referring again to Figure 7.3a, entrepreneurs are willing and able to supply different amounts of the output depending on their production system and cost structure. This is summarized by the profit searching rule described in Chapter 5 and given in equation 7.1.

$$MR = P = Supply = MC \qquad (7.1)$$

How to produce?

Reviewing Chapters 2 and 3, the answer to "How to produce?" is straightforward. Entrepreneurs should produce in a technically efficient (Chapter 2) and production cost efficient (Chapter 3) manner. The relationship between technical efficiency, production cost efficiency, and the supply curve is given by Figure 7.4.

Each entrepreneur within a market is a profits maximizer (i.e. Architectural Plan for Profit Figure 7.1 and illustrated by *Max* $\Pi = TR - TC$). Profits are maximized by entrepreneurs simultaneously making total revenue (TR) as large as possible while making total costs (TC) as small as possible. The producer surplus illustrated in Figure 7.4 shows this as the supply curve is both technically efficient and production cost efficient. Thus, the producer surplus in Figures 7.3a and 7.3c is also technically efficient and production cost efficient.

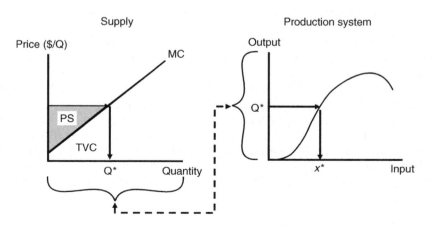

Figure 7.4 The relationship between supply, producer surplus, and the
production cost efficiency.[‡]

[‡]PS denotes producer surplus and MC and TVC denotes marginal cost and total variable cost, respectively (Chapter 6).

For whom to produce or provide?

Entrepreneurs produce or provide commodities for consumers. This statement seems obvious and simple. However, how do we model consumers' preferences for commodities? This is done using demand (Chapter 6). Entrepreneurs produce or provide commodities for whoever is willing and able to pay for them. As shown by Figure 7.3b, not every consumer purchases the same amount of output and some consumers will no longer purchase your output if the market price rises above a given level.

Final thoughts on markets and market equilibrium

The apparently simple model illustrated in Figure 7.3c is a culmination of all the economic information presented in the previous chapters. The Architectural Plan for Profit (Figure 7.1) presented and the beginning of this chapter summarized the economic information that an entrepreneur should obtain to help them in their profit searching behavior. Markets are efficient mechanisms for entrepreneurs to respond to consumers demand and for allocating commodities among consumers. The implication is that entrepreneurs' profits and consumers' net benefits are as large as possible. What markets are not especially good at are equitable allocations of resources.[4] In addition, the market structure that I used to develop Figures 7.3a, 7.3b, and 7.3c was of workable competition. In the next section, I will examine briefly the implications of modifying this market structure. Finally, there are potential spillover or external effects that may affect individuals not involved in market transactions between entrepreneurs and consumers. The classic example of a negative externality is pollution created by entrepreneurs in the process of producing or providing outputs to consumers. The pollution negatively impacts individuals (e.g. property damage, health, and well-being, etc.) and this negative impact is not reflected in the market price (i.e. P* of Figure 7.3c) nor are these individuals compensated for the negative impacts. In these cases, as pollution is tied to output production, entrepreneurs may be producing too much output. A classic example of a positive externality is my getting a flu shot every fall. You benefit from the reduced probability of me passing on the flu. Again, market price will not reflect this positive externality nor do you compensate me for the benefit you receive from my actions. In this case, too few people will probably get flu shots. A more in-depth discussion of positive and negative externalities, their impacts on market equilibriums and potential solutions are beyond the scope of this book. If readers are interested there is a vast literature within the areas of resource economics, environmental economics, and ecological economics on this topic.

Market structures

As defined previously, a market is the sum of all entrepreneurs and all consumers (Figures 7.3a, 7.3b, and 7.3c). However, market structures depend on a number of factors: (1) number and size of buyers, sellers, and potential entrants; (2) the degree of product differentiation; (3) amount and cost of information about product price and quality; and (4) conditions for entry and exit (Brickley *et al.* 2007). I will examine briefly six different market structures. The following discussion is not meant to be all encompassing. I would encourage readers to research these topics further if they are interested.

Perfect competition

The conditions required for perfect competition are: (1) many buyers and many sellers; (2) no barriers to entry to or exit from the market; (3) perfect information with respect to price and products are available to all; (4) transactions costs are zero; and (5) homogenous products. The resulting market can be described by Figure 7.3c and market price only reflects how scarce a good or service is relative to other goods and services. While the market demand curve has a negative slope (as does each consumer's demand curve), individual entrepreneurs face a demand curve that is perfectly elastic; that is, the price line defined by the market equilibrium price – P* in Figures 7.3a, 7.3b, and 7.3c – defines an entrepreneur's demand curve. Thus, an entrepreneur's average revenue is identically equal to its marginal revenue, which is equal to market price (Chapter 4). Entrepreneurs are described as taking the price given by the market (i.e. price takers). In terms of the profit searching rule as given by equation (7.1), entrepreneurs' decisions are limited only to how much to produce or provide given the current market price.

As you can observe, the conditions required for perfect competition are extremely strict and one would probably be hard pressed to point to a market that satisfies all these conditions. Thus, perfect competition is often viewed as one endpoint (a theoretical endpoint) on a continuum of potential different market structures.

Workable competition

Workable competition is defined as the case where no one buyer or group of buyers and no one seller or group of sellers can influence price (Chapter 1).[5] This is the default market structure I have used for the book. The implication of this market structure is that there are many substitutes that are relatively easy to obtain. Thus, entrepreneurs and consumers have very little, if any, market power in terms of price setting. Entrepreneurs are described as taking the price given by the market (i.e. price takers). Therefore, the resulting market can be also described by Figure 7.3c, and market price reflects how scarce a good or service is relative to other goods and services. While the market demand curve has a negative slope (as does each consumer's demand curve), individual entrepreneurs face a demand curve that is nearly elastic; that is, the price line defined by the market equilibrium price – P* in Figures 7.3a, 7.3b, and 7.3c – for all practical purposes defines an entrepreneur's demand curve. While an entrepreneur may have a very limited ability to set price, with respect to the market as a whole their average revenue approximates marginal revenue and both may be defined by market price (Chapter 4). In terms of the profit searching rule given by equation (7.1), an entrepreneur's decisions are basically limited to how much to produce or provide given the current market price.

Monopolistic competition

Monopolistic competition is characterized by a structure in which there are many sellers of similar products that can be differentiated; for example, breakfast cereal, toothpaste, laundry, and dish detergent. What about maple syrup or dimension lumber sawn from loblolly pine or other pine species? This differentiation can be the result of brand loyalty, advertising, and effective marketing. The more effective an entrepreneur differentiates their products from those of their competitors, the more market power they have allowing them to set their price. Because there is competition, their price-setting behavior does not affect the market as a whole. No barriers keep entrepreneurs from entering or exiting the market.

The market an entrepreneur faces given monopolistic competition is shown in Figures 7.5a and 7.5b.

While Figures 7.5a and 7.5b look complicated I will examine each, paying close attention to the economic information these models provide. The entrepreneur's average and marginal cost curves in Figure 7.5a are the same as those discussed in Chapter 3; namely, equations (3.5) and (3.7), respectively. Thus, the economic information contained in the cost components has already been examined.[6] The unique feature of Figure 7.5a is there is a consumer's demand curve which is different than the entrepreneur's marginal revenue curve. Instead of the entrepreneur

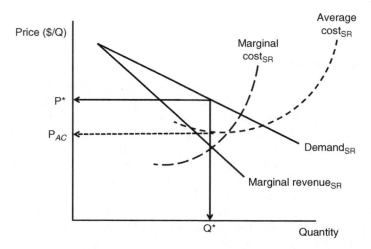

Figure 7.5a Market equilibrium given monopolistic competition in the short-run as viewed by an entrepreneur.[‡]

[‡]SR denotes short-run.

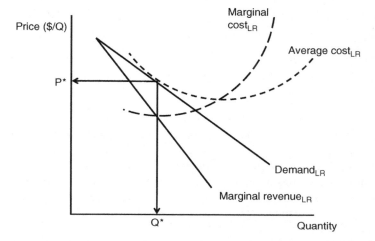

Figure 7.5b Market equilibrium given monopolistic competition in the long-run.‡

‡LR denotes long-run.

facing a horizontal demand curve (as with perfect and workable competition), the demand curve has a negative slope. According to Chapter 6, a consumer's demand curve has a negative slope or there is an inverse relationship between price and quantity due to their opportunity costs. From an entrepreneur's point of view, they want to turn a consumer's demand into revenue. Therefore, what is the relationship between a consumer's demand curve and an entrepreneur's revenue?

The consumer's demand curve defines an entrepreneur's average revenue curve.[7] Because a consumer's demand curve has a negative slope, the entrepreneur must lower price to get the consumer to buy additional units of output. As an entrepreneur lowers their price, the average revenue per unit sold would decrease as would the additional revenue per unit, or marginal revenue.[8] Because of the ability to differentiate among similar outputs, the entrepreneur now has two decisions; namely, how much to produce and at what price. Entrepreneurs are still profit maximizers and the profit searching rule as given by equation (5.5) still holds. Thus, the most profitable level of output to produce or provide given Figure 7.5a would be Q^*. The price the entrepreneur would charge would be what the customer is willing and able to pay; namely, P^* given Figure 7.5a.

According to Figure 7.5a, P_{AC} defines the average "total" cost of producing the most profitable level of output, Q^*. In addition, Figure 7.5a shows that P^* is greater than P_{AC} or average total cost (ATC) which is illustrated by equation (7.2a):

$$P^* > P_{AC} = ATC = \frac{TC}{Q^*}$$

$$P^* > \frac{TC}{Q^*}$$

(7.2a)

Multiplying both sides of equation (7.2a) by Q* shows that profits are positive:

$$P^* \cdot Q^* > \frac{TC}{Q^*} \cdot Q^*$$
$$TR > TC \tag{7.2b}$$
$$TR - TC = \Pi > 0$$

where *TR* denotes total revenue, *TC* denotes total cost, and Π denotes profit. Thus, in the short-run, the entrepreneur would be making a positive economic profit.[9] Again, this is because entrepreneurs can differentiate their products.

One of the conditions of monopolistic competition is that there are no barriers to entry or exit. As other entrepreneurs see that there are economic profits to be made, this will entice more entrepreneurs to enter this market, increasing competition and reducing the share of the market held by any one entrepreneur in the long-run. Figure 7.5b illustrates the long-run market model. Due to the reduced market share, the demand curve faced by any entrepreneur will shift left as will the marginal revenue curve. The interpretation of the demand (average revenue) and marginal revenue curves do not change. The cost curves now represent the concept that there are no fixed inputs and no fixed costs. For example, in the long-run you can build a new plant to produce addition ISO beams, purchase a larger sugarbush, or buy more acres for your loblolly pine plantation. In the long-run, economic profits are driven to zero as P* will equal P_{AC}. Entrepreneurs would still be making positive accounting profits, however.

How would you describe the market for maple syrup? If you produced New York maple syrup, would you be able to differentiate it from that produced in Vermont, New Hampshire, or Canada? How would you describe the market for dimension lumber sawn from a pine tree? If you produce dimension lumber from your loblolly pine plantation would you be able to differentiate your dimension lumber from that produced in Northeastern, Midwestern, or Pacific Northwest areas of the United States? What about Canada, Latin America, or Australia?

Oligopoly

An oligopoly is a market structure characterized by a few sellers who control most, if not all, of the market. Potential new entrepreneurs find there are significant barriers to entry (e.g. economies of scale, technology, patents). An oligopolist has two decisions to make; how much to produce and at what price. However, the oligopolist not only faces each consumer's negatively sloped demand curve, they must also recognize that fellow oligopolists will react to any pricing and production decisions just as they would react to any pricing and production decisions made by their fellow oligopolists. It is the move–countermove strategic nature of the pricing and production decisions that makes modeling these markets interesting. The modeling approaches described by game theory have been used to analyze the interactions among oligopolists. While the technical nature of these models is beyond the scope of this book, I would point interested readers to Kreps (1990), Gibbons (1992), and Bierman and Fernandez (1993).

Monopoly

A monopoly is a market structure characterized by one seller and many buyers. There are significant barriers to entry for any potential new entrepreneurs (e.g. economies of scale, technology, patents, or regulation). A monopolist has two decisions to make; how much to produce and at what price. The market structure for a monopoly can be illustrated by Figure 7.6a.

As can be seen, Figure 7.6a is similar to Figure 7.5a. The interpretations of demand, marginal revenue, marginal cost, and average cost curves are the same as in Figure 7.5a. In a monopoly market, the entrepreneur does not have to worry about distinguishing their products from other similar products nor think strategically in terms of prices and production amounts as there are no similar products or sellers to contend with, respectively. That said, monopolists can still be characterized as a profit maximizer and will still employ the profit searching rule, equation (5.5). Examining Figure 7.6a shows that the production decision follows the profit searching rule; namely, produce at the output level, Q*, such that $MR = MC$. The pricing decision depends on the consumer's willingness and ability to pay. Even though there is a single seller in a monopoly market, the monopolistic entrepreneur may not charge as much as they want because however inelastic the demand curve may be (see Chapter 6), it still has a negative slope or there is a negative relationship between price and quantity reflecting the consumer's opportunity costs. Thus, a monopolist must lower the price of the commodity in order to induce the consumer to buy more. Charging any price lower (greater) than P* in Figure 7.6a would imply that MR is greater (less) than MC and the monopolist's profits would increase by lowering (increasing) price.[10]

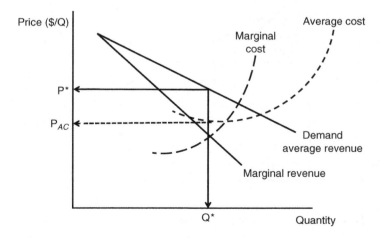

Figure 7.6a Market equilibrium given a monopoly.

Comparing the production and pricing decisions with those described for monopolistic competition, the reader will note that they are virtually the same. A further comparison of Figures 7.5a and 7.5b with Figure 7.6a is that there is no distinction between the long- and short-run for a monopolist. This is due to the fact that there are barriers to entry that keep out any potential competitors. The classic example of a monopoly is the spatial monopolies for utilities (e.g. water and sewage) that are sanctioned by many municipal governments within urban areas. Electricity is a unique case in the United States with many states deregulating certain operations of utilities that sell to consumers. Thus, while consumers may choose from many different producers of electricity, the delivery structure (i.e. power lines and towers, etc.) are still monopolies sanctioned by governments.

The relationship between the market solutions given a monopoly versus workable competition market are illustrated in Figure 7.6b.

Figure 7.6b illustrates the market power that a monopolist has in terms of pricing and production decisions as compared to workable competition. The more inelastic the consumer's demand curve, the greater the control the monopolist has over setting price. In Figure 7.6b, this is illustrated by the difference between P* and P_{WC}, with P* greater than P_{WC}. Again, the monopolist does not have unlimited market power with respect to pricing as the consumer's demand curve has a negative slope. The pricing decision depends on the monopolist's knowledge and modeling the consumer's demand relative to the monopolist's marginal revenue. As illustrated by Figure 7.6b, the monopolist will produce or provide less output than given workable competition; namely, Q* is less than Q_{WC}. Production decisions are based on the marginal costs relative to the marginal revenues given that marginal revenue depends on pricing. Determining the degree of market power requires the entrepreneur to have a working knowledge of consumer's demand

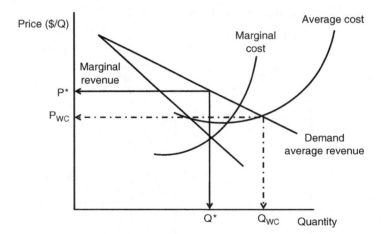

Figure 7.6b Market equilibrium: monopoly versus workable competition.[‡]

[‡]P_{WC} and Q_{WC} denote the market equilibrium price and quantity given workable competition.

(Chapter 6) and their production process, costs, and revenues (Chapter 2, Chapter 3, and Chapter 4). Figure 7.6b shows a level of detail probably not available to entrepreneurs; however, it reflects the required economic information an entrepreneur must collect and the relationships that should be analyzed continuously. This is the value of Figure 7.6b.

Monopsony and oligopsony

A monopsony is a market structure characterized by one buyer and many sellers. The classic example of a monopsony is a "One Mill Town." This mill is the only purchaser of labor and natural resource inputs that are available in a given region. An oligopsony is a market structure characterized by few buyers and many sellers. The presence of spatial constraints, for example, the high costs associated with transporting logs from where they are harvested to a pulp or paper mill, may allow these mills to exhibit monopsony or oligopsony power over the price they pay for these inputs (Murray 1995a and 1995b).

Like monopolistic competition, monopoly, and oligopoly market structures, a monopsony market structure gives the entrepreneur market power. In the case of a monopoly, the market power is reflected in output production and pricing decisions as illustrated in Figures 7.6a and 7.6b. In the case of a monopsony, the market power is reflected in the price or wage the monopsonist will pay for inputs – such as labor – and the amount they will purchase. Like Figure 7.6a, Figure 7.7a models the monopsonist's decisions concerning input price and amount purchased.

At first glance, Figure 7.7a may appear intimidating; however, we can actually build on the knowledge gained from Figures 7.5a, 7.5b, 7.6a, and 7.6b. Because monopolist's are profit maximizers, it would be logical that the profit searching rule would still be a tenable guide to examining their decision behavior. However, the profit searching rule as laid out in equation (5.5) and Figures 5.5a and 5.5b of

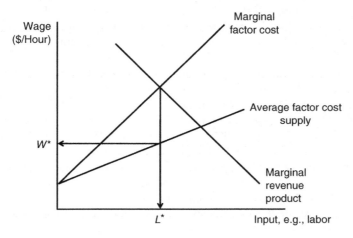

Figure 7.7a Market equilibrium given a monopsony.

Chapter 5, is described with respect to output. There is an analogous profit search-ing rule that was developed in Appendix 4 and Appendix 5 – profits with respect to inputs and summarized in equation (7.3).

$$\frac{\Delta\Pi}{\Delta x_j} = MRP_j - MFC_j; \text{ for all inputs} \tag{7.3}$$

where *MRP* denotes marginal revenue product or the additional revenue gener-ated from selling the additional output produced using more of the *j*th input (e.g. labor); *MFC* denotes marginal factor cost or the price or wage the entrepreneur must pay for the input.[11] Profits increase (decrease) if MRP is greater than (less than) MFC. This is modeled in Figure 7.7a by the monopsonist's choice of L^*. To the left of L^*, MFC is less than MRP and to the right of L^*, MFC is greater than MRP.

Figure 7.7a shows that the marginal factor cost curve is above the average fac-tor cost curve; that is, the amount the monopsonist is willing to pay for the input depends on the number of inputs that they purchase. The more they purchase, the more they must pay for all of the inputs purchased not just the additional unit. This is analogous to the marginal revenue being below the average revenue curve in Figure 7.6a.[12] In addition, the average factor cost describing the monopsonist's supply curve is analogous to average revenue describing the monopolist's demand curve.[13] Thus, the monopolist would only pay a wage of W^* per hour for labor, as shown in Figure 7.7a, for the labor L^* hours of labor.

An analogy can be drawn between an oligopsony and an oligopoly in terms of price to pay for inputs and the amount of inputs to purchase and the price to charge for outputs and the amount of outputs to produce, respectively. The oli-gopsonists' average factor cost curve defines their supply curve, but they must also recognize that their fellow oligopsonists will react to any pricing and input purchase decisions as they would react to pricing and production decisions made by their fellow oligopsonists. If ologopsonists compete against each other, the result will be to bid up the price paid for the input. However, if ologopsonists join together in input pricing decisions, each ologopsonist will benefit from a lower input price.

Figure 7.7b illustrates the market power that a monopsonist has in terms of input pricing and input purchasing decisions as compared to workable competition.

The more inelastic the consumer's demand curve the greater the control the monopolist has over setting price. In Figure 7.6b, this is illustrated by the differ-ence between W^* and W_{WC}, with W^* less than W_{WC} and L^* less than L_{WC}. The degree of market power – as illustrated by the difference between W^* and W_{WC} – depends on the elasticity of the supply curve. In the case of a single input buyer, the supply curve would be relatively inelastic and the difference between the mar-ginal factor cost and the average factor cost curves would be large as would the buyer's market power. If the supply curve is relatively elastic, there would be more buyers in the market and the power would be reduced for any one buyer to set input price.

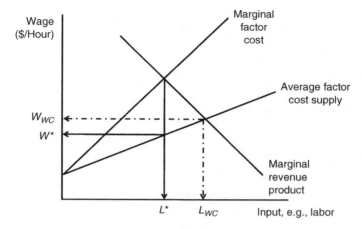

Figure 7.7b Market equilibrium: monopsony versus workable competition.[‡]

[‡]W_{WC} and L_{WC} denote the market equilibrium wage and labor (e.g. Hours) given workable competition.

Market power

The essential part of the above discussion on market structure is identifying if you as an entrepreneur have market power and what are the implications with respect to profit maximization. Successful entrepreneurs know the market structure for the outputs they produce or provide. As an entrepreneur your objective is to turn a consumer's willingness and ability to buy the commodities you produce into your revenue. Market structure defines the market power you have to capture this revenue given your ability to price your commodities relative to your competitors and the amount you produce relative to your competitors. Few sellers in the market are able to charge more and produce less than would result in a market with workable competition. This is the result of markets characterized by monopoly, oligopoly, or monopolistic competition. The elasticity of the demand curve, the number of sellers, and the interactions among the sellers defines the market power.

Successful entrepreneurs know the market structure for the inputs they purchase. As an entrepreneur, your objective is to pay as little for inputs as possible. Market structure defines the market power you have to affect the price you pay for inputs. In a market characterized by few buyers, they have the market power to pay less and purchase fewer inputs than would result in a market described by workable competition. This is the result of markets characterized by monopsony and oligopsony. The elasticity of the supply curve, the number of buyers, and the interactions among the buyers defines the market power.

Product price searching strategies

Profit is equal to total revenue ($P \times Q$) minus total cost. To make profits as large as possible, you must make total revenue as large as possible and total costs as small as possible for any given input(s)–output(s) combination. Examining this relationship reveals that total revenue is half of profit and product price, *P*, is half of total revenue. Relatively small changes in product price may imply relatively large changes in total revenue, as price changes are magnified by the amount sold (Longenecker *et al.* 2006). Thus, a key to making total revenue as large as possible is to develop a sound pricing strategy. As an entrepreneur, your pricing strategy depends on your knowledge of the market structure for your products. Your pricing strategy is to (1) set price to cover explicit and implicit measures of total costs; (2) reflect your consumers' willingness and ability to pay; and (3) recognize any competition.[14] Simply put, the goal of your pricing strategy is to charge as much as you can regardless what it costs you to produce or provide the outputs.

Average cost, cost-plus, or full cost pricing strategies

According to Silberberg and Suen (2001: 181–2) "the cost of producing the last unit of output is the same as the cost of producing the first or any other unit of output and is, in fact, the average cost of the output." This defines the logic of average cost, cost-plus, or full cost pricing strategies; namely, set output price to cover total accounting production costs plus an additional amount for markup or profit. Equations (7.4a) and (7.4b) illustrate this

$$P = \frac{TVC}{Q} + \frac{TFC}{Q} + m \cdot \left(\frac{TVC}{Q} + \frac{TFC}{Q} \right)$$
$$= ATC + m \cdot ATC = ATC \cdot (1 + m) \tag{7.4a}$$

$$P = \frac{TVC}{Q} + \frac{TFC}{Q} + \frac{\Pi_Q}{Q} \tag{7.4b}$$

where *TVC* denotes total variable cost, *TFC* denotes total fixed cost, *ATC* denotes average variable cost, *AFC* denotes average fixed costs, *m* in equation (7.4a) defines the entrepreneur's per unit markup above average total costs, and Π_Q in equation (7.4b) denotes the entrepreneur's target profit for producing a given output level, *Q*.[15] The target profit is often defined as a percentage of the assets; for example, buildings, equipment, and land.

Table 7.1 displays the annual and average production cost information for six maple syrup operations.

Based on Table 7.1 for example, if the maple syrup entrepreneur of 1,530-, 2,550-, or 5,100-gallon per year operations charged $19.79, $18.28, and $16.70 respectively, they would cover their fixed and variable annual production costs. Table 7.2 illustrates using these strategies for estimating an output price given the six different sized maple syrup operations given varying percentage of asset markup values.

Table 7.2 shows that the owners of an operation producing 127.5 gallon of maple syrup per year would price between $48.32 and $68.23 per gallon of maple syrup. The owners of an operation producing 1,530 gallon of maple syrup per year

Table 7.1 Total annual and average production costs of six different maple syrup operations[‡]

Cost item	127.5 (gal/yr)	255 (gal/yr)	765 (gal/yr)	1,530 (gal/yr)	2,550 (gal/yr)	5,100 (gal/yr)
		Annual Production Costs				
Variable Costs						
Labor	$1,601.79	$2,347.50	$4,641.17	$8,259.34	$13,083.57	$25,144.14
Supplies	$1,208.08	$2,271.45	$5,068.54	$9,204.94	$14,812.56	$28,748.85
Other	$471.12	$689.85	$1,261.92	$2,120.02	$3,667.98	$6,528.32
Fixed Costs	$2,245.20	$3,449.24	$7,184.52	$10,688.95	$15,048.79	$24,726.19
Total Cost	$5,526.19	$8,758.05	$18,156.14	$30,273.25	$46,612.90	$85,147.50
		Average Production Costs				
Average Variable Costs	$25.73	$20.82	$14.34	$12.80	$12.38	$11.85
Average Fixed Costs	$17.61	$13.53	$9.39	$6.99	$5.90	$4.85
Average Total Costs	$43.34	$34.35	$23.73	$19.79	$18.28	$16.70

[‡]Source of cost data CFBMC (2000). Costs are in 1999 US dollars using an exchange rate of 1.49 Canadian dollar per US dollar. (gal/yr) denotes gallons of maple syrup per year. These costs do not include the cost of establishing a larger sugarbush or the costs purchasing the additional equipment.

Table 7.2 Average cost, cost-plus-markup, and full cost pricing for six different maple syrup operations[‡]

Cost item	127.5 (gal/yr)	255 (gal/yr)	765 (gal/yr)	1,530 (gal/yr)	2,550 (gal/yr)	5,100 (gal/yr)
Total Variable Cost ($/gal)	$43.34	$34.35	$23.73	$19.79	$18.28	$16.70
Assets[†] ($/gal)	$248.84	$205.49	$157.12	$142.41	$114.97	$102.82
2%	$4.98	$4.11	$3.14	$2.85	$2.30	$2.06
4%	$9.95	$8.22	$6.28	$5.70	$4.60	$4.11
6%	$14.93	$12.33	$9.43	$8.54	$6.90	$6.17
8%	$19.91	$16.44	$12.57	$11.39	$9.20	$8.23
10%	$24.88	$20.55	$15.71	$14.24	$11.50	$10.28
Pricing ($/gal)						
2%	$48.32	$38.46	$26.88	$22.63	$20.58	$18.75
4%	$53.30	$42.56	$30.02	$25.48	$22.88	$20.81
6%	$58.27	$46.67	$33.16	$28.33	$25.18	$22.86
8%	$63.25	$50.78	$36.30	$31.18	$27.48	$24.92
10%	$68.23	$54.89	$39.45	$34.03	$29.78	$26.98

[‡]Source of cost data CFBMC (2000). Costs are in 1999 US dollars using an exchange rate of 1.49 Canadian dollar per US dollar. (gal/yr) denotes gallons of maple syrup per year. ($/gal) denotes 1999 US dollars per gallon.
[†]Assets include structures, equipment, and land. ($/gal) denotes 1999 US dollars per gallon using an exchange rate of 1.49 Canadian dollar per US dollar.

would price between $22.63 and $34.03 per gallon of maple syrup. As a point of comparison, in the *New Hampshire Forest Market Report, 1998–1999* published by the University of New Hampshire Cooperative Extension, the retail price was $33.10 per gallon of maple syrup (http://extension.unh.edu/resources/resource/267/NH_Forest_Market_Report,_1998-1999 accessed on 17 February 2010).

The advantage of these strategies is that they are relatively simple and the entrepreneur should have all the necessary data to develop equivalent tables if they follow the Architectural Plan for Profit. There are two concerns with using these methods. First, if actual sales are less than predicted sales or production, then average cost per unit sold will be higher than estimated and thus, profit per unit will be smaller. Second, according to equations (7.4a) and (7.4b), fixed costs are simply allocated on a per unit basis to the product. For example, in Table 7.1 the fixed costs are allocated to the production of maple syrup. While the primary output of the Maple Syrup Operation case study was sap collection and production of maple syrup, these businesses often produce other maple syrup related products (e.g. maple sugar, maple candies, etc.). In addition, they may also be tourist destinations. Thus, not all the variable and fixed costs are associated with just maple syrup production. As allocating these truly joint production costs among all the outputs is purely arbitrary (Carlson 1974 and Blocher *et al.* 2002), caution should be taken when using average cost, cost-plus, or full-cost pricing strategies in the case of joint production.

Marginal (incremental) cost pricing

The concept of marginal cost price can be traced to the profit searching rule given in equation (7.1). The basic idea is that the price should cover the change in production cost of producing a change in output. The application of marginal cost pricing is illustrated by Figure 7.8.

As can be seen by Figure 7.8, the marginal cost – and consequently the price – of output Q_1 is different than the marginal cost (and price) of outputs Q_2 and Q_3. Based on this pricing strategy, each individual unit of output would be priced at its incremental production costs.[16] Thus, the price associated with Q_1 is $P_{Min\ AVC}$ or the price associated with the minimum point on the average variable cost curve. The prices associated with Q_2 and Q_3 are $P_{Min\ ATC}$ and P, respectively.

Using $P_{Min\ AVC}$ and $P_{Min\ ATC}$ as reference pricing points illustrates a potential problem with marginal cost pricing. Incremental production costs for output levels less than Q_1 would result in prices less than $P_{Min\ AVC}$ and will not be large enough to cover fixed production costs in the short-run. Incremental production costs for output levels between Q_1 and Q_2 would result in prices greater than $P_{Min\ AVC}$ but less than $P_{Min\ ATC}$ and will by definition only be large enough to cover a portion of fixed production costs in the short-run.[17] In addition, if the production process is described as non-continuous (e.g. need to add a new machine to increase production or a series of sequential steps must be completed before the input can move to the next step in the production process; Chapter 2), the incremental costs will be enormous at the point of discontinuity but not elsewhere.

Price searching revisited

There are many other types of price searching strategies that an entrepreneur may use; for example, block pricing, peak-load pricing, price discrimination, commodity

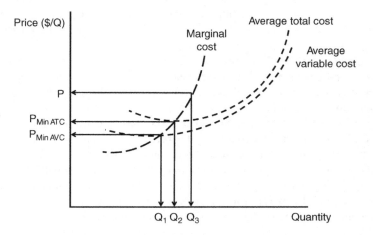

Figure 7.8 Marginal cost pricing.

bundling, and price matching, etc. (Baye 2009). The strategies described above deal primarily with setting price to cover explicit and implicit measures of total costs. However, as described in Chapter 6 and the introduction to this chapter, market price is defined by supply and demand, where supply represents the entrepreneur's opportunity cost of producing or providing an output. What is missing from these strategies is an explicit examination of the demand (i.e. the opportunity cost of the consumer) and any possible market power that an entrepreneur may possess. Thus, your ability to set price depends on the own-price elasticity of demand (i.e. the price responsiveness of your consumers; Table 6.7, Chapter 6), reflected by the appropriate market structure described above.[18] If the market demand curve is relatively elastic, then you have very little market power to set price. If the market demand curve is relatively inelastic, they you have some degree of market power to set price.

Derived demand

The demand for any productive resource is derived from the demand for the commodities they produce. For example, students have a demand for paper for taking notes. Their demand for paper gives rise to a demand for trees to make the paper. The market for car loans is derived from the market for cars and that most people need a loan to buy a car. The same can be said for a house mortgage. The Hancock Timber Resource Group, a Timber Investment Management Organization, examined the softwood lumber market and found that housing is the

> largest end-use for softwood lumber, a change in housing starts translates fairly directly into a change in lumber usage and … that over the past thirty years, 86 percent of the variation in the volume of lumber produced in the U.S. can be explained by changes in the demand for housing as measured by single family housing starts.
>
> (Hancock Timberland Investor, Q3 2006,
> http://www.htrg.com/research_archives_2006.htm
> accessed on 21 February 2010)

Rideout and Hesseln (2001) illustrate this relationship using three separate markets: a natural resource market, intermediate product market, and a final product market. A two market model of derived demand can be illustrated by Figure 7.9a.

In addition, for productive resources (e.g. natural resources) to flow through intermediate markets to a final consumer good market there must be a sufficient price difference among the market. The price difference illustrates the value added by each market. For example, there must be a sufficient price difference between the softwood stumpage market and pulp market for pulp mills to make a profit producing pulp. There must be a sufficient price difference between the pulp market and the market for notebook paper for paper mills to make a profit producing notebook paper. This is illustrated in Figure 7.9b.

As illustrated, the demand for natural resources, for the most part, is derived from the commodities they produce. As described by McGuigan and Moyer

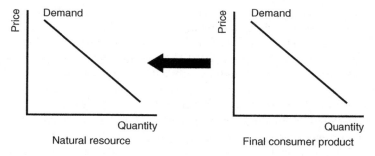

Figure 7.9a Derived demand.

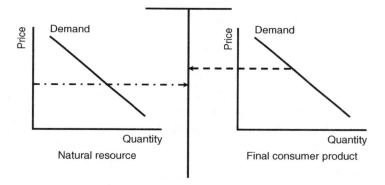

Figure 7.9b Derived demand.

(1993), two sets of factors must be examined for these types of resources. First, we must identify the criteria or specifications used by firms purchasing the resources as inputs into their production process (e.g. grade, defect, length, available quantity of a defined quality, etc.). For example, many state agencies publish stumpage prices by some measure of tree quality. Links to this information have been compiled by the United States Forest Service and can be found at http:// www.srs.fs.usda.gov/econ/data/prices/index.htm (accessed on 21 February 2010). In the Northeastern United States, similar publications are the *Sawlog Bulletin* and the *Sawlog Bulletin Magazine*, commonly called the "Log Street Journal", (http://sawlogbulletin.com/ and http://www.sawlogbulletin.org/ accessed on 21 February 2010). Forestland owners who derive revenue from the sale of timber are in essence trying to predict stumpage prices decades out in the future (Smith 1988). Wagner and Sendak (2005) examined the trends in regional stumpage prices from 1961 to 2001 for the Northeastern United States. In addition, Smith (1988) speculated that species considered as not very commercially viable today may become commercially viable in the future. Second, and perhaps the most important, we must identify the significant factors affecting the demand for the ultimate consumer good. For example, the Hancock Timberland Resource Group (http://www.htrg.com/ accessed on 21 February 2010), Forest Research Group (http://www.forestresearchgroup.com/ accessed on 21 February 2010), James W. Sewall Company (http://www.sewall.com/about/newsinfo/digital-library/news-letters.php accessed on 12 March 2010), and RMK Timberland Group (http://www.rmktimberland.com/index.html accessed on 21 February 2010) publish reports describing the impacts on stumpage prices and timberland values given changes in regional, national, and international economic factors.

The impacts of the final consumer products market on the derived demand of the natural resource can be illustrated by Figure 7.9c.

Figure 7.9c shows that demand shifts originating in the final consumer product market, filter through any intermediate markets, and finally impact the market for the natural resource. The magnitudes of the demand shifts moving from final consumer product to the market for the natural resource depend on the market structure associated with each of the markets, the economic variables and socio-demographic factors of demand given in equation (7.5), and the elasticities of supply and demand:

$$Q_D = f(P;\ P_S,\ P_C,\ Y,\ E_p,\ E_Y,\ SocDem,\ N) \tag{7.5}$$

where the economic variables and socio-demographic factors of demand are defined in Chapter 6, and N denotes the number of consumers in the market.

It is worth repeating the caution described by McGuigan and Moyer (1993): Entrepreneurs dealing with natural resources need to be keenly aware of the criteria or specifications used by firms purchasing these resources as inputs into their production process and the significant factors affecting the demand for the ultimate consumer good.

Derived supply

Analogous to the concept of derived demand is derived supply. For example, softwood and hardwood stumpage are used as inputs in the production process for pulp, and pulp is an input into producing various different types and grades of paper. The paper is then packaged and sold to students who demand it for taking

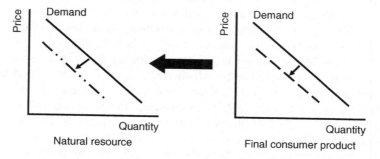

Figure 7.9c Derived demand: shifts move from final consumer product to natural resource.

notes. This linking of production processes (i.e. the output of one firm is the input into another until a final consumer product is manufactured) describes the concept of derived supply and is illustrated by Figure 7.10a.

With derived demand, demand shifts started in the final consumer product market and filtered down through any intermediate markets to the market for natural resources. In the case of derived supply, the shifts start in the market for natural resources, filter through any intermediate markets, and end in the final consumer product market. For example, Prestemon *et al.* (2008) examining the response of trading partners for the United States given an invasive defoliator such as the Asian Lymantria moth illustrates the concept of derived supply:

> A ban on the importation of United States logs by our trading partners would have a much larger effect on the forest sector than the loss of affected tree species directly caused by the [Asian Lymantria] moth invasion. Such a ban would lower the price of industrial roundwood in the United States, and simultaneously induce drops in processed wood product prices. The most affected industries would be plywood where the average price would be 2.3 percent lower, chemical wood pulp (3.2 percent lower) and printing paper (2.3 percent lower).
>
> (Prestemon *et al.* 2008: 413)

The impacts of the changes in the market for natural resources on final consumer products market can be illustrated by Figure 7.10b.

Figure 7.10b shows that a supply shift originating in the natural resources market filters through any intermediate markets, and impacts the final consumer products market. The magnitudes of the supply shifts as they move from natural resources market to final consumer product to the market depend on the market structure associated with each of the markets, the economic variables of supply given in equation (7.6), and the elasticities of supply and demand:

$$Q_S = f(P;\ ProdSys,\ P_I,\ E_P,\ N) \tag{7.6}$$

where the economic variables of supply are defined in Chapter 6, and N denotes the number of suppliers in the market. The connection between the natural

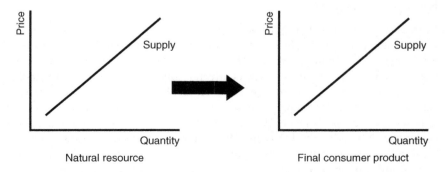

Figure 7.10a Derived supply.

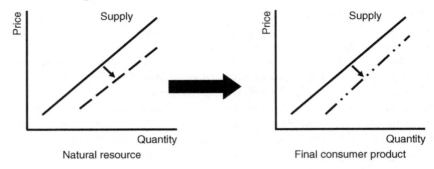

Figure 7.10b Derived supply: shifts move from natural resource to final consumer
product.

resources and intermediate production markets are that the output of the natural
resource market is the input into an intermediate production market. Thus, the
market equilibrium price in the natural resources market plus any transportation
and transactions costs would describe the price of an input in the intermediate
production market. The same can be argued for the connection between intermediate production and final consumer product market.

While McGuigan and Moyer (1993) described derived demand, a similar caution could be made for derived supply. Entrepreneurs dealing with natural
resources need to be keenly aware of the criteria or specifications used by firms
purchasing these resources as inputs into their production process and the significant factors affecting the supply of these inputs. Finally, overlaying the concept of
derived demand on derived supply you will get a picture of how economic information is transmitted from final consumer product market to the natural resources
market through derived demand, then back to the consumer product market
through derived supply. This constant flow of information back and forth among
the markets means that McGuigan and Moyer's (1993) cautionary statements to
entrepreneurs involved in any of these markets should not be forgotten.

Know your markets

Markets are not static, but dynamic over time due to changes in economic and
socio-demographic factors. Successful entrepreneurs are not blind to these changes.
Again borrowing from McGuigan and Moyer (1993), entrepreneurs should be
keenly aware of the significant factors affecting final product and input markets.
This was illustrated in the examples given in the sections entitled "Derived
demand" and "Derived supply." Continuing the example taken from the forestry
profession started in Chapter 6, Hagan *et al.* (2005) examine the changes in timberland ownership during the 1980s and 1990s in the Northeastern United States and
the implications for management regimes. They determined a significant decrease
in the number of acres owned by the forest industry and a corresponding increase
in the number of acres owned by other types of landowners. As the number of

forestland owners increased, the parcel size owned decreased. The implications for management are moving from managing commodity outputs, primarily timber, to managing forested ecosystems to produce a suite of ecosystem goods (e.g. renewable – timber and nonrenewable – minerals) and services (e.g. water and air purification, carbon sequestration). Arano and Munn (2006) examined the management behavior of various ownership categories. They found that in Mississippi, Timber Investment Management Organizations and industrial landowners generally invest three times as much in management activities related directly to timber production as compared to nonindustrial private forest landowners. However, nonindustrial private forest landowners control approximately 72 percent of the forestland. Belin *et al.* (2005) examined the attitudes of nonindustrial private forest landowners of Vermont, New Hampshire, and western Massachusetts toward an ecosystem approach to managing their forestlands. Even though the concept of an ecosystem-based management approach is not yet well defined, this concept seemed more consistent with their stated goals for owning forestland (Butler and Leatherberry 2004; Hagan *et al.* 2005; Butler 2008 and 2010).

The question is where to obtain information on these economic and socio-demographic factors and how they may impact your business. For the examples used here, the information was available through various sources, such as professional organization or society publications, extension publication, applied research publications, and newsletters. As an entrepreneur, you must actively search out and read these types of publications.

How to use economic information – *market equilibrium and structure* – to make better business decisions?

Markets are the primary means through which our economy determines what and how much of each product to produce, how to produce, and for whom to produce? Answering these questions also depends on the structures of any market. Successful entrepreneurs know the market structure for the commodities they produce or provide and for the inputs they purchase. As an entrepreneur, your objective is to turn a consumer's willingness and ability to buy your commodities into revenues and to pay as little for inputs as possible. Market structure defines the market power you have to capture this revenue given your ability to price your commodities relative to your competitors and to affect the price you pay for inputs. The Architectural Plan for Profit provides an outline for you to collect economic information to analyze market structure and price searching.

Markets are dynamic. Profit searching behavior includes being aware of the economic and socio-demographic factors that affect your input and output markets. If you focus only on what is in front of you, you will potentially miss changes in downstream input and upstream output markets that might affect your business. The goal of obtaining economic information about your markets and their structure is to increase the percentage of proactive decision making relative to reactive decision making.

8 Capital theory
Investment analysis

In the previous seven chapters, we examined the economic information used to build the Architectural Plan for Profit, develop the profit searching rule, and examine economic models of producers, consumers, and markets. The purpose was to help in answering the question "How to use economic information to make better business decisions?" An assumption that was made, although not explicitly, was that time – as a relevant input – is considered to have a zero opportunity cost. All other relevant inputs (e.g. labor) were examined in terms of their opportunity costs with respect to profit. While we included time (i.e. plantation age) as an input in the production process for the Loblolly Pine Plantation case study, we did not examine its impact on the Architectural Plan for Profit explicitly (Figure 8.1).

How does time affect the Pillars of Cost, Price, and Value? How is profit modified to reflect the economic information used to build these pillars? It is now time to rectify this omission in this and the next two chapters.

As a preview to examining how time affects the Architectural Plan for Profit, think about the following example. College students often obtain loans to help finance their college education. Table 8.1a contains an example set of loans obtained by a college student.

The student loans have the following conditions: (1) the annual interest rate is 7.43 percent; (2) no interest accrues while you are in school; (3) payment starts six months after graduation; and (4) you are permitted up to ten years to repay the loan through monthly payments. Based on Table 8.1a, over a four-year college career the amount borrowed is $17,125 and based on a 7.43 percent interest rate, the monthly payments would be $202.65.[1] Table 8.1b shows that at the end of the ten-year repayment period the amount paid in interest payments is $7,193.00.

According to Table 8.1b, the interest payment is more than the loan for the senior year (e.g. 1.3 times). As a point for comparison, if no interest was charged the monthly payments would be $142.71 (= 17,125/120).

So what is the economic interpretation of the interest rate and why is it positive? To examine the effects of time on the Architectural Plan for Profit, I will have to first address these two questions. I will then describe the mechanics of compounding and discounting or accounting for time. I will next examine six common tools that are used to analyze the profitability of investments and end with a discussion on capital budgeting.

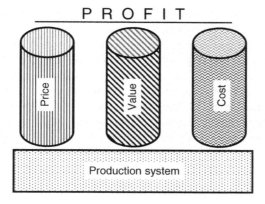

Figure 8.1 The Architectural Plan for Profit.

Table 8.1a Example student loan amounts

Year	Loan amount[‡]
Freshman	$2,625.00
Sophomore	$3,500.00
Junior	$5,500.00
Senior	$5,500.00

[‡]Loan amounts are in 2010 US dollars.

Table 8.1b Student loan interest payments

Total payments[‡]	$24,318.00
Principle borrowed	$17,125.00
Interest payments	$7,193.00

[‡]$24,318.00 = 202.65 × 120.

Case studies

The case studies used in this chapter will include the Inside-Out (ISO) Beams (Patterson *et al.* 2002 and Patterson and Xie 1998), Mobile Micromill (Becker *et al.* 2004), Maple Syrup Operation (Huyler 2000 and CFBMC 2000), Great Lakes Charter Boat Fishing (Lichtkoppler and Kuehn 2003), Select Red Oak (Hahn and Hansen 1991), and Loblolly Pine Plantation (Amateis and Burkhart 1985 and Burkhart *et al.* 1985). They can be found on the Routledge website for this book and their summaries will not be repeated here.

The interest rate

Previous choices that were analyzed dealt with production and consumption holding time constant. Now the fundamental question is consumption today or consumption in the future. Using the definition of an opportunity cost – the value of the next best alternative forgone – we can rephrase the question as "What is the opportunity cost of consumption today?". Answer: I am forgoing consumption in the future. Or, "What is the opportunity cost of consumption in the future?" Answer: I am forgoing consumption today. The opportunity cost described here is the opportunity cost of consumption at different points in time or the opportunity cost of time. The interest rate enumerates the opportunity cost of time. If production is the focal point of the discussion, the interest describes the opportunity cost of capital or the fact that you have alternative investment opportunities. As described by Hirshleifer (1970), the objective is to determine a preferred time-pattern of consumption or production; that is, balancing consumption or production now versus sometime in the future. Irving Fisher (1930) postulated that people are impatient given the choice of consuming now versus the possibility of greater income sometime in the future from not consuming but investing today. It is this tension or time preference that gives rise to positive interest rates.

In addition to time preference, the interest rate is also defined as a function of risk and inflation (Brealey *et al.* 2008; Silberberg and Suen 2001). I will describe each of these factors and the difference between the nominal and real interest rates.

Time preference

As stated above, the concept of time preference was described by Irving Fisher (1930) as people being impatient given the choice of current versus future consumption. Based on this impatience, time preference is defined as

> Time Preference – We value current consumption or income more highly than that accruing in the future. Therefore, we must be compensated if we are to put off current consumption.[2]

Hirshleifer (1970) defines that an individual's time preference depends on endowment, market opportunities, and productive opportunities. The endowment of most college students is low relative to their parents or Mr Bill Gates, the founder of Microsoft. For example, it would be tenable to postulate that $1,000 would constitute a large percent of a college student's endowment but a relatively smaller percent of their parent's endowment and would be inconsequential to Mr Gates' endowment. Consequently, a college student's rate of time preference for that $1,000 would be positive and larger than that of their parents. While their parents would still probably have a positive rate of time preference for the $1,000, Mr. Gates' time preference for $1,000 would likely be relatively low if not zero.[3] As an additional point of comparison, how would you describe the time preference of a subsistence farmer for that same $1,000 compared to the typical college student?

The concept of time preference is not universally accepted as the principle factor of a positive interest rate (Silberberg and Suen 2001). These authors argue that if production technology is increasing over time, then current consumption reduces future consumption by the rate of production technology change or conversely if current consumption is forgone, then future consumption would increase by the rate of production technology change. Thus the rate of production technology change would be sufficient to imply that the interest rate is positive.[4]

Risk and uncertainty

Not all future events are known with 100 percent certainty. The greater the gamble, the greater the reward we would require for forgoing current consumption. This relationship is captured by the concept of risk and uncertainty.

> Risk and Uncertainty – Exposure to the chance of loss. If the chance can be described in statistical terms, this is commonly described as risk. If the chance cannot be described in statistical terms, this is commonly described as uncertainty.

Risk and uncertainty can be categorized as market, biological, or political. While market risk is not unique to natural resources, they may have production processes that often take a long time. For example, in the Loblolly Pine Plantation case study the process could take up to 40 years (Chapter 2). The owners of this loblolly pine plantation are gambling that there will be a market for loblolly pine in the future and the future value of the loblolly pine will cover all production costs. The possibility that the owners are growing the loblolly pines for 40 years subjects them to many different biological hazards such as losing the trees due to drought, fire, insects, or diseases. Finally, many local, regional, state, and national regulations can limit or restrict the types of management activities that can be preformed (e.g. Ellefson *et al.* 1995). This describes the nature of political risk. For example, the Endangered Species Act of 1973 (16 U.S.C §§ 1531–1544) prohibits any person from "taking" a listed species (16 U.S.C §§ 1538(a)(1)(B)). Takings are actions that "harass, harm, pursue, hunt, shoot, kill, trap, capture, collect, or to attempt to engage in such conduct" (16 U.S.C §§ 1532(19)). Kennedy *et al.* (1996) describe such a case with the red-cockaded woodpecker.[5]

Inflation

Inflation is related directly to the sum of money required to purchase goods and services. If it takes more money to purchase the same goods and services this year than it did last year, then this describes inflation.

> Inflation (Deflation) – A fall (rise) in the *purchasing power of money*, often experienced as a rise (fall), on average, of the monetary value of goods and services.
>
> (Heyne *et al.* 2006: 537, emphasis added)

Simply, inflation means each dollar will buy fewer goods and services than before. The problems created by inflation are caused almost entirely by uncertainty of the future purchasing power of money. Thus, inflation distorts the signals that are provided through market prices observed by consumers and entrepreneurs. If the fall in the purchasing power of money was certain and constant over time, consumers and entrepreneurs would factor this constant into all economic decisions.

If inflation affected all goods and services equally, then while the absolute price levels would increase, the tradeoff relationships would not change and therefore economic decisions would remain the same. However, inflation affects goods and services differently; thus, it impacts the relative prices.

Nominal, real, and risk-free interest rates

Interest rates are most often defined as real and nominal. Fisher (1930) developed the relationship between the nominal and real interest rate given in equation (8.1):

$$
\begin{aligned}
i &= (1+r)\cdot(1+\omega) - 1 \\
&= r + \omega + r\cdot\omega
\end{aligned}
\tag{8.1}
$$

where i denotes the nominal interest rate, r denotes the real interest rate, and ω denotes the anticipated inflation rate ($\omega \geq 0$). Basically the difference between the nominal and real interest rates is accounting for inflation.[6]

In addition, Silberberg and Suen (2001) and Brealey *et al.* (2008) note that market interest rates have historically included a risk premium to account for external factors other than inflation. Let the risk adjusted real interest rate, $\tilde{r}$, be given by equation (8.2a):[7]

$$
\tilde{r} = r_f + \eta
\tag{8.2a}
$$

where r_f denotes a risk-free real interest rate (e.g. the return on 1-year US Treasury bills) and η denotes a risk premium ($\eta \geq 0$). If anticipated inflation is low, equation (8.2a) can be modified to include a risk premium:

$$
\tilde{i} = \tilde{r} + \omega = r_f + \eta + \omega
\tag{8.2b}
$$

Equations (8.2a) and (8.2b) can be used as models for the real and nominal opportunity costs of capital as defined by the capital markets.[8] While it is not accurate to do so, for practical purposes I will drop the distinction between the risk adjusted real interest rate and the real interest rate, and the risk adjusted nominal interest rate and the nominal interest rate. In addition, for practical purposes, I will use following definitions and relationships defined in equations (8.3a) and (8.3b) between the real and nominal interest rates in this and the next two chapters:

Real interest rate

$$
\begin{aligned}
r &= r_f + \eta \\
&\cong tp + \eta
\end{aligned}
\tag{8.3a}
$$

Nominal interest rate

$$
\begin{aligned}
i &= (1+r)\cdot(1+\omega) - 1 \\
&= r + \omega + r\cdot\omega \\
&= r_f + \eta + \omega + (r_f + \eta)\cdot\omega \\
&= tp + \eta + \omega + (tp + \eta)\cdot\omega
\end{aligned}
\tag{8.3b}
$$

where the risk-free interest rate is a proxy for time preference, tp.

As can be seen by equations (8.3a) and (8.3), the difference between the nominal and real interest rate is accounting for inflation. It would be reasonable to ask: Should you choose the nominal or real interest rate? As stated previously, the problems created by inflation are caused almost entirely by uncertainty of the future purchasing power of money. Forecasting the rate of inflation tomorrow can probably be done with a high level of confidence. However, the further you predict into the future, the less confident you would be concerning your prediction about inflation; thus, my preference for using the real interest rate. The general rule is that if you are using nominal cash flows, or cash flows that have not been adjusted for inflation, use the nominal interest rate.[9] If you are using real cash flows, or cash flows that have been adjusted for inflation, use the real interest rate.

The interest rate – revisited

Just as there are markets for goods and services, there are also markets for consumers and entrepreneurs to exchange funds for current versus future consumption or production (i.e. borrowing and lending). In the market for goods and services, market equilibrium will result in a market equilibrium price and quantity (Figure 7.3c). The markets for borrowing and lending are called capital markets. Market equilibrium in the capital markets would result in equilibrium prices for borrowing and lending and the amount of funds available at that price. The price for borrowing and lending is the interest rate. The interest rate can be used to describe the opportunity cost of time, but it also can be used to describe the opportunity cost of capital or the fact that there are alternative investment opportunities. In general, borrowers of capital will try to obtain the lowest interest rate as possible so the amount paid in interest will be as small as possible (e.g. Table 8.1b). Conversely, if a person is lending capital, they will try to obtain the largest interest rate as possible so the amount they receive in interest payments will be as large as possible, *ceteris paribus*. If capital markets function perfectly, the opportunity cost of time would be equal to the opportunity cost of capital (Price 1993). Unfortunately, this is not the case (Hirshleifer 1970; Johansson and Löfgren 1985; Price 1993; Klemperer 1996; Luenberger 1998; Silberberg and Suen 2001); that is, there is often a difference between their borrowing and lending rates, with the rate charged for borrowing capital being usually greater than the rate they receive for lending or investing.[10] Nonetheless, if the price of borrowing capital (the opportunity cost of capital) is less than what the individual would be willing to pay (the opportunity cost of time), the individual would borrow the capital. And, if the price of loaning capital (the opportunity cost of capital) is greater than what the individual would have been willing to accept (the opportunity cost of time), the individual would loan or invest their capital.

Fisher (1930), Henderson and Quandt (1980), Luenberger (1998), Silberberg and Suen (2001), Copeland *et al.* (2005), and Brealey *et al.* (2008) describe two broad categories of information capital markets use to define the opportunity cost of capital. The first can be described as internal factors or the amount and timing of the cash flows associated with a specific investment. For example, the ISO Beams, Mobile Micromill, Maple Syrup Operation, and Great Lakes Charter Boat

Fishing case studies all describe in varying levels of detail the timing of specific revenues and costs associated with each activity.

The second information category can be described as external factors; for example, the purchasing power of the cash flows in the different periods, the risk, derived demand and supply concepts, and other relevant business information such as patents, research and development, and entrepreneurial history. I will examine each briefly. The purchasing power of the cash flow in different periods addresses whether revenues and costs are appreciating in real terms. As Klemperer (1996), Davis *et al.* (2001), Silberberg and Suen (2001), and Brealey *et al.* (2008) point out, there is no universal or unique risk premium. Each investment reflects its own unique combination of market, biological, and political risk factors. Chapter 7 discusses that the demand for any productive resource is derived from the demand for the commodities they produce (Figures 7.9a, 7.9b, and 7.9c). In addition, the linking of production processes (i.e. the output of one firm is the input into another until a final consumer product is manufactured) describes the concept of derived supply (Figures 7.10a and 7.10b). The relative strengths of these forward and backward linkages among the markets will impact the opportunity cost of capital for any investment. Finally, entrepreneurs engage in activities to develop new cost structures, and more efficient ways to produce and deliver goods and services to consumers (Heyne *et al.* 2010). The capital markets consider these innovations, research and development progress, and patents in assessing the opportunity cost of capital. Thus, there is no unique real or nominal interest rate that could be used to examine all investments and the opportunity cost of capital is set by the capital markets using both internal and external factors of each investment (Brealey *et al.* 2008; Johnstone 2008).

What is the appropriate interest rate?

The conclusion from the previous discussion was that there is no unique interest rate (i.e. opportunity cost of time and the opportunity cost of capital) that can be used to examine all investments. So where does this leave us with respect to analyzing investments especially in managing natural resources sustainably with production systems that may take 20 to 100-plus years?

The basic assumption from Chapter 1 that landowners are maximizers and that they maximize their net benefits is still tenable even if accounting for time is included in the analysis. Consequently, for any given risk level, the investment chosen should increase an investor's net benefit no less than the next best alternative investment. The interest rate used to analyze investments would depend on a number of factors unique to each landowner, including the landowner's time preference, wealth, objectives, alternative uses of capital (e.g. mutual funds, stocks and bonds), degree of risk aversion, length of investment, and variability of returns associated with each investment. For example, Bullard *et al.* (2002) illustrates this by examining different interest rates required by nonindustrial private forest landowners in Mississippi. The nominal before tax interest rates were 8.0 percent for forestry investments lasting 5 years, 11.3 percent for forestry investments lasting 15 years, and 13.1 percent for forestry investments lasting 25 years. The real interest rates for the same investment horizons were 5.7 percent, 8.9 percent, and 10.7 percent respectively. According to a 2006 report by James W. Sewall Company, Institutional Investor, Timber Investment Management Organizations, and Real Estate Investment Trusts real interest rates range from 6 to 9 percent (Timberland Report. 2006. *Discount Rates and Timberland Investment*. Vol. 8, No. 3. http://www.sewall.com/about/newsinfo/digital-library/newsletters.php accessed on 12 March 2010).

Present value and future value

Time is a relevant input and people place different weights on cash flows that occur in different periods. In order for this economic information to be portrayed correctly and used accurately in decision making, various formulae have been developed to find the equivalence of these many revenues and costs at a common reference point. These formulae were developed when analysts did not have ready access to personal computers. The two most important ones are: future value and present value. With these two formulae and spreadsheets, you should be able to handle most problems.[11]

The use of the future and present value formulae are based on the following definitions. The chronological date of the reference point is set by the analyst. All costs and revenues in the *t*th period occur at the end of the period. Periods are one year long. Thus, all cash flows occur on December 31st of any given year. Compounding is the process of carrying a revenue or cost forward in time at a defined interest rate. Discounting is the process of carrying a revenue or cost backward in time at a defined interest rate. Capitalizing is the process of discounting a series of cash flows to one point in time.[12]

Analyses involving compounding and discounting can become intricate and thus keeping track of revenues and costs can be problematic and capitalizing them frustrating. For this reason, I will present a simple tool that can assist in working with these problems before I talk about future and present value formulae. This tool is called a cash flow diagram.

Cash flow diagrams

For all practical purposes, the interest rate entrepreneurs face will be positive. This interest rate will be used to place weights on *any* revenue or cost. The mechanics of accounting for revenues and costs in different time periods start with drawing a representation of the cash flows associated with the investment. This representation is a cash flow diagram and is illustrated by Figure 8.2a.

The cash flow diagram starts with a horizontal line called the time line that represents the duration of the investment and is divided into periods. For example, Figure 8.2a could represent a seven-year investment horizon with seven annual periods, or a 70-year investment horizon with ten year periods. Costs and revenues are pictured with vertical arrows; revenues or cash flows generated by the

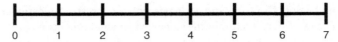

Figure 8.2a The time line of a cash flow diagram.

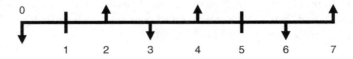

Figure 8.2b The cash flow diagram.

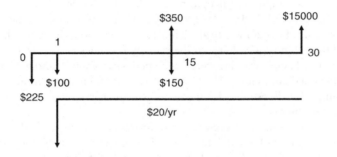

Figure 8.3 Example cash flow diagram.

investment are depicted by arrows pointing upward from the time line where the transaction occurred; cost or cash flows paid out by the investors are depicted by arrows pointing down (Figure 8.2b).

As can be seen by Figure 8.2b, there are costs at years 0, 3, and 6; and there are revenues at years 2, 4, and 7. The power of a cash diagram is that it can take a relatively complex situation and portray it in a concise format. For example, a management plan calls for $225 per acre to prepare the site for planting pine. Planting pine seedlings costs $100 per acre in year 1. A pulpwood thinning is planned at age 15. The expected cost of this thinning is $150 per acre and the expected pulpwood revenue is $350 per acre. You expect that yearly administrative expenses will be $20 per acre. The management plan calls for a clearcut at age 30 with an expected revenue of $15,000 per acre.[13] Figure 8.3 illustrates the cash flow diagram for the series of cash flows related to this management plan.

A direct relationship can also be drawn between a cash flow diagram and the production system. In the case of the Loblolly Pine Plantation case study, Figure 2.6 illustrated the total product curve. Figures 8.4a and 8.4b illustrate this relationship.

The horizontal axis on both Figures 8.4a and 8.4b is the same as Figures 8.2a, 8.2b, and 8.3. Figures 8.5a and 8.5b show a similar relationship can be developed for the Wisconsin oak given in Figure 2.9 and Table 2.7.

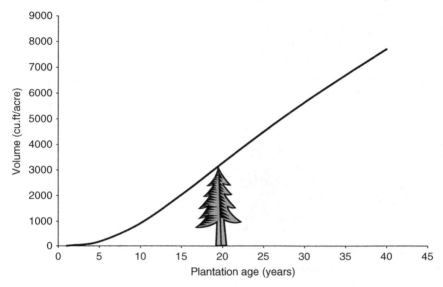

Figure 8.4a Loblolly pine total product curve.[‡]

[‡]Volume is measured as cubic feet per acre (cu.ft/ac) given $A_I = 25$, $S = 66$, $N_p = 1210$; Figure 2.6.

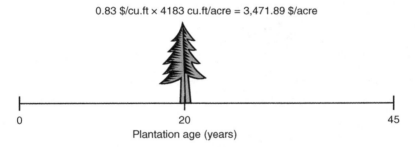

Figure 8.4b Loblolly pine cash flow diagram.[‡]

[‡]Volume is measured as cubic feet per acre (cu.ft/ac) given $A_I = 25$, $S = 66$, $N_p = 1210$; Table 2.6. Price is based on a 2009 southern United States wide pine sawtimber average.

The flexibility and utility of a cash flow diagram cannot be overstated: the power of a cash diagram is that it can take a relatively complex situation and portray it in a concise format. The cash flow diagram should be drawn before calculations are made. The cash diagram can be shown to a landowner or client and agreed upon and will save you from most misunderstandings that will cost you time and money later on. A student in 2000 gave me the following poem which summarizes all these ideas.

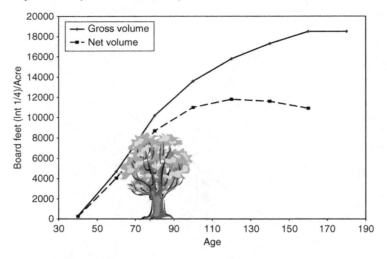

Figure 8.5a Unthinned oak stand in Southwestern Wisconsin (medium site) total product curve.[‡]

[‡]Volume is measured as board feet (International ¼) per acre (bf/ac); Figure 2.9.

0.17 \$/bf × 4050 bf/acre = 688.5 \$/acre

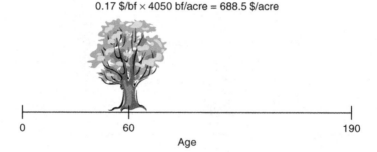

Figure 8.5b Unthinned oak stand in Southwestern Wisconsin (medium site) cash flow diagram.[‡]

[‡]Volume is measured as net board feet (International ¼) per acre (bf/ac); Table 2.7. Price is a 2010 northern United States wide oak sawtimber average.

It's So Simple

Investment solutions are hard to find
Miscalculations will put you in a bind
Keeping time and money together
Can prove to be rough

What do I do?

When the going gets tough
If you find yourself

Stuck in a jam
Don't lose your calm
Draw the Cash Flow Diagram
Sean Lyman
College of Environmental Science
and Forestry
B.S. 2001, A.A.S. 1999

Future value formula

The basic concept of a future value is to find the equivalence of a revenue or cost in the future. This is illustrated by the cash flow diagram in Figure 8.6.

The future value formula is given by equation (8.4):

$$V_t = V_0(1+r)^t \tag{8.4}$$

where V_t defines the equivalent value of a cash flow t periods in the future, V_0 defines the value of a cost or revenue in time 0, r denotes the real interest rate from equation (8.3a), and t denotes the number of 1-year periods that V_0 is compounded.[14]

Examining Figure 8.6 and equation (8.4) reveals the following three features. First, Figure 8.6 illustrates the finding of the equivalence of a cash flow, V_0, t-periods in the future, V_t. That is, an individual would be indifferent between the value V_0 now or the value V_t t-periods in the future. Second, the procedure for finding the equal values is given by equation (8.4). The economic information required to calculate V_t are the initial cash flow, V_0, the interest rate, r, in decimal form, and the number of 1-year periods, t. For example, if I invest $20,000 at 6 percent, what is its value nine years from today? The solution is illustrated in Figure 8.7.

The interpretation of Figure 8.7 is that given the economic information of an initial value of $20,000, an interest rate of 6 percent which reflects both time preference and risk, and a period of nine years, an individual would be indifferent between $20,000 now and $33,789.58 nine years from today. Finally, Figure 8.6 and equation (8.4) represent the fundamental concept of compensation required to forgo current consumption. This will be illustrated using equations (8.5a) through (8.5c).

$$\begin{aligned} V_1 &= V_0 + V_0 \cdot r \\ &= V_0(1+r) \end{aligned} \tag{8.5a}$$

The right-hand side of equation (8.5a) shows that the difference between consumption now and in period 1 is the amount of compensation $V_0 \cdot r$. As the interest rate r is defined as a percent, the amount of compensation is a percent of what was forgone. Equation (8.5b) illustrates the compensation required to forgo consumption between periods 1 and 2:

$$\begin{aligned} V_2 &= V_1 + V_1 \cdot r \\ &= V_1(1+r) \end{aligned} \tag{8.5b}$$

Figure 8.6 Future value cash flow diagram.

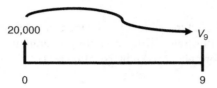

Figure 8.7 Future value example.

$$V_9 = 20,000 \cdot (1 + 0.06)^9$$
$$= 20,000 \cdot (1.06)^9$$
$$= 33,789.58$$

Equation (8.5c) illustrates the compensation required to forgo consumption between now and in period 2 by using the definition of the compensation given in equation (8.5a).

$$V_2 = V_1(1 + r)$$
$$= V_0(1 + r) \cdot (1 + r)$$
$$= V_0(1 + r)^2 \tag{8.5c}$$

If this process is repeated many times, the end result is equation (8.4). The compensation is modeled using a geometric progression. The economic implications of this will be examined after a discussion of present value.

Present value formula

The basic concept of a present value is to find the equivalence of a future revenue or cost. This is illustrated by the cash flow diagram in Figure 8.8.

The future value formula is given by equation (8.6):

$$V_0 = V_t (1 + r)^{-t}$$
$$= \frac{V_t}{(1 + r)^t} \tag{8.6}$$

where V_0 defines the equivalent value of a cash flow in time 0, V_t defines the value of a cost or revenue t periods in the future, r denotes the interest rate from equation (8.3a), and t denotes the number of 1-year periods that V_t is discounted.

Figure 8.8 Present value cash flow diagram.

Examining Figure 8.8 and equation (8.6) reveals the following three features. First, Figure 8.8 illustrates the finding of the equivalence now, V_0, of a cash flow, V_t, t-periods in the future, V_t; that is, an individual would be indifferent between the value V_t t-periods in the future or the value V_0 now. Second, the procedure for finding the equal values is given by equation (8.6). The economic information required to calculate V_0 are the initial cash flow, V_t, the interest rate, r, in decimal form, and the number of 1-year periods, t. For example, what is the value today of receiving $100,000 20 years from now given a 6 percent interest rate? The solution is illustrated in Figure 8.9.

The interpretation of Figure 8.9 is that given the economic information of a future value 20 years from now of $100,000 and an interest rate of 6 percent which reflects both time preference and risk, an individual would be indifferent between $100,000 and $31,180.47 today.

Future and present value formulae – revisited

Equations (8.4) and (8.6) define the future and present respectively and are repeated here

Future value formula

$$V_t = V_0(1 + r)^t$$

Present value formula

$$V_0 = V_t(1 + r)^{-t}$$
$$= \frac{V_t}{(1 + r)^t}$$

In both formulae, the compensation is defined by the geometric progression $(1 + r)^t$ or $(1 + r)^{-t}$. Thus, for any given interest rate $(1 + r)^t$ grows very quickly in magnitude as time increases and conversely $(1 + r)^{-t}$ becomes small in magnitude very quickly as time increases. Consequently, the weight placed on any future value, $(1 + r)^{-t}$, in terms of impact on an economic decision is very small. The choice of the interest rate has a similar effect. The economic implications of these simple algebraic relationships are important when dealing with production systems that may

Figure 8.9 Present value example.

$$V_0 = 100,000 \cdot (1 + 0.06)^{-20}$$
$$= 100,000 \cdot (1.06)^{-20}$$
$$= \frac{100,000}{(1.06)^{20}}$$
$$= 31,180.47$$

take decades to result in a good such as stumpage that can be sold to produce lumber or furniture. The entrepreneur or a consultant working for a landowner or stakeholders must be very thoughtful in determining the appropriate cash flow diagram that represents all the relevant costs and revenues and the appropriate interest rate.[15]

The Mobile Micromill case study (Becker et al. 2004)

I would caution readers to exercise due diligence when reviewing published analyses. I will use the Mobile Micromill case study as an example. Table 8.2a contains the cost data for purchasing the mill.

Table 8.2b contains the calculated monthly payments, the total amount paid, and the interest payments using a seven-year loan period and an interest rate of 7 percent. The calculations used here are similar to those used to determine the monthly payments from the student loan example.

The monthly payments calculated by Becker *et al.* (2004; 8) are $5,884.00. The difference is due to the study's authors miscalculating the monthly payments using the following procedure:

$$5,884.07 = \frac{303,229.50 \cdot \left(1 + \dfrac{0.07}{12}\right)^{(12 \times 7)}}{12}$$

Using this monthly payment, the total interest paid would be $191,032.78. This incorrect analysis resulted in an overstatement of the interest paid by $109,832.43. The monthly payments and finance charges for the tractor-loader, forklift, and pickup were calculated in the same way.

Investment analysis tools

The concept of forgoing current consumption generally implies that you are investing capital in some opportunity such that at future date you will be able to

Table 8.2a Mobile Micromill example[‡]

Item	Cost (2003 US$)
Micromill	426,185.00
Sales tax	0.00
Mill delivery, setup, and training	7,000.00
Total delivered price	433,185.00
Down payment (30%)	129,955.50
Amount financed	303,229.50

[‡]Cost data are taken from Becker *et al.* (2004: 8). Costs are in terms of United States 2003 dollars (2003 US$).

Table 8.2b Monthly payments and total interest paid

Monthly payments	$4,576.55
Total payment	$384,429.85
Interest paid	$81,200.35

increase your consumption. A similar argument could be made with respect to production, in that you are investing in a technology that will allow you to produce more in the future. In any case, we need some way to examine these alternative investments reliably that is consistent with the assumption that individuals are maximizers. I will examine seven common investment analysis tools: (1) net present value, (2) benefit cost ratio, (3) profitability index, (4) internal rate of return, (5) return on investment, (6) payback period, and (7) breakeven analysis.

Net present value

Net present value (NPV) is the sum of the present value of the revenues minus the present value of the costs. This is shown in equation (8.7):

$$NPV_T = \sum_{t=1}^{T} R_t (1+r)^{-t} - \sum_{t=0}^{T} C_t (1+r)^{-t}$$

$$= \sum_{t=1}^{T} \frac{R_t}{(1+r)^t} - \sum_{t=0}^{T} \frac{C_t}{(1+r)^t}$$

(8.7)

where R_t denotes the revenue at time t, r denotes the interest rate, and C_t denotes the cost at time t. The economic information required to use NPV are: (1) the amount and timing of all cash flows and (2) an appropriate opportunity cost of capital given as the interest rate. The interpretation of NPV is tied directly to its general decision criterion.[16] If $NPV > 0$, this implies the investment is returning the desired discount rate *plus* a present value of additional net revenue. The implication is the wealth of an investor increases and the investment is acceptable. If $NPV = 0$, this implies the investment is returning the desired interest rate. The investment does not increase or decrease wealth. The investor is indifferent as

192   *Capital theory: investment analysis*

their wealth does not change. If *NPV* < 0 the investment is rejected as the investor's wealth decreases.

Table 8.3 illustrates calculating the NPV of the management plan with the cash
flow diagram in Figure 8.3 given a discount rate of 9 percent.

I would like to highlight three aspects of the information given in Table 8.3.
First, based on the results of the NPV calculations, this management plan would
increase the wealth of the landowner by 663.27 $/acre. In other words, investing
in this management plan returns the landowner the required 9 percent interest rate
plus 663.27 $/acre of additional present value of net revenue. While calculating
the NPV can be somewhat onerous and the result should be calculated correctly,
interpreting the result accurately is paramount to providing useful economic
information to an investor. Second, the present value of the annual administrative
costs can be calculated by summing the present value of each individual yearly
cost. Or, they can be determined by using the present value of a terminating every-
period series formula given in Appendix 7. Finally, Table 8.3 shows the calculations comprised of rows, columns, and cells. Each row represents a specific cash
flow from the cash flow diagram (Figure 8.3). The column headings denote if the
cash flow represents a revenue, cost, or present value. The cells under the present
value column headings contain equation (8.6) with the appropriate cell refer-
ences to the specific cash flow, timing, and interest rate. Basically, Table 8.3 is a
spreadsheet.

As Luenberger (1998: 27) states: "It is widely agreed (by theorists, but not
necessarily by practitioners) that, overall, the best [investment analysis] criterion
is that based on net present value." NPV is the standard by which all other investment analysis tools are evaluated. NPV is consistent with the assumption that

Table 8.3 Net present value example based on the cash flow diagram given in
Figure 8.3[‡]

Year	Revenue	Cost	Present value revenues	Present value costs
0		225.00		225.00
1		100.00		91.74
15	350.00	150.00	96.09	41.18
Annual Administrative Cost		20.00		205.47
30	15,000.00		1,130.57	
Total			1,226.66	563.39

[‡]The units on all revenues and costs are dollars per acre ($/acre). The discount rate is 9 per-
cent or 0.09.
Annual administrative costs

$$205.47 = \sum_{t=1}^{30} 20 \cdot (1.09)^{-t} = 20 \cdot \frac{(1.09)^{30} - 1}{0.09 \cdot (1.09)^{30}}$$

$NPV = 1,226.66 - 563.39 = \$663.27$ per acre.

individuals maximize their wealth (Brealey *et al.* 2008; Copeland *et al.* 2005; Luenberger 1998). This does not imply that NPV is perfect as an investment analysis tool. As larger NPV values are preferred to smaller ones, NPV is biased toward projects with large numbers for revenues and costs; as the scale of an investment increases so does the NPV. This criticism can be ameliorated by calculating revenues and costs on a per unit bases; for example, per acre as in Table 8.3. I will discuss more about this topic in the section entitled "Capital budgeting." As described in equation (8.7), the amount and timing of all cash flows and the interest rate are all known with certainty. In Chapter 10, I will examine approaches that can be used if the cash flows are not known with certainty.

Benefit cost ratios

Benefit cost ratios (BCR) calculate an investment's (discounted) benefits per dollar of (discounted) costs. This is shown in equation (8.8):

$$BCR_T = \frac{\sum\limits_{t=1}^{T} R_t \,(1+r)^{-t}}{\sum\limits_{t=0}^{T} C_T \,(1+r)^{-t}} = \frac{\sum\limits_{t=1}^{T} \dfrac{R_t}{(1+r)^t}}{\sum\limits_{t=0}^{T} \dfrac{C_t}{(1+r)^t}} \tag{8.8}$$

where the variables are defined as before. The economic information required to use BCR are the same as NPV: (1) the amount and timing of all cash flows and (2) an appropriate opportunity cost of capital given as the interest rate. It should be noted that the BCR will calculate a result with no units. This is because the units of the numerator or the present value of the revenues must equal the units on the denominator or the present value of the costs. The interpretation of the BCR is tied directly to its general decision criterion. If $BCR > 1$, this implies the investment is returning greater per dollar revenues than per dollar costs in present value terms. The implication is the wealth of an investor increases, that is $NPV > 0$, and the investment is acceptable. If $BCR = 1$, this implies the investment's per dollar revenues are equal to its per dollar costs in present value terms. The investment does not increase or decrease wealth; that is $NPV = 0$. The investor is indifferent as their wealth does not change. If $BCR < 1$ the investment is rejected. The wealth of the investor decreases as the per dollar costs are greater than the per dollar revenues in present value terms; that is $NPV < 0$.

Using the cash flow diagram of the management plan given in Figure 8.3, the BCR is given by equation (8.9):

$$BCR = \frac{1,226.66 \,(\$/\text{acres})}{563.39 \,(\$/\text{acres})} = 2.18 \tag{8.9}$$

Three aspects of the economic information resulting from equation (8.9) should be noted. First, the BCR is a ratio 2.18:1 or for every \$2.18 of present value of revenue, there is $1.00 of present value of cost. Second, the BCR provides information on how much the discounted costs (revenue) could increase (decrease) before $NPV < 0$. In this case, the discounted cost would have to increase by 2.18 before $NPV = 0$. Conversely, the discounted revenues would have to decrease by 2.18 before $NPV = 0$. Third, the interpretation of BCR seems simpler than NPV; a BCR greater than one and the project is acceptable.

There are, however, three concerns with using BCRs. Examining equation (8.8) shows that BCR is a fraction; thus, as the denominator of a fraction decreases faster than any changes in the numerator, the value of the fraction will decrease. In terms of the BCR, this is related directly to how costs are defined and treated; for example, the treatment of initial investment cost versus operating costs. In the cash flow diagram given in Figure 8.3 for the management plan, if the year 1 planting costs of $100 per acre, the year 15 thinning costs of $150 per acre, and the yearly administrative expenses of $20 per acre are defined as normal operating expenses and included as part of the investments net cash flow, as part of the numerator in equation (8.8). While the initial investment cost of $225 is defined as the denominator in equation (8.8), the BCR would be 1.95. If only positive net revenues (e.g. $R_t - C_t > 0$, for all t) are included in the numerator and all other costs in the denominator, the BCR would be 4.06. However, the NPV does not change given these various cost definitions. Second, the BCR is biased towards investments with small discounted cost elements. Finally, BCR is not related directly to economic efficiency because wealth is not typically maximized when the BCR is maximized. Thus, the BCR is not consistent with the assumption that humans maximize their wealth.

Profitability index

Profitability index (PI) calculates the investment's NPV per dollar of initial cost and is given by equation (8.10a):

$$PI_T = \frac{\sum_{t=1}^{T} R_t (1 + r)^{-t} - \sum_{t=0}^{T} C_T (1 + r)^{-t}}{C_0}$$

$$= \frac{\sum_{t=1}^{T} \frac{R_t}{(1 + r)^t} - \sum_{t=0}^{T} \frac{C_t}{(1 + r)^t}}{C_0} \tag{8.10a}$$

$$= \frac{NPV_T}{C_0}$$

An alternative form is given by equation (810b) and calculates the investment's present value of all future net cash flows per dollar of initial cost:

$$PI_T = \frac{NPV_T - C_0}{C_0} \tag{8.10b}$$

For either equations (8.10a) or (8.10b), the decision criteria is if *PI* > 1, the investment is acceptable as the *NPV* > 0. If equation (8.10b) is used, a *PI* = 1 defines a breakeven point.

Using the cash flow diagram of the management plan given in Figure 8.3, the PI calculated using equation (8.10a) is 2.95. This means every $1.00 per acre of initial costs generates $2.95 per acre of NPV. The PI calculated using equation (8.10b) is 1.95; that is, every $1.00 per acre of initial costs generates $1.95 per acre of present value of net revenue beyond the initial cost.

As can be seen, the PI is related directly to BCR. In fact, equation (8.10a) is a variant of a BCR described in the previous section. The PI tries to address the problem of scale associated with NPV. However, there are better procedures to account for scale that will be discussed in the section entitled "Capital budgeting." Because the PI is a variant of the BCR, it suffers from the same criticisms and is not consistent with the assumption that humans maximize their wealth.

Internal rate of return

The internal rate of return (*irr*) is generally attributed to Fisher (1930). However, Boulding (1935) and Keynes (1936) also published manuscripts discussing the *irr* at roughly the same time. I will use Boulding (1935) as my primary reference as he most clearly lays out the foundation of the *irr* as used in forestry. Based on Boulding (1935), the *irr* is commonly defined as the "interest rate" or "discount rate" that sets the NPV equal to zero as illustrated by equation (8.11):

$$\sum_{t=0}^{T} \frac{R_t}{(1+irr)^t} - \sum_{t=0}^{T} \frac{C_t}{(1+irr)^t} = \sum_{t=0}^{T} \frac{(R_t - C_t)}{(1+irr)^t} = 0 \qquad (8.11)$$

where R_t denotes revenues at time t with $R_t \geq 0$, C_t denotes costs at time t with $C_t \geq 0$, $(R_t - C_t)$ denotes the net cash flow at time t, with $(R_0 - C_0) < 0$ and at least one $(R_t - C_t) > 0$ for $t > 0$; and *irr* is the internal rate of return. The economic information required to determine the *irr* consists only of the amount and timing of all cash flows and the result is the "internal rate of return." The decision criteria; however, requires that an appropriate opportunity cost of capital obtained from the capital markets be provided. If the *irr* is greater (less) than an appropriate opportunity cost of capital the investment is acceptable (not acceptable). If the *irr* is greater (less) than an appropriate opportunity cost of capital, there is a greater likelihood that the investment's NPV is greater than (less than) 0. Using the cash flow diagram of the management plan given in Figure 8.3, the *irr* is approximately 0.1253.

There are three main problems with the *irr*, the least of which is its calculation. Mathematically, the *irr* is a function of the "root" or "zero" of a polynomial equation. The degree of the polynomial defines the number of potential roots or zeros. As some of the roots may be positive, negative, or zero, the *irr*s may be positive, negative or zero. If the net cash flow does not change sign at least once, you will not be able to calculate an *irr* that makes economic sense (Descartes' Rule of

Signs). The calculation of the *irr* becomes exceedingly complicated when the project involves positive and negative net cash flows scattered though time. If the net cash flow changes signs more than once, there may be multiple positive *irr*s (Descartes' Rule of Signs). Given the complexity of many cash flows, the *irr* can only be determined iteratively.

These are the steps that I would advise to estimating the *irr*. (1) Set up the spreadsheet just like you are going to calculate NPV. (2) Determine if the net cash flow changes signs at least once. (3) Change the interest rate until NPV is approximately equal to zero.[17]

The second problem is interpreting the *irr*.[18] Some people confuse the *irr* as a rate of return similar to the interest rate because both "appear" in the NPV formula; for example, equation (8.7) versus equation (8.11). A classic example of this misinterpretation is given by the following set of articles: Webster *et al.* (2009), Wagner (2009), and Pickens *et al.* (2009). Unfortunately, the *irr* can only be interpreted as a rate of return under very strict circumstances (Wagner 2009; also see Appendix 9). Interpreting the *irr* similar to an interest rate drawn from the capital market does not follow from the mathematics of its estimation. The *irr* depends solely on the amount and timing of a project's cash flow. The interest rate is established in the capital markets, is the expected rate of return offered by other investments of equivalent in risk to the project being evaluated, and is based on the capital markets assessment of the internal and external factors of each investment (Brealey *et al.* 2008; Johnstone 2008).

Finally, the *irr* favors projects that have low capital costs and relatively early returns and uses a single "interest rate" over the life of the project. It is also well documented in the finance and forest economics literature (e.g. Brealey *et al.* 2008; Copeland *et al.* 2005; Davis *et al.* 2001; Rideout and Hesseln 2001; Luenberger 1998; Klemperer 1996; Gregory 1987; Johansson and Löfgren 1985; Leuschner 1984; Clutter *et al.* 1983) that if the objective is to maximize an individual's wealth, *irr* is not always consistent with NPV to analyze investments. As the decision criteria of the *irr* requires defining an appropriate market derived opportunity cost of capital, why not use NPV?

Return on investment

Return on Investment (ROI) is a non-discounted ratio of revenues to costs. There are many similar variations of ROI that use different measures of revenues to costs. Two of the common ones are return on assets and return on equity. The various variations are given by equation (8.12a) through (8.12e)

Return on Investment

$$\frac{Book\ Value_{t+1} - Book\ Value_t}{Book\ Value_t} \quad (8.12a)$$

$$\frac{(Net\ Income + Interest) \cdot (1 - tax\ rate)}{Book\ Value} \quad (8.12b)$$

$$\frac{Income}{Total\ Assets} \tag{8.12c}$$

$$\frac{Net\ Income}{Book\ Value} \tag{8.12d}$$

$$\frac{Book\ Value_T - Book\ Value_t}{Book\ Value_t} \tag{8.12e}$$

The ratio is commonly used as a measure of efficiency. The greater the ratio, the greater the measure of per dollar revenue or income is to per dollar asset value. The definition of ROI depends directly on the definitions of the terms in the numerator and the denominator. The problems with ROI are due to accounting measures of value versus economic cash flow (Chapter 2). The larger value of T in equation (8.12e), the more the overstatement of efficiency is due to the non-discounted analysis. Different rules of accounting (e.g. depreciation) would cause different measurements of ROI. Neither the merits of an investment nor efficiency depend on how accountants classify cash flows. As a result, ROI is not necessarily consistent with the assumption that individuals maximize their wealth.

Payback period

The payback period of an investment is found by counting the number of years it takes before the cumulative forecasted net cash flow equals the initial investment. This is a non-discounted summation. Using the cash flow diagram of the management plan given in Figure 8.3, the payback period is 30 years. The decision criterion is the shorter the payback period the better. Payback periods are most often used for investments similar to considering the replacement of incandescent light bulbs with compact fluorescent light bulbs. In these cases, the goal is to reduce operating costs.

There are two problems with using the payback period. First, this is a non-discounted sum, therefore it does not account for the opportunity cost of capital, risk, or any other measures of opportunity cost. Second, it ignores all net cash flows after the payback period. As a result the payback period is not consistent with the assumption that individuals maximize their wealth.

Breakeven analysis

The breakeven analysis of an investment is found by counting the number of years it takes before the present value of the revenues is greater than the present value of the costs. The Mobile Micromill case study (Becker *et al.* 2004; Table 8, p. 30) is an example of a breakeven analysis. While there is no set decision criterion for breakeven analysis, a short payback period would be preferred over a longer one. Breakeven analysis does not ignore the opportunity cost of capital, it does ignore all net cash flows after the breakeven period. As a result breakeven analysis is not consistent with the assumption that individuals maximize their wealth.

Investment analysis tools – revisited

As Brealey *et al.* (2008), Kierulff (2008) and Graham and Harvey (2001) describe, NPV and *irr* are the most common investment analysis tools used in business. This result is consistent with the conclusion reached by Luenberger in 1998. While none of the investment analysis tools examined for determining acceptable investments is infallible, NPV is the best criterion to use and is the standard by which all other investment analysis tools are compared (Brealey *et al.* 2008; Copeland *et al.* 2005; and Luenberger 1998). Consequently, my recommendation is that NPV should always be used in addition to and regardless of other invest-ment analysis tools requested.

Patterson *et al.* (2002) used NPV to examine the feasibility of producing ISO beams using softwoods and hardwoods. What is interesting about Patterson *et al.*'s investment analysis is the use of three production levels and five interest rates; as the authors note, this was done to account for risk. This approach will be examined in more detail in Chapter 10. The Maple Syrup Operation and Great Lakes Charter Boat Fishing case studies do not explicitly use an investment anal-ysis tool like NPV nor calculate annual loan payments, but their authors recognize the importance of these capital costs by including them in their analyses. The Loblolly Pine Plantation case study was mentioned at the beginning of this chap-ter to introduce time as a relevant input. The timing of cutting trees based on an appropriate silvicultural system is often the principal tool used in sustainably managing a forested ecosystem. The investment analysis of these silvicultural systems will be covered in Chapter 9. Finally, while the production system described in the Loblolly Pine Plantation case study included time as an input, the Select Red Oak case study does not. Thus, the description of the production proc-ess for Select Red Oak does not facilitate any type of investment analysis. Additional research and information would be required to answer any questions concerning the opportunity cost of time in this case study. I would encourage the reader to review these case studies given the above discussion.

Capital budgeting

In the previous section, I examined tools that can be used to determine if a single investment is acceptable. However, an entrepreneur is often faced with choosing between a number of alternative investments. In addition, the choice is made more difficult by having a fixed budget with which to make these investments. The purpose of this section is to determine how to choose among investments.

Investments can be classified by attributes, not unlike classifying a tree into family, genus, and species. I will use three attributes to classify investments. The primary attributes are:

Mutually exclusive – Only one investment can be chosen; e.g. using an acre of land to plant loblolly pine for timber, or scotch pine for Christmas trees.

Independent – Investments are not dependent on one another with respect to adoption; e.g. a precommerical thinning on one stand does not preclude a commercial thinning in a different stand.

Interdependent – the feasibility of one investment is dependent on whether others are undertaken.

The secondary attributes are:

Divisible – If you can invest in part of a project; e.g. adding money to a savings account.

Indivisible – If you cannot invest in part of a project; e.g. buying a truck or pulp mill is indivisible, it is an all-or-nothing proposition.

The tertiary attributes are

Repeating – If you can replicate the exact project; e.g. a management regime that calls for repeating patterns of regeneration and harvest of the same area.

Non-repeating – If you cannot replicate the exact project.

Approximately 90 percent of the investments you will deal with will be either: mutually exclusive, indivisible, and non-repeating or independent, indivisible, and non-repeating. The main exception is including repeating. I will examine these two examples.

Mutually exclusive, indivisible, and non-repeating investments

I will use the following two projects in this section.

Parcel A

Parcel A is 30 acres, has just been cut, and will be managed with a prescription calling for site preparation and planting followed by a commercial thinning at age 50 and a harvest at age 80. The site prep and planting will cost $3,000; the commercial thin will net $36,000; and the final harvest will net $240,000. The project's discount rate is 5 percent. The cash flow diagram and spreadsheet are given in Figure 8.10.

Parcel B

Parcel B is 10 acres and has a 30-year old stand to be precommercially thinned immediately followed by a commercial thinning in 20 years, which includes improving the road system and other infrastructure, and a final harvest in 50 years. The precommercial thin at age 30 will cost $1,000; the commercial thin will net $12,000; the road improvements will cost $8,000; and the final harvest will net

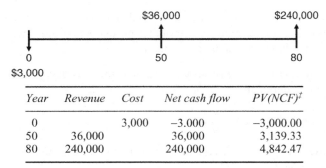

Year	Revenue	Cost	Net cash flow	PV(NCF)[‡]
0		3,000	−3.000	−3,000.00
50	36,000		36,000	3,139.33
80	240,000		240,000	4,842.47

Figure 8.10 Parcel A cash flow diagram and spreadsheet.

[‡]PV(NCF) denotes present value of the cash flow calculated using equation (8.6).
Net Present Value = $4,981.81.

$60,000. The project's discount rate is 5.5 percent. The cash flow diagram and spreadsheet are given in Figure 8.11.

Overview

From Figures 8.10 and 8.11, the NPVs of parcel A and parcel B are $4,981.81 and $4,496.91, respectively. Based on the NPVs, investment in the management of parcel A would be preferred as it results in the highest NPV. However, as the entrepreneur only has $3,000 in capital to spend on either investment they are mutually exclusive. In both parcels the investments are to manage the entire parcel. Thus, they are indivisible. Finally, neither investment will be repeated.

A closer examination of these projects reveals that they not only differ in the initial cost (Scale) but they also differ in the length of time (Horizon) the entrepreneur's capital is tied up. For parcel A, the investment horizon is 80 years and for parcel B it is 50 years. If parcel B is chosen, there is the opportunity to reinvest for an additional 30 years and would then match the investment horizon of parcel A. If the entrepreneur chooses parcel A, they use all their capital; however, with parcel B there is the opportunity cost of $2,000. These differences are important. Simply, there is an opportunity cost of the different investment scale and horizon that must be included in the analysis. To illustrate how to account for each difference, I will examine each separately and then combine them.

Investment horizon

The differing investment horizons imply there is an opportunity cost of reinvesting any proceeds. To account for any potential reinvestment opportunities, normalize different investment horizons to the investment horizon of the longest investment. Logic would dictate that only positive net cash flows or positive net revenues have the potential to be reinvested. Comparing parcels A and B shows

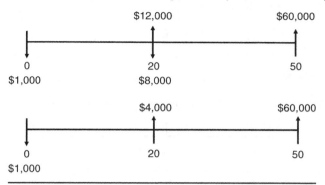

Year	Revenue	Cost	Net cash flow	PV(NCF)‡
0		1,000	−1.000	−1,000.00
20	12,000	8,000	4,000	1,370.92
50	60,000		60,000	4,125.99

Figure 8.11 Parcel B cash flow diagrams and spreadsheet.

‡PV(NCF) denotes present value of the cash flow calculated using equation (8.6).

Net Present Value = $4,496.91.

that: (1) parcel A has a positive net cash flow of $36,000 at year 50 that could be reinvested for 30 years; and (2) parcel B has a positive net cash flow of $4000 at year 20 that could be reinvested for 60 years and a positive net cash flow of $60,000 at year 50 that could be reinvested for 30 years. In addition to identifying any net cash flow that could be reinvested, the entrepreneur must also define a potential reinvestment interest rate. For purposes of illustration, the reinvestment rate for both investments will be 6 percent.

The cash flow diagram and spreadsheet for parcel A's reinvestment is given in Figure 8.12.

Equation (8.13) illustrates accounting for the reinvestment for parcel A:

$$\$4,171.91 = \frac{\$36,000 \cdot (1.06)^{30}}{(1.05)^{80}}$$
$$= \frac{\$206,775.68}{(1.05)^{80}}$$

(8.13)

The numerator of equation (8.13) calculates the future value of the $36,000 (see equation (8.4)) and the denominator calculates the present value the result (see equation (8.6)). The NPV of parcel A adjusting for investment horizon is $6,014.38.

The cash flow diagram and spreadsheet for parcel B's reinvestment is given in Figure 8.13.

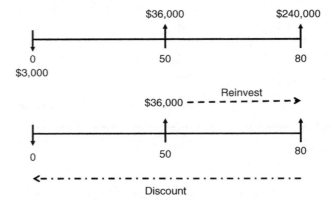

Year	Revenue	Cost	$NCF^{\ddagger}$	$FV(NCF_{80})^{\ddagger}$	$PV(NCF)^{\ddagger}$
0		3,000	−3.000		−3,000.00
50	36,000		36,000	206,765.68	4,171.91
80	240,000		240,000		4,842.47

Figure 8.12 Investment horizon – parcel A's reinvestment cash flow diagram and spreadsheet.

$\ddagger$NCF denotes net cash flow. FV(NFC$_{80}$) denotes the future value of 36,000.
PV(NFC) denotes present value of the net cash flow.
Net Present Value = $6,014.38.

Equations (8.14a) and (8.14b) illustrate accounting for the reinvestment for parcel B:

$$\$1{,}820.60 = \frac{\$4{,}000\cdot(1.06)^{30}}{(1.05)^{80}}$$
$$= \frac{\$131{,}950.76}{(1.05)^{80}} \tag{8.14a}$$

$$\$4754.78 = \frac{\$60{,}000\cdot(1.06)^{30}}{(1.05)^{80}}$$
$$= \frac{\$344{,}606.47}{(1.05)^{80}} \tag{8.14b}$$

The numerators of equations (8.14a) and (8.14b) calculate the future value of the $4,000 and $60,000, respectively. The denominators calculate the present value. The NPV of parcel B adjusting for investment horizon is $5,575.38.

Normalizing for investment horizon in this example did not change the preferred alternative; that is, investing in the management of parcel A would increase the entrepreneur's wealth more than investing in the management of parcel B. How would the analysis, results, and interpretation change if the reinvestment rate

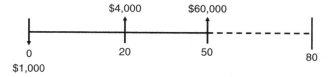

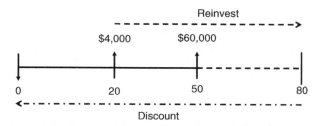

Year	Revenue	Cost	NCF‡	FV(NCF$_{80}$)‡	PV(NCF)‡
0		1,000	−1.000		−1,000.00
20	12,000	8,000	4,000	131,950.76	1,820.60
50	60,000		60,000	344,606.47	4,754.78

Figure 8.13 Investment horizon – parcel B's reinvestment cash flow diagram and spreadsheet.

$\ddagger$NCF denotes net cash flow. FV(NFC$_{80}$) denotes the future value. PV(NFC) denotes present value of the net cash flow.
Net Present Value = $5,575.38.

was the same as the interest rate? How would the analysis, results, and interpretation change if the reinvestment rate was less than the interest rate?

Investment scale

The entrepreneur only has $3,000 in capital; thus, if the entrepreneur chooses parcel A, they use all their capital but with parcel B there is the opportunity cost of $2,000. Again, there are reinvestment opportunities due to scale. To account for any potential reinvestment opportunities, normalize different investment scales to the investment horizon of the largest initial investment. The potential reinvestment opportunities are any interest income that can be earned.

Parcel B is the only alternative that has the potential to earn interest income. The cash flow diagram and spreadsheet for parcel B's reinvestment is given in Figure 8.14.

Equation (8.15) illustrates the interest income earned by accounting for scale:

$$\$919.46 = \frac{\$2,000\cdot(1.06)^{80}}{(1.055)^{80}} - \$2,000$$
$$= \frac{\$211,591.99}{(1.055)^{80}} - \$2,000$$

(8.15)

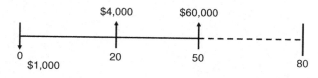

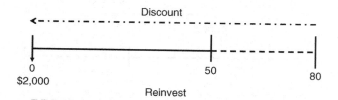

Year	Revenue	Cost	NCF‡	FV(NCF₈₀)‡	PV(NCF)‡
0		1,000	−1.000		−1,000.00
		2,000		211,591.99	919.46
20	12,000	8,000	4,000		1,370.92
50	60,000		60,000		4,125.99

Figure 8.14 Investment scale – parcel B's reinvestment cash
flow diagram and spreadsheet.

‡NCF denotes net cash flow. FV(NFC₈₀) denotes the future value. PV(NFC)
denotes present value of the net cash flow.
Net Present Value = $5,416.37.

The numerator of equation (8.15) calculates the future value of the $2,000. The denominators calculate the present value the result. Finally, subtracting the initial investment of $2,000 provides the interest income earned by accounting for scale. The NPV of parcel B adjusting for investment scale is $5,416.37.

Normalizing for investment scale in this example changed the preferred alternative; that is, investing in the management of parcel B would increase the entrepreneur's wealth more than investing in the management of parcel A. How would the analysis, results, and interpretation change if the reinvestment rate were the same as the interest rate? How would the analysis, results, and interpretation change if the reinvestment rate were less than the interest rate?

Investment horizon and scale

As parcel A did not need to be normalized for investment scale, Figure 8.15 illustrates the cash flow diagrams and spreadsheet to adjust parcel B for investment horizon and scale.

Normalizing for investment horizon and scale in this example changed the preferred alternative; that is, investing in the management of parcel B would increase the entrepreneur's wealth more than investing in the management of parcel A.

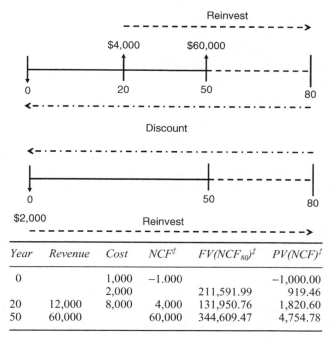

Figure 8.15 Investment horizon and scale – parcel B's reinvestment cash flow diagram and spreadsheet.

‡NCF denotes net cash flow. FV(NFC₈₀) denotes the future value. PV(NFC)
denotes present value of the net cash flow.
Net Present Value = $6,494.84.

How would the analysis, results, and interpretation change if the reinvestment rate
were the same as the interest rate? How would the analysis, results, and interpretation change if the reinvestment rate were less than the interest rate?

Interest versus reinvestment rates

Brazee (2003), Klemperer (1996), and Price (1993) provide a discussion as to
why interest rates may be different than reinvestment rates. A common assumption used in the corporate finance literature is that they are the same (Brealey *et al.*
2008; Copeland *et al.* 2005; and Luenberger 1998). As an entrepreneur, you will
need to understand the implications of dealing with investment horizon and scale
given interest and reinvestment rates that are the same or are different.

Independent, indivisible, and non-repeating

With independent investments, the concept of reinvestment is meaningless because,
by definition, independent investments are not exclusive and consequently do not

require accounting for project scale or investment horizon. The rule for independent investments is to determine all projects whose $NPV \geq 0$; and find the combination of projects with the largest total NPV within the given budget.

Table 8.4 gives a list of ten different independent investments.

If the entrepreneur has a budget of $40,000, $30,000, or $20,000, which investments would be acceptable? Deciding which investments are acceptable can be cumbersome as all feasible combinations must be determined and from those that are feasible the one that generates the greatest wealth for the investor must be chosen. Luenberger (1998) illustrates how the use of the use of Linear Programming with a binary decision variable – a variable that can only have a value of 0 (reject) and 1 (accept) – can solve these types of problems. However, a discussion of binary Linear Programming is beyond the scope of this book.

Capital budgeting – final thoughts

I have examined: (1) mutually exclusive, indivisible, non-repeating and (2) independent, indivisible, non-repeating investments. While approximately 90 percent of the investments you encounter will be either of the two described above, this is 2 out of 12 ($= 3 \times 2 \times 2$) possible combinations of the primary, secondary, and tertiary attributes describing investments. When capital and other resources are limited and you have various permutations of the 12 possible combinations of the primary, secondary, and tertiary attributes describing investments, then the capital budgeting decision can become tremendously complex. In these cases, get help.

How to use economic information – *capital theory: investment analysis* – to make better business decisions?

There is an opportunity cost associated with time. As a student, the opportunity cost of going to an 8:00am class is time not spent sleeping. While there has been

Table 8.4 Independent, indivisible, non-repeating investments

Project	Initial cost ($)	Horizon	Net present value ($)
1	763	9	456
2	4,687	20	273
3	5,995	11	−93
4	8,666	7	217
5	1,829	8	495
6	9,895	11	−82
7	790	9	157
8	6,112	17	81
9	1,227	14	−86
10	8,614	6	208

no money assigned to this opportunity cost, it is a cost nonetheless. In business decisions where time is a relevant input, the interest rate can be used to describe the opportunity cost of time; but, it also can be used to describe the opportunity cost of capital or the fact that there are alternative investment opportunities. The opportunity cost of capital represents the internal and external factors of investments as assessed by the capital markets. If the price of borrowing capital (the opportunity cost of capital) is less than what the individual would be willing to pay (the opportunity cost of time), the individual would borrow the capital. And, if the price of loaning capital (the opportunity cost of capital) is greater than what the individual would have been willing to accept (the opportunity cost of time), the individual would loan or invest their capital.

Many different formulae have been developed to find the equivalence at a common reference point of revenues and costs that may occur at different points in time. The two most important ones are: future value and present value. With these two formulae and the flexibility of spreadsheets, you should be able to handle most of your problems. The key to working with investments that have a variety of revenues and costs throughout time is to draw a cash flow diagram of the potential investment.

As Luenberger (1998: 27) states: "It is widely agreed (by theorists, but not necessarily by practitioners) that, overall, the best [investment analysis] criterion is that based on net present value." My advice is to follow the recommendations of theorists and put theory into practice by using NPV. This tool provides the flexibility to analyze investments described as mutually exclusive, independent, and interdependent. In addition, it is consistent with the assumption that individuals maximize their wealth. NPV is also consistent with the Architectural Plan for Profit, and the same economic information used to build it are used in calculating NPV.

Finally, a few concluding thoughts.[19] The time horizons of investment alternatives or projects in natural resource management may be very long. For example, in a northern hardwood stand an uneven-age management plan may define 50 to 100 years between entries that generate revenue by harvesting trees. How certain are the estimates of product prices, costs, and harvest volumes for 50 to 100 years in the future? One approach to dealing with potential uncertainty is to evaluate pessimistic, realistic, and optimistic scenarios for each alternative.[20] Second, when reporting results of a cost-benefit analysis to a client, use appropriate and logical levels of precision. For example, reporting a management alternative for an uneven-aged forest as $1,456.17 per acre from an infinite series of 25-year harvest cycles invites questions: Were management actions priced to the nearest one cent per acre? Are you sure of this precision? Are you sure of this precision for forever? Rounding to the nearest $1 (or $10) is usually more honest. Third, lowball your revenues and highball your costs, or round your revenues down and your costs up. It is always easier to explain to your client why costs were under budget and revenues over budget (your wonderful management of the project) than the opposite situation; that is, plan for the worst, hope for the best. Finally, use common sense; use it early and often. For example, if two alternatives return $1,235 and $1,245 respectively, is the second one clearly superior to the first? Even if the level of certainty is the same between the two alternatives, the $10 difference is less than a 1 percent change. Which alternative provides returns you would be comfortable implementing if it were your own money or your grandmother's money? Professional honesty and integrity is the best policy.

9   The forest rotation problem

As discussed in Chapter 6, the primary reasons for family forest landowners to own forestland are esthetics, privacy, recreation, and protecting nature.[1] Near the bottom of reasons for forestland ownership is the generation of income from timber sales (Butler and Leatherberry 2004; Belin *et al.* 2005; Hagan *et al.* 2005; Butler 2008 and 2010). Nonetheless, these owners can still be considered as maximizers; that is, their behavior is consistent with maximizing the net value of *their* forestland ownership that includes timber and non-timber net values.[2] One of the primary tools a landowner can use to manipulate forested ecosystems in a sustainable manner to produce timber and non-timber goods and services (e.g. water quality and quantity, recreation, wildlife, carbon storage, and mushrooms, etc.) is to cut trees according to a sustainable forest management plan.[3] To analyze a landowner's decision of when to cut their trees, I will start with a simple case assuming the primary reason for forestland ownership is income from timber sales. I will then relax this assumption to include other non-timber ownership goals and examine the impact on a landowner's cutting decision.

In Chapter 8, time was identified as a relevant input. Its impact on Profit and the Pillars of Price, Value, and Cost of the Architectural Plan for Profit (Figure 9.1) was discussed with respect to the opportunity cost of time or the discount rate.

In addition, time was identified as a relevant input in the Loblolly Pine Plantation case study's production process. As hardwood or softwood trees grow, they increase in volume (quantity) and quality (grade or product class). For the Loblolly Pine Plantation case study, the value of the output increases with greater diameter and height. Smaller loblolly pines are used for pulpwood and as the trees increase in diameter and height, they can be used as Chip-n-Saw, sawtimber, plywood logs, or power poles. The lowest valued output is pulpwood and the highest valued output is power poles. This brief description reiterates the importance of answering the three fundamental questions used to systematically analyze any production system (Chapter 2) and their impact on building the Pillars of Price, Value, and Cost.

While the first part of this chapter will focus on a forest landowner whose primary goal is generating profit from the sale of timber, I want to reiterate the importance of understanding the production system especially when overlaying non-timber goals and objectives and their production systems on a production system for producing timber.

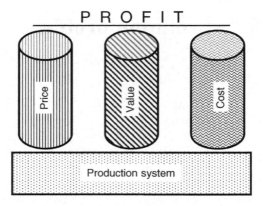

Figure 9.1 The Architectural Plan for Profit.

The chapter will be divided into four sections. The first section will analyze a landowner's cutting decision given an even-aged forest. The second section will analyze a landowner's cutting decision given an uneven-aged forest. The third section will reintroduce the non-timber goals and objectives of many forest land-owners and how these objectives may affect cutting decisions. I will end with a discussion of the economic information examined and how it can be used to make better business decisions.

Case studies

The case studies used in this chapter will include the Select Red Oak (Hahn and Hansen 1991) and Loblolly Pine Plantation (Amateis and Burkhart 1985 and Burkhart *et al.* 1985). They can be found on the Routledge website for this book and their summaries will not be repeated here.

The even-aged forest rotation problem

An even-aged forest contains a series of even-aged stands and the spread of ages in an even-aged community does not differ by more than 20 percent of the intended rotation age (Nyland 1996). I started building the Architectural Plan for Profit with the Production System (Chapter 2); hence, I will start by re-examining the "even-aged production system" with respect to developing cutting decision rules and the economic information they contain. Next, I will develop an economic analysis of a single and multiple rotation problems. I will end by examining the concept of forest rent.

Biological rotations

The mathematical description of the Loblolly Pine Plantation is given in equations (2.6) to (2.8). Table (2.6) and Figure 2.6 describe the total product. Table (2.9) and Figure 2.11 describe the average and marginal product. This information can be used to describe three different biological rotation ages. The first is the maximum of total product or where marginal product equals zero. A rotation age defined by this point would give the maximum volume production for a single rotation. In the Loblolly Pine Plantation this is at a stand age of 40 years as limited by equations (2.6) to (2.8). The second is the maximum of average product or where marginal product equals average product. In forestry terminology, this is called the culmination of mean annual increment (CMAI) and occurs at a stand age of 28. A rotation age defined by CMAI would give the maximum volume or biomass production per acre over multiple rotations. This rotation is defined as maximum sustained yield (Johansson and Löfgren 1985; Samuelson 1995) or optimal biological rotation age. The final is the maximum of marginal product. In forestry terminology, marginal product defines periodic annual increment (PAI). A rotation aged defined by maximizing PAI would give the maximum growth of the stand and occurs at a stand age of 16.

Examining the above definitions of rotation ages relative to the Architectural Plan for Profit shows they contain no economic information about the Pillars of Price (price of the output – sawtimber, for example), Value (do consumers demand the output? Do sawmills want these loblolly pines as an input to produce sawtimber?, for example), and Cost (any wage payments for inputs including the opportunity cost of time – the cost of establishing the plantation and an appropriate discount rate, for example). They contain very limited economic information; namely, only the production system and its descriptors. The rotation ages defined above optimize some concept of biological production, but no statements can be made concerning optimizing profits.

The single rotation problem

Assume that the forest landowner's sole objective is to maximize the net present value (NPV) of the final harvest. To analyze this objective, I will use a management regime defined by an initial cost for establishing a stand and then determining when to harvest the trees. This is illustrated in Figure 9.2.

Equation (9.1) is used to search for the largest NPV from all the potential rotation ages as illustrated in Figure 9.2:

$$NPV_t = \left. \begin{array}{l} P \cdot Q(t) \cdot (1 + r)^{-t} - C_0 \\[2mm] \dfrac{P \cdot Q(t)}{(1 + r)^t} - C_0 \end{array} \right\} ; \quad t = 0, 1, 2, \ldots \tag{9.1}$$

where P denotes output price, $Q(t)$ denotes the production process, $P \cdot Q(t)$ denotes the value of the harvest at time t, r denotes the real interest rate, and C_0 denotes the costs of establishing the stand.[4] Continuing with the Loblolly Pine Plantation case study, $Q(t)$ defines the cubic feet per acre yield given by equations (2.6) to (2.8) and illustrated in Table (2.6) and Figure 2.6. The cost of establishing the

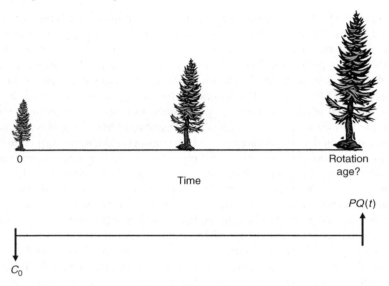

Figure 9.2 Single rotation problem with cash flow diagram.

stand includes shearing, raking, piling, chopping, burning, and herbicide. This removes whatever material may be on the site and killing any herbaceous plants that may hinder the growth of the loblolly pines. The site will then be planted with 1,210 seedlings per acre or a 6-foot by 6-foot spacing using 1-0 stock.[5] The establishment costs were estimated at $305.54 per acre (Dubois *et al.* 1997). The 1997 stumpage for sawtimber or the output price was $1.81 per cubic feet (Timber Mart-South http://www.tmart-south.com/tmart/index.html accessed on 12 April 2010). Finally, the real interest rate is 8 percent. Bullard *et al.* (2002) examined the hurdle rates of return for nonindustrial private forest landowners in Mississippi. They found for timberland investments ranging from 5 to 25 years, the real interest rates ranged from 6 to 11 percent.

The landowner must decide when to cut the trees to maximize the net present value of the final harvest. Table 9.1 illustrates calculating the NPV of various different potential rotation ages.

Focusing on the first three columns of Table 9.1 shows that the financially optimal single rotation age for the Loblolly Pine Plantation case study is at a plantation age of 18, or $T = 18$.[6] This can also be illustrated by graphing the range of NPV values as shown in Figure 9.3.

Once the range of NPV values has been calculated, determining the financially optimal single rotation age for the Loblolly Pine Plantation case study seems very straightforward. However, care must be taken with interpreting the results in Table 9.1 and Figure 9.3 economically. I will examine five concerns associated with interpreting the financially optimal single rotation age for the Loblolly Pine Plantation case study is at a plantation age of 18, or $T = 18$.

Table 9.1 Single rotation problem for the Loblolly Pine Plantation case study[‡]

Plantation age	Yield (cu.ft/ac)	NPV ($/ac)	MRP ($/ac/yr)	r · PQ(t) ($/ac/yr)
15	2,675.73	1,221.20	515.99	387.45
16	2,960.80	1,258.72	510.73	428.72
17	3,242.97	1,280.88	502.66	469.58
18	3,520.69	*1,289.16*	492.34	509.80
19	3,792.70	1,285.11	480.29	549.18
20	4,058.05	1,270.33	466.96	587.61
21	4,316.04	1,246.36	452.77	624.96

[‡]Yield is measured as cubic feet per acre (cu.ft/ac). NPV denotes net present value and is measured as dollars per acre ($/ac). MRP denotes marginal revenue product and is measured as dollars per acre per year ($/ac/yr). r denotes the real interest rate. P denotes the output price. Q(t) denotes the yield at plantation age t. For example:

$$1289.16 \text{ \$/acre} = 3520.69 \cdot 1.81 \cdot (1.08)^{-18} - 305.54$$

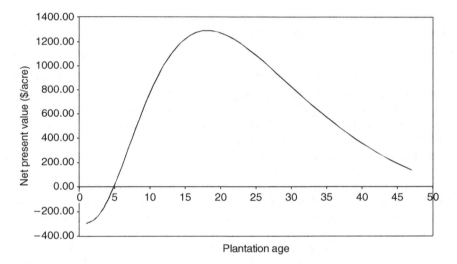

Figure 9.3 Net present value – Loblolly Pine Plantation case study.

First, there is a difference between the age of the trees (i.e. stand age) and the time used to calculate the NPV (i.e. investment or plantation age). The site was planted with 1-0 stock or seedlings that were 1 year old. Therefore if the plantation is 18 years old the trees are actually 19 years old. Calculating the NPV is based on the length of the investment not, in this case, the actual age of the trees as illustrated by the cash flow diagram in Figure 9.2.

Second, equation (9.1) defines a NPV calculation. Therefore if *NPV* > 0, the investment is returning the desired discount rate *plus* a present value of additional net revenue (Chapter 8). In the Loblolly Pine Plantation example, the entrepreneur is earning the required 8 percent plus a present value of additional net revenue of $1,289.16 per acre.

Third, while this may seem like a trivial question: How does the entrepreneur evaluate if they have maximized the NPV? Using Figure 9.2, the maximum NPV would occur at the top of the curve. Drawing on similarities from maximizing total product (Chapter 2) and profit (Chapter 5), a maximum was defined if a marginal condition, marginal product, or marginal profit were set equal to zero, respectively. The entrepreneur's management decision is whether to cut the plantation today or wait for next year. This requires defining the opportunity costs in terms of revenues and costs. If the entrepreneur decides to let the plantation grow for an additional year there will be an increase in net revenue as the trees grow in quantity (volume per acre; e.g. cubic feet per acre) and quality (product class; e.g. pulpwood, Chip-n-Saw, sawtimber, plywood logs, or power poles); this describes the opportunity cost in terms of revenues. If the entrepreneur decides to let the plantation grow for an additional year they will lose one year's worth of interest income on the net revenue from the sale of the timber; this describes the opportunity cost in terms of costs.

Fisher (1930) employed the same reasoning to derive the optimality conditions for equation (9.1) given in equation (9.2a):[7]

$$MRP_T = P \cdot MP_T = r \cdot [P \cdot Q(T)] = MC_T \qquad (9.2a)$$

where MRP_T denotes the marginal revenue product (see equation (4.9)) at the optimal rotation age, T; P denotes output price; MP_T denotes marginal product (see equations (2.11) and (2.12)) at the optimal rotation age, T; $Q(T)$ denotes the yield at the optimal rotation age T; and MC_T denotes the marginal cost at the optimal rotation age T (Chapter 3). A closer examination of equation (9.2a) shows it is analogous to the profit searching rule describe in Chapter 5; that is, the left-hand side of equation (9.2a) defines a marginal revenue concept and the right-hand side defines a marginal cost concept. If the entrepreneur decides to let the plantation grow for an additional year, there will be an increase in net revenue as the trees grow; MP_T describes current annual increment or how volume changes if the trees are allowed to grow for an additional year. Consequently $MRP_T = P \cdot MP_T$ defines the opportunity cost in terms of revenue. For the Loblolly Pine Plantation case study, MRP is given by the fourth column in Table 9.1. The MRP at the optimal single rotation age, $T = 18$, is given by equation (9.2b):

$$\begin{aligned} MRP_{18} &= P \cdot MP_{18} \\ &= 1.81 \ (\$/cu.ft) \cdot 272.01 \ (cu.ft/ac/yr) = 492.34 \ (\$/ac/yr) \end{aligned} \qquad (9.2b)$$

where price is 1.81 dollars per cubic feet ($/cu.ft); marginal product is 272.01 cubic feet per acre per year (cu.ft/ac/yr), and marginal revenue product is 492.34 dollars per acre per year ($/ac/yr) or the opportunity cost in terms of revenue.

If the entrepreneur decides to let the plantation grow for an additional year, they will lose one year's worth of interest income on the net revenue from the sale of the timber. The right-hand side term in equation (9.2a), $r \cdot [P \cdot Q(T)]$, is the interest rate times total revenue that would have been received had the timber been cut and is given in equation (9.2c).

$$MC_{18} = r \cdot [P \cdot Q(18)]$$
$$= 0.08 \cdot [1.81 \ (\$/cu.ft) \cdot 3520.69 \ (cu.ft/ac)] = 509.80 \ (\$/ac/yr) \qquad (9.2c)$$

where r denotes the 8 percent real interest rate, price is as defined in equation (9.2b), yield at plantation age, $Q(18)$, is 3520.09 cubic feet per acre (cu.ft/ac); and the interest income lost if the trees are not harvested is 509.80 dollars per acre for a year ($/ac/yr) or the opportunity cost in terms of cost and marginal cost. In addition, $r \cdot [P \cdot Q(T)]$ is defined as stand rent or the opportunity cost of capital in the trees.

Table 9.2 summarizes the optimality condition and the corresponding cash flow diagram given in equation (9.2a).

Before plantation age 18, MRP is greater than stand rent, $r \cdot [P \cdot Q(T)]$, and the entrepreneur's decision would be to let the plantation grow as their net value growth is greater than what could be made in an alternative investment. If the entrepreneur grows the plantation after plantation age 18, stand rent is greater than MRP and the entrepreneur's decision would be to cut the trees as net value growth in the trees is less than what could be made in an alternative investment. At plantation age 18, the opportunity cost in terms of revenue, MRP, is approximately balanced with the opportunity cost in terms of cost, stand rent. The reason that MRP does not equal stand rent exactly as described in equation (9.2a) is that the production system defines annual yields. If yields were defined in terms of days, weeks, or months, then we could pinpoint the precise date that would satisfy

Table 9.2 Optimality conditions and cash flow diagram for the single rotation problem for the Loblolly Pine Plantation case study[‡]

Plantation age	Yield (ft^3/Ac)	NPV ($/Ac)	MRP ($/Ac/PA)	iPQ(t) ($/Ac/PA)	
15	2,675.73	1,221.20	515.99	387.45	
16	2,960.80	1,258.72	510.73	428.72	MRP > iPQ(t)
17	3,242.97	1,280.88	502.66	469.58	
18	*3,520.69*	*1,289.16*	*492.34*	*509.80*	
19	3,792.70	1,285.11	480.29	549.18	
20	4,058.05	1,270.33	466.96	587.61	MRP < iPQ(t)
21	4,316.04	1,246.36	452.77	624.96	

[‡]Yield is measured as cubic feet per acre (cu.ft/ac). NPV denotes net present value and is measured as dollars per acre ($/ac). MRP denotes marginal revenue product and is measured as dollars per acre per year ($/ac/yr). r denotes the real interest rate. P denotes the output price. Q(t) denotes the yield at plantation age t.

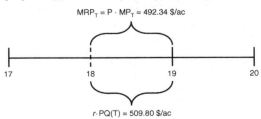

$MRP_T = P \cdot MP_T = 492.34 \ \$/ac$

$r \cdot PQ(T) = 509.80 \ \$/ac$

equation (9.2a). However, yield tables are generally not that precise and are often defined using stand age increments of greater than one year; for example, the unthinned oak stand in Southwestern Wisconsin yields are given every 20 years (see Table 2.7). Equation (9.2a) defines a search rule to estimate the financially optimal single rotation age.

Equation (9.2a) can be rewritten to express the optimality condition as given in equation (9.3)

$$r = \frac{P \cdot MP_t}{P \cdot Q(t)} = \frac{MRP_t}{TR_t} = \frac{P \cdot \dfrac{\Delta Q(t)}{\Delta t}}{P \cdot Q(t)} = \frac{\dfrac{\Delta Q(t)}{\Delta t}}{Q(t)} \tag{9.3}$$

which describes that the trees should be cut when their value growth (the right-hand side term) is equal to the interest rate (the left-hand term) (Johansson and Löfgren 1985). Table 9.3 derives the right-hand side of equation (9.3). As the interest rate was defined as 8 percent, according to Table 9.3 and equation (9.3) the financially optimal single rotation age is 18.

As with Table 9.2, the reason equation (9.3) is not illustrated with the data provided in Table 9.3 is that the yields are not given in smaller time increments.

A close examination of equation (9.3) shows that output price would cancel out of the right-hand side term leaving $\dfrac{\dfrac{\Delta Q(t)}{\Delta t}}{Q(t)}$ as shown in Table 9.3. In equation (9.1), I assumed a constant price; namely, the output was assumed to be sawtimber thus the price reflected sawtimber. If the price is a function of product class and product class changes as the trees grow in height and diameter, then the price would not be constant. Nonetheless, the rule described in equation (9.3) seems reasonable. Allow the trees to grow if their annual net value growth is greater than the interest rate. If not, cut as you could take the net revenue put it in a bank for one year at 8 percent and earn a greater net return than leaving the trees grow an additional year.

Table 9.3 Annual value growth of Loblolly Pine Plantation[‡]

Plantation age	Yield, $Q(t)$ (cu.ft/ac)	$P \cdot Q(t)$ ($/ac)	$MP, \dfrac{\Delta Q(t)}{\Delta t}$ (cu.ft/ac/yr)	$P \cdot MP$ ($/ac/yr)	$\dfrac{P \cdot \dfrac{\Delta Q(t)}{\Delta t}}{P \cdot Q(t)} = \dfrac{\dfrac{\Delta Q(t)}{\Delta t}}{Q(t)}$
15	2,675.73	4,843.06	285.08	515.99	0.107
16	2,960.80	5,359.05	282.17	510.73	0.095
17	3,242.97	5,869.78	277.71	502.66	0.086
18	3,520.69	6,372.44	272.01	492.34	0.077
19	3,792.70	6,864.78	265.35	480.29	0.070
20	4,058.05	7,345.07	257.99	466.96	0.064
21	4,316.04	7,812.03	250.15	452.77	0.058

[‡]Yield is measured as cubic feet per acre (cu.ft/ac). P denotes 1.81 dollars per cubic feet ($/cu.ft). MP denotes marginal product and is measured as cubic feet per acre per year (cu.ft/ac/yr).

Four, how do the financially optimal single rotation age compare to the optimal biological rotation age (CMAI).[8] The not-so-simple answer is derived using equation (9.4) which starts with equation (9.3a).

$$P \cdot MP_T = r \cdot [P \cdot Q(T)]$$

$$\frac{\Delta Q(T)}{\Delta T} = r \cdot Q(T)$$

$$\frac{\Delta Q(T)}{\Delta T} = r \cdot Q(T) \cdot \frac{T}{T} \qquad (9.4)$$

$$CAI = \frac{\Delta Q(T)}{\Delta T} = rT \cdot \frac{Q(T)}{T} = rT \cdot MAI_T$$

$$CAI = rT \cdot MAI_T$$

where *CAI* denotes current annual increment (Chapter 2) and *MAI* denotes mean annual increment (Chapter 2). The interpretation of the result of equation (9.4), $CAI = rT \cdot MAI_T$, depends on the relationship between CAI or marginal product and MAI or average product. This basic relationship is illustrated in Figure 9.4a.

Interpreting $CAI = rT \cdot MAI_T$ is divided into three arguments:

1   If $r > 0$, $T > 0$, and if $rT > 1$, then the financially optimal single rotation age is defined where $CAI > MAI$ or a rotation age shorter than the CMAI.

2   If $rT < 1$, then the financially optimal single rotation age is defined where $CAI < MAI$ or a rotation age longer than the CMAI.

3   If $rT = 1$, then the financially single optimal rotation age is defined where $CAI = MAI$ or a rotation age equal to the CMAI.

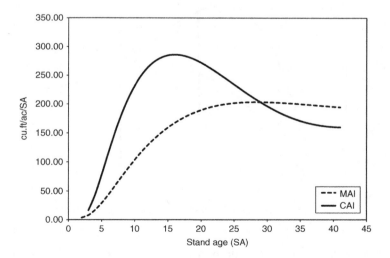

Figure 9.4a Average product (mean annual increment – MAI) and marginal product (current annual increment – CAI) for the Loblolly Pine Plantation.

Comparing these arguments with the Loblolly Pine Plantation case study, at the financially optimal single rotation age $rT = 0.08 \times 18 = 1.44 > 1$ and CAI and MAI at plantation age 18 or stand age 19 are 272.01 and 185.30 cubic feet per acre per year, respectively. Thus, the financially optimal single rotation age is less than the optimal biological rotation age as shown by Figure 9.4b.

The above four arguments were made based on a very simple single rotation scenario; i.e. establishing the plantation and then harvesting. Examining the cash flow diagram given in Figure 9.2, the NPV calculated using equation (9.1) determines the rotation age that maximizes the NPV of the final harvest and nothing else. The importance of this observation is that no value can be attributed to the land before or after the rotation. Johansson and Löfgren (1985) describe this implication as land is so abundant that its opportunity cost is zero. Bentley and Teeguarden (1965), Uys (1990), and Hseu and Buongiorno (1997) address this problem by modifying equation (9.1) to include the purchase and sale of the land. Equation (9.5) illustrates this as well as any other intermediate net cash flows (e.g. precommercial or commercial thinning).

$$NPV_t = P \cdot Q(t) \cdot (1+r)^{-t} + \sum_{k=1}^{K}\sum_{j=0}^{t} B_{kj} \cdot (1+r)^{-t} + L_t \cdot (1+r)^{-t}$$

$$- \sum_t C_t \cdot (1+r)^{-t} - L_0; \quad t = 0, 1,... \tag{9.5}$$

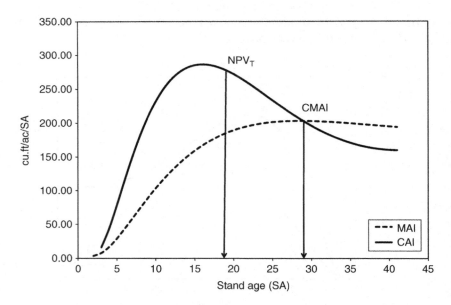

Figure 9.4b Optimal financial single rotation age versus biological rotation age for the Loblolly Pine Plantation.[‡]

[‡]MAI denotes mean annual increment. CAI denotes current annual increment. NPV_T denotes the financially optimal single rotation age. CMAI denotes the optimal biological rotation age.

where B_{kj} denotes the net cash flow of the kth intermediate action and L_t and L_0 denote the sale and purchase price of the land at end and beginning of the rotation, respectively. The objective is to determine the optimal rotation age, T, that maximizes present value of net cash flow resulting from a single rotation scenario (e.g. establishment costs, any timber stand improvements such as herbicides, thinning, and a final harvest) including the purchase and sale of the land.[9] Equation (9.5) looks intimidating and determining potential solutions is not as simple to display as those in equation (9.2a), and is beyond the scope of this book. However, the concepts of the opportunity cost in terms of revenue (i.e. marginal revenue) – What additional revenues may be gained by letting the plantation grow for an additional year? – and the opportunity cost in terms of costs (i.e. marginal cost) – What is being given up to keep the plantation in place? – are still useful in examining this more complex problem.

Finally, even as simple or complex as the problem given in equations (9.1) or (9.5) may describe, they still represent a single even-aged rotation problem. If the entrepreneur's goal is to replace the current plantation after it is cut with another plantation and so on, these formulations do not account for the extra cost of not cutting the plantation now which is delaying the net revenue from all future rotations. The next section examines this issue.

The multiple rotation problem

The issue of determining a financially optimal multiple rotation age is the focus of this section. To develop a model to examine the multiple rotation problem, I will again use the Loblolly Pine Plantation case study and describe the management regime to include the cost of establishing the plantation and revenue generated from a harvest. As was discussed at the end of the previous section, the harvest decision now must include the additional cost of delaying the net revenue from all future rotations. This is illustrated in Figure 9.5.

As illustrated by Figure 9.5, the timing of establishing future plantations depends on when the previous one is harvested. Calculating the present value of the two rotations illustrated in Figure 9.5 is shown in Figure 9.6.

Equation (9.6) extends this to an infinite number of rotations:

$$LEV_t = P \cdot Q(t) \cdot (1+r)^{-t} - C_0 + [P \cdot Q(t) \cdot (1+r)^{-t} - C_0] \cdot (1+r)^{-t}$$
$$+ [P \cdot Q(t) \cdot (1+r)^{-t} - C_0] \cdot (1+r)^{-2t} + \cdots$$
$$= NPV_t + NPV_t (1+r)^{-t} + NPV_t (1+r)^{-2t} + \cdots \qquad (9.6)$$

There are three observations from examining equation (9.6). First, the left-hand term has the notation of LEV_t which denotes land expectation value at time t to distinguish the solution of equation (9.6) from equation (9.1).[10] Second, the parameters of equation (9.6) – P, $Q(t)$, r, and C_0 – are held constant for an infinite number of rotations. I will revisit the implications of this observation later in this chapter. Third, solving for the optimal rotation age, T, that maximizes LEV is very problematic as there are an infinite number of rotation ages. Fortunately, this

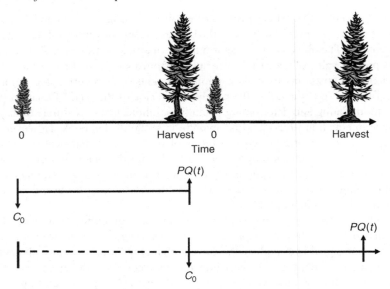

Figure 9.5 Multiple rotation problem with cash flow diagram.

problem formulation and its solution was described by Faustmann (1849), Pressler (1860), and Ohlin (1921) and is given in equation (9.7):

$$LEV_t = \left.\begin{array}{c} \displaystyle\sum_{\theta=0}^{\infty} (PQ(t)\cdot(1+r)^{-t} - C_0)\cdot[(1+r)^{-t}]^{\theta} \\[2ex] \dfrac{PQ(t)\cdot(1+r)^{-t} - C_0}{1-(1+r)^{-t}} \\[2ex] \dfrac{NPV_t}{1-(1+r)^{-t}} \end{array}\right\} \; ; \quad t = 0, 1, 2, \ldots$$

(9.7)

Equation (9.7) is often called the Faustmann Formula or Faustmann Model.[11]

The landowner must decide when to cut the trees to maximize the LEV.[12] Table 9.4 illustrates calculating the LEV of various different potential rotation ages.

Table 9.4 shows that the financially optimal multiple rotation age for the Loblolly Pine Plantation case study is at a plantation age of 15, or $T = 15$.[13] This can also be illustrated by graphing the range of LEV values as shown in Figure 9.7.

What does an LEV of 1,783.40 dollars per acre ($/ac) at a plantation age 15 mean? LEV deals with more than just the current rotation, it takes into account *all* future rotations. An LEV of 1,783.40 $/ac at plantation age 15 is the present net value of the current and all future rotations. The assumption is we started with bare land that is regenerated and then harvested. This exact management regime is going to be repeated forever. LEV calculates the value of that bare land to produce a periodic net revenue stream. Therefore, an LEV of 1,783.40 $/ac is the

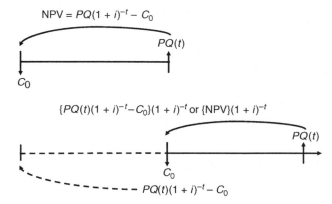

Figure 9.6 Present value of the first two rotations for the Loblolly Pine Plantation case study.

Table 9.4 Multiple rotation problem for the Loblolly Pine Plantation case study[‡]

Plantation age	Yield (cu.ft/ac)	NPV ($/ac)	LEV ($/ac)
13	2,104.73	1,095.23	1,732.12
14	2,389.62	1,167.03	1,769.46
15	2,675.73	1,221.20	*1,783.40*
16	2,960.80	1,258.72	1,777.57
17	3,242.97	1,280.88	1,755.28
18	3,520.69	1,289.16	1,719.45
19	3,792.70	1,285.11	1,672.70
20	4,058.05	1,270.33	1,617.33

[‡]Yield is measured as cubic feet per acre (cu.ft/ac). NPV denotes net present value and is measured as dollars per acre ($/ac). LEV denotes land expectation value and is measured as dollars per acre ($/ac).

$$1783.40 \ \$/acre = \frac{2675.73 \cdot 1.81 \cdot (1.08)^{-15} - 305.54}{1 - (1.08)^{-15}}$$

$$= \frac{2675.73 \cdot 1.81 - 305.54 \cdot (1.08)^{15}}{(1.08)^{15} - 1}$$

value of the bare land given the Loblolly Pine Plantation case study. You would not pay more than 1,783.40 $/ac given you were going to follow the above management regime forever.

Examining Figure 9.7 reveals that it is very similar to Figure 9.3, which is logical because the numerator in equation (9.7) is the same as equation (9.1). Thus, the analysis of equation (9.7) and Figure 9.7 can build on what has been already discussed. The maximum LEV would occur at the top of the curve in Figure 9.7. The entrepreneur's management decision is "should the plantation be cut today or wait for next year?" This again requires defining the opportunity

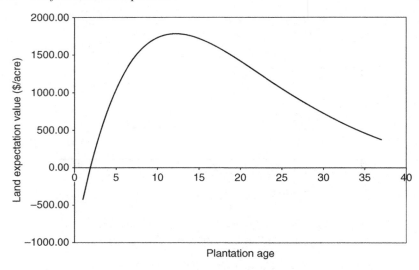

Figure 9.7 Land expectation value – Loblolly Pine Plantation case study.

costs in terms of revenues and costs. If the entrepreneur decides to let the plantation grow for an additional year, net revenue will increase as the trees grow in
quantity and quality. This describes the opportunity cost in terms of revenues. If
the entrepreneur decides to let the plantation grow for an additional year, they will
lose one year of interest income on the net revenue from the sale of the timber *plus*
delaying the net revenue from all future rotations; this describes the opportunity
cost in terms of costs.

The opportunity costs in terms of revenue and cost are given in equation (9.8a):[14]

$$MRP_T = P \cdot MP_T = r \cdot [P \cdot Q(T)] + r \cdot LEV_T = MC_T \tag{9.8a}$$

The left-hand side of equation (9.8a) is exactly the same as the left-hand side of
equation (9.2a) and has the same interpretation; it defines the increase in net
revenue as the trees grow and the opportunity cost in terms of revenue. The right-
hand side of equation (9.8a) denotes marginal cost at the financially optimal multiple rotation age, T, and is given in equation (9.8b):

$$r \cdot [P \cdot Q(T)] + r \cdot LEV_T = MC_T \tag{9.8b}$$

The first term on the left-hand side of equation (9.8b) defines stand rent,
$r \cdot [P \cdot Q(T)]$, and is the interest rate times total revenue that would have been
received if the timber would have been cut or the opportunity cost of the stand.
The concept of stand rent given in equation (9.8b) is the same as used in determining the financially optimal single rotation age and has the same interpretation. The
second term on the left-hand side of equation (9.8b) defines land rent, $r \cdot LEV_T$,
and is the cost – delaying the net revenue from all future rotations – due to length-
ening the rotation age by one year (Johansson and Löfgren 1985).

The information contained in equation (9.8a) is illustrated in Table 9.5.

The opportunity cost in terms of revenue at the financially optimal multiple rotation age, $T = 15$, is $MRP_{15} = 515.99$ dollars per acre per year ($/ac/yr). If the plantation is not cut at plantation age 15 it will generate 515.99 dollars per acre of additional revenue. Stand rent, $r \cdot [P \cdot Q(T)]$, and land rent, $r \cdot LEV_T$, at the financially optimal multiple rotation age, $T = 15$, are 387.45 and 142.67 $/ac/yr respectively. The stand rent of 387.45 $/ac is the interest income that could be earned if the plantation is cut and the revenue invested at 8 percent. If the plantation is not cut and allowed to grow for an additional year, the entrepreneur would forgo 142.67 $/ac of potential income due to not establishing the next rotation. Comparing Tables 9.2 and 9.3 or the financially optimal single and multiple rotation ages, respectively, shows that the economic information contained in MRP and stand rent is the same.

Comparing equations (9.8a) and (9.2a) shows that the opportunity cost in terms of cost for LEV contains two components (stand rent and land rent), while for NPV the opportunity cost in terms of cost only contains stand rent. If land rent is positive, $r \cdot LEV_T > 0$, then marginal revenue product must increase to cover this additional cost. Given that stumpage prices are constant at 1.81 dollars per cubic feet, this implies shortening the rotation age. This is illustrated in Table 9.6 and Figure 9.8.

Table 9.5 Optimality conditions and cash flow diagram for the multiple rotation problem for the Loblolly Pine Plantation case study[‡]

Plantation age	Yield (cu.ft/ac)	LEV ($/ac)	MRP ($/ac/yr)	$r \cdot PQ(t)$ ($/ac/yr)	$r \cdot LEV$ ($/ac/yr)	MC ($/ac/yr)
13	2,104.73	1,732.12	515.66	304.76	138.57	443.33
14	2,389.62	1,769.46	517.85	346.02	141.56	487.57
15	2,675.73	*1,783.40*	515.99	387.45	142.67	530.12
16	2,960.80	1,777.57	510.73	428.72	142.21	570.93
17	3,242.97	1,755.28	502.66	469.58	140.42	610.00
18	3,520.69	1,719.45	492.34	509.80	137.56	647.35
19	3,792.70	1,672.70	480.29	549.18	133.82	683.00
20	4,058.05	1,617.33	466.96	587.61	129.39	716.99

[‡]Yield is measured as cubic feet per acre (cu.ft/ac). LEV denotes land expectation value and is measured as dollars per acre ($/ac). MRP denotes marginal revenue product and is measured as dollars per acre per year ($/ac/yr). r denotes the real interest rate. P denotes the output price. $Q(t)$ denotes the yield at plantation age t. MC denotes marginal cost and is the sum of $r \cdot PQ(t)$ plus $r \cdot LEV$ and is measured as dollars per acre per year ($/ac/yr). $r \cdot PQ(t)$ denotes stand rent and $r \cdot LEV$ denotes land rent.

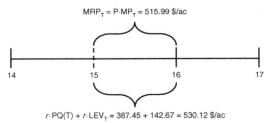

$MRP_T = P \cdot MP_T = 515.99$ $/ac

14   15   16   17

$r \cdot PQ(T) + r \cdot LEV_T = 387.45 + 142.67 = 530.12$ $/ac

Table 9.6 Comparison of the financially optimal single and multiple rotation ages for the Loblolly Pine Plantation case study[‡]

Plantation age	Yield (cu.ft/ac)	NPV ($/ac)	LEV ($/ac)	MRP ($/ac/yr)	$r \cdot PQ(t)$ ($/ac/yr)	$r \cdot LEV$ ($/ac/yr)
13	2,104.73	1,095.23	1,732.12	515.66	304.76	138.57
14	2,389.62	1,167.03	1,769.46	517.85	346.02	141.56
15	2,675.73	1,221.20	*1,783.40*	515.99	387.45	142.67
16	2,960.80	1,258.72	1,777.57	510.73	428.72	142.21
17	3,242.97	1,280.88	1,755.28	502.66	469.58	140.42
18	3,520.69	*1,289.16*	1,719.45	492.34	509.80	137.56
19	3,792.70	1,285.11	1,672.70	480.29	549.18	133.82
20	4,058.05	1,270.33	1,617.33	466.96	587.61	129.39

[‡]Yield is measured as cubic feet per acre (cu.ft/ac). NPV denotes net present value and is measured as dollars per acre ($/ac). LEV denotes land expectation value and is measured as dollars per acre ($/ac). MRP denotes marginal revenue product and is measured as dollars per acre per year ($/ac/yr). *r* denotes the real interest rate. P denotes the output price. Q(*t*) denotes the yield at plantation age *t*.

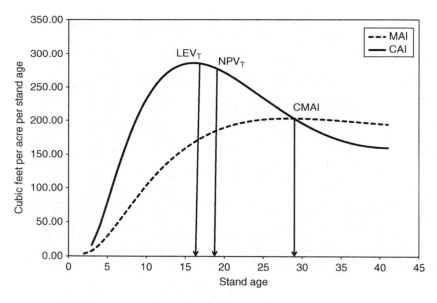

Figure 9.8 Comparison of land expectation value, net present value, and optimal biological rotation ages for the Loblolly Pine Plantation case study.[‡]

[‡]MAI denotes mean annual increment. CAI denotes current annual increment. LEV_T denotes the financially optimal multiple rotation age. NPV_T denotes the financially optimal single rotation age. CMAI denotes the optimal biological rotation age.

Figure 9.8 also compares the financially optimal multiple and single rotation ages with the optimal biological rotation age. As shown, the optimal biological rotation age is the longest, followed by the financially optimal single rotation age with financially optimal multiple rotation age the shortest. This also follows the fact that calculating the optimal biological rotation age uses almost no economic information, while calculating the financially optimal multiple rotation age uses the most economic information.[15]

Equation (9.6) and (9.7) describe the LEV of a very simple management regime. However, just as NPV can describe more complicated cash flows, so can LEV. Equation (9.9a) illustrates the net present value of a management regime with intermediate net cash flows (e.g. precommercial or commercial thinning, etc.):

$$NPV_t = P \cdot Q(t) \cdot (1 + r)^{-t} + \sum_{k=1}^{K} \sum_{j=0}^{t} B_{kj} \cdot (1 + r)^{-t}$$
$$- \sum_t C_t \cdot (1 + r)^{-t}; \quad t = 0, 1, ... \tag{9.9a}$$

where NPV_t denotes the net present value of a management regime with intermediate net cash flows; B_{kj} denotes the net cash flow of the kth intermediate action and the other variables are defined as before. The LEV of this management regime's more complex cash flow is given in equation (9.9b):

$$LEV_t = \frac{NPV_t}{1 - (1 + r)^{-t}} \tag{9.9b}$$

Examining equation (9.9b) reveals two observations. First, the objective is to determine the optimal rotation age, T, that maximizes the value of the land asset. Second, it is virtually identical to equation (9.7). Consequently, LEV can be used to determine the financially optimal multiple rotation age of simple to complex management regimes.[16]

Forest rent

Gregory (1972) and later Johansson and Löfgren (1985) define forest rent as the maximum average net revenue or a net sustainable yield concept, respectively. Equation (9.10) defines forest rent for the Loblolly Pine Plantation case study

$$FR_t = \frac{P \cdot Q(t) - C}{t}; \quad t = 1, 2, 3, ... \tag{9.10}$$

where FR_t defines the forest rent at time t and all the other variables are defined as before. The entrepreneur's objective would be to determine the rotation age that maximizes forest rent. Table 9.7 shows the forest rent calculations for the Loblolly Pine Plantation case study.

Examining equation (9.10) and Table 9.7 reveals two observations. First, equation (9.10) and the corresponding calculations in Table 9.7 show that the opportunity cost of time or the discount rate is set equal to zero. As described by Gregory

Table 9.7 Forest rent for the Loblolly Pine Plantation case study[‡]

Plantation age	Yield (cu.ft/ac)	Total revenue ($/ac)	Forest rent ($/ac)
24	5,042.07	9,126.15	367.53
25	5,267.78	9,534.69	369.17
26	5,485.57	9,928.88	370.13
27	5,695.73	10,309.27	***370.51***
28	5,898.67	10,676.60	370.40
29	6,094.87	11,031.71	369.87
30	6,284.84	11,375.56	369.00
31	6,469.15	11,709.16	367.86

[‡]Yield is measured as cubic feet per acre (cu.ft/ac).

$$370.51 = \frac{1.81 \cdot 5695.73 - 305.54}{27}$$
$$= \frac{10309.27 - 305.54}{27}$$

(1972), the logic of forest rent follows from the idea in that a fully regulated even-aged forest has the same number of acres harvested and regenerated every year.[17] Given the structure of a fully regulated forest, the entrepreneur's objective would be to maximize the net revenue flow. However, as Gregory (1972) points out, this logic is flawed. As shown in Chapter 8, the opportunity cost of time or the discount rate is not equal to zero. In addition, the mathematical formulation of equation (9.10) ties a cost and revenue together that are not related; that is, the cost of the acres being regenerated is not tied to the revenue generated from a harvest. They are not the same trees. The trees being regenerated will not be ready for harvesting until later. While Gregory (1972: 295) states that forest rent "deserves no serious consideration" as a result of this grave mistake, it serves as an important object lesson. In investment analysis, cost should be tied directly to the revenues they generate. This is the importance of reviewing the concept of forest rent.[18]

Second, the rotation age that maximizes forest rent will, in general, be longer than the financially optimal single and multiple rotation ages (Johansson and Löfgren 1985; Hyytiäinen and Tahvonen 2003).[19] For the Loblolly Pine Plantation case study, given that the plantation is established on bare land, and the afforestation and reforestation costs and output prices are constant, then maximum net revenue or net sustainable yield would occur at a rotation age longer than CMAI.[20] Finally, Johansson and Löfgren (1985), Möhring (2001), and Hyytiäinen and Tahvonen (2003) show that the rotation age given by LEV – equations (9.6) and (9.7) – is the same as forest rent if the discount rate is set equal to zero.

Optimal financial rotation age given an existing forest

The optimal single and multiple rotation ages described above assumed that the entrepreneur started with bare land, established, and then harvested the forest.

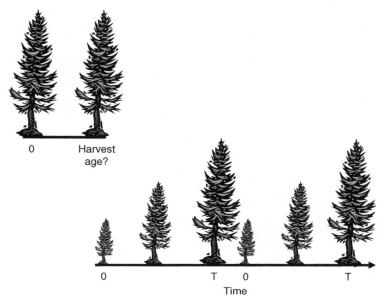

Figure 9.9 Financially optimal rotation age given an existing forest.

What if the land had an existing forest? I will use the Loblolly Pine Plantation case study to demonstrate the investment analysis of this scenario. Figure 9.9 illustrates this situation.

The existing forest is illustrated by the upper left-hand corner of Figure 9.9. Once this existing forest is harvested, a continuous regeneration–harvest pattern can be established which is illustrated by the lower right-hand corner of Figure 9.9. This portion of Figure 9.9 is taken from Figure 9.5.

The investment analysis model to analyze this problem, based on that developed by Rideout (1985), is given by equation (9.11)

$$FV = P \cdot Q(N + t) \cdot (1 + r)^{-t} + LEV_t \cdot (1 + r)^{-t} \qquad (9.11)$$

where *FV* denotes the forest value; $Q(N + t)$ denotes the per acre volume of the existing forest at plantation age *N* with $t = 0, 1, 2, 3,...$; LEV_t denotes the land expectation value of the continuous regeneration–harvest management regime at a rotation age of *t*. The entrepreneur's objective is to maximize the FV. If equation (9.11) is to be maximized, each component must be maximized. As Rideout (1985) points out, this is problematic as the term $P \cdot Q(N + t) \cdot (1 + r)^{-t}$ would dominate the analysis. The term LEV_t denotes a steady state condition; that is, for any given rotation age, *t*, the regeneration–harvest pattern is constant. The optimal steady state condition is given by LEV_T from equations (9.8a) and (9.9b). Substituting LEV_T into equation (9.11) gives:

$$FV_t = P \cdot Q(N + t) \cdot (1 + r)^{-t} + LEV_T \cdot (1 + r)^{-t}; \quad t = 0, 1, 2, 3,... \qquad (9.12)$$

Table 9.8 Investment analysis of a 12-year-old loblolly pine plantation[‡]

Plantation age	Yield (cu.ft/ac)	Total revenue ($/ac)	FV ($/ac)
12	1,823.64	3,300.79	5,084.19
13	2,104.73	3,809.56	5,178.66
14	2,389.62	4,325.22	5,237.15
15	2,675.73	4,843.06	5,260.30
16	2,960.80	5,359.05	5,249.91
17	3,242.97	5,869.78	5,208.62
18	3,520.69	6,372.44	5,139.56

[‡]Yield is measured as cubic feet per acre (cu.ft/ac). Total revenue is measured as dollars per acre ($/ac). Forest value (FV) is measured as dollars per acre ($/ac). LEV_T is defined as 1783.40 dollars per acre ($/ac) from Table 9.4.

$$5260.30 = 1.81 \cdot 2675.73 \cdot (1.08)^{-3} + 1783.40 \cdot (1.08)^{-3}$$

The entrepreneur's objective is to determine the t that maximizes the FV given the plantation age of the existing stand. Table 9.8 illustrates this analysis for the Loblolly Pine Plantation case study given the existing plantation is 12 years old.

Based on the information contained in Table 9.8, the entrepreneur should wait three years before cutting the existing plantation and regenerating the next rotation.[21]

The analysis presented here assumed the only revenue was from the sale of the timber without other costs or revenues. Equation (9.12) can be modified to include additional costs and revenues within the existing forest and the land expectation value terms. However, the analysis approach would be similar to that described above.

The production system – revisited

The foundation of the Architectural Plan for Profit is the Production System (Chapter 2). This still holds, even if you are calculating financially optimal rotation ages. The financially optimal single and multiple rotation age were determined to be 1,289.16 dollars per acre at a plantation age of 18 and 1,783.40 dollars per acre at plantation age of 15, respectively. These solutions were determined assuming a 1.81 dollar per cubic foot price for sawtimber. So the question is: Are 19- or 16-year-old loblolly pine trees sawtimber? Table 9.9 defines loblolly pine product classes by diameter at breast height (DBH).

A review of the Loblolly Pine Plantation production system shows that DBH is not one of the relevant inputs used in answering the question from Chapter 2 "What are the inputs?" While the description of the production system is complex, equations (2.6) to (2.8), it is not sufficient for determining if in fact 19- or 16-year-old loblolly pine trees are sawtimber and if the interpretations of the calculated NPV and LEV values are accurate. This point needs to be re-emphasized. It is not the mathematical calculations nor the profit searching rule developed and

Table 9.9 Loblolly pine product class by diameter at breast
height (DBH) range[‡]

Product class	DBH range
Pulpwood	4.6-inches ≤ DBH ≤ 9-inches
Chip-n-saw	9-inches ≤ DBH ≤ 12-inches
Sawtimber	12-inches ≤ DBH

[‡]Product class ranges are based on a paper entitled: "Economics of growing slash and loblolly pine to a 33-year rotation – impact of thinning at various stumpage prices" by E. David Dickens, Coleman W. Dangerfield, Jr., and David J. Moorhead from the Warnell School of Forest Resources at the University of Georgia, Athens (http://warnell.forestry.uga.edu/service/library/for05-06/econ6.pdf accessed on 26 April 2010).

described in equations (9.2a) and (9.8a) that are of concern; these mathematical calculations are done correctly. It is the economic interpretation of the results that is important. Because the production system does not provide volume by DBH class, the economic interpretation should be questioned.

Tasissa *et al.* (1997) and Sharma and Oderwald (2001) provide mathematical relationships between volume, height, and DBH for loblolly pine. These relationships can be used to estimate DBH for the Loblolly Pine Plantation case study. Based on the information provided by these authors and the description of the management regime at the beginning of this chapter, it would take approximately 30 to 40 years to grow a 12-inch loblolly pine.[22] The land expectation value using a plantation age of 30 is 916 dollars per acre.

In contrast, the Select Red Oak case study provides explicit information on the DBH of individual select red oak trees. However, the inputs are defined as DBH and site index. What is missing to determine a financially optimal rotation age is stand age. Thus, a secondary source of information would again be required to link DBH and site index to tree age. Then, a financially optimal rotation age could be estimated for individual select red oak trees.

I purposefully developed and presented the financially optimal single and multiple rotation ages. First, the mechanics of calculating and the economic implications of the financially optimal single and multiple rotation ages can be overwhelming when introduced. Thus, the first part of this chapter focused on that component.

Second, an entrepreneur can get lost in the mechanics of estimating rotation ages and forget the Architectural Plan for Profit. The Pillars of Price, Value, and Cost are built on the economic information generated from the Production System (Figure 9.1). If inaccurate or incomplete information is used to build the Pillars of Price, Value, and Cost then Profits (or in this case land expectation value) will not convey what the entrepreneur thinks it does. In this case, make sure you are producing the output you believe you are by answering the three fundamental questions to systematically examine any production system sufficiently. Conclusion: Do not get so lost in or enamored by the mechanics of what you are doing that you forget if the economic interpretation of what you are doing makes sense. After all, the reason you do these calculations is to generate accurate economic information to help make better business decisions.

The even-aged rotation problem – revisited

While the development of the financially optimal rotation ages used the Loblolly Pine Plantation case study, the use of the Faustmann Formula, equation (9.7), has a long and rich history and the general results described in Table 9.6 and Figure 9.8 have been applied worldwide and with many different species managed using even-aged and uneven-aged silvicultural systems.[23] The number of articles and manuscripts using variations of the Faustmann Formula with respect to sustainable forest management shows no signs of slowing down. Newman (2002 and 1988) provides a review of this vast literature. In addition, many journals have had special issues devoted to research based on the Faustmann Formula (e.g. *Journal of Forest Economics* 1995 1(1) and 2000 6(3), *Forest Science* 2001 47(4), *Forest Policy and Economics* 2001 2(2)).

Table 9.6 and Figure 9.8 demonstrate that the financially optimal single and multiple rotation ages, the optimal biological rotation age, and the forest rent rotation age do not appear to be *all* that different. Thus, would it matter in practice which one was chosen? Hyytiäinen and Tahvonen (2003) examined this question and concluded that rotation ages based on either the optimal biological or forest rent criteria lead to major losses in economic value. They did not examine the financially optimal single rotation age as the landowners were managing their lands for multiple rotations. However, if the lands are managed for multiple rotations and the financially optimal single rotation age was chosen, this would also result in the loss in economic value.

The Faustmann Formula, equations (9.6) and (9.7), describes the present value of a perpetual periodic series (Appendix 7). Forever is a very long time for the parameters and variables that describe the market conditions, production system, and the management regime to be constant. That is why using equation (9.8a) is a useful tool to estimate the financially optimal multiple rotation age. The managerial implications of risk and uncertainty associated with these parameters and variables also have a rich historical record in the literature; for example, Hool (1966) and Lembersky and Johnson (1975) are two early papers examining these issues. Chapter 10 will cover a few methods that entrepreneurs can use to incorporate risk into their economic analyses. As forever is a long time and biological and market systems are dynamic, an entrepreneur should develop a strategy to revisit their analyses periodically given new information for the parameters and variables that describe market conditions, production system, and the management regime.

The Faustmann Formula describes the perpetual periodic generation of net revenues from forest management. For the forest to generate this periodic net revenue stream perpetually, an argument could be made that the Faustmann Formula implies managing the forest sustainably – as an ecosystem – so it can produce this net revenue stream. Unfortunately, there is no single accepted definition of sustainable forest management, and I will not attempt to address that issue here; however, common themes emerge from these definitions that include maintaining the forest system, use of forest-based ecosystem goods and services that does not impact future generations, multiple stakeholders, and scale. The Faustmann Formula would seem to address one of the tenets of sustainable forest management. However, the determination of the financially optimal multiple rotation age requires that the exact same management regime be repeated forever and the land, as a biological and ecological production system, also remains constant forever. The latter condition is important to revisit briefly. The continual harvesting and regenerating of a stand will probably impact the productive ability of the land (Erickson *et al*. 1999; Navarro 2003). In addition to making sure you are producing the output you expect, you need to be aware of how your management actions affect the production system – in this case the forested ecosystem. While the Faustmann Formula is a conceptual improvement over the single rotation model, equation (9.1), entrepreneurs should be aware of all its economic as well as biological and ecological implications.

The uneven-aged cutting cycle problem (selection system)

An uneven-aged forest contains one or more uneven-aged stands that are comprised of at least three distinct age classes or cohorts (as measured by some combination of age, diameter at breast height, and tree height) irregularly mixed within the area. Diameter distributions are commonly used to manage uneven-aged stands and thus the forest. The structure of an uneven-aged forest is such that cutting in a diameter class will impact the growth in other diameter classes. This is in contrast to an even-age forest where age classes are spatially distinct such that cutting one age class does not impact the growth of any other age class. It is this particular interrelationship among the diameter classes of an uneven-aged forest that make it more interesting and difficult to analyze economically.

Harvesting an uneven-aged stand never removes the growing stock completely. In contrast, regenerating an even-aged stand was based on harvest or removing the growing stock completely; i.e. a clearcut.[24] The uneven-aged stands ability to regenerate and produce volume is dependent on the diameter distribution of the tree species left in the reserve growing stock after a harvest. As there is not a distinct harvest that removes all the growing stock, the amount of time between the periodic harvests is defined as a cutting cycle. The above description of the production system of an uneven-aged stand is illustrated in Figure 9.10.

Thus, the management decisions are (1) to determine amount of reserve growing stock, (2) the appropriate diameter distribution and species composition for the reserve growing stock, and (3) the length of the cutting cycle that maximize the entrepreneur's wealth.

The economic analysis will be conducted at the individual uneven-aged stand level rather than at the uneven-aged forest level due to the complexities of uneven-aged forest-level problems.[25] The entrepreneur's wealth objective can be described by equation (9.13) as described by Chang (1982) and Rideout (1985)

$$
\begin{aligned}
USV_{t,G} &= P \cdot Vol + \sum_{\theta=0}^{\infty} \{P \cdot [Q(t,G) \cdot (1+r)^{-t} - G]\} \cdot [(1+r)^{-t}]^{\theta} \\
&= P \cdot Vol + \frac{P \cdot [Q(t,G) \cdot (1+r)^{-t} - G]}{1 - (1+r)^{-t}} \\
&= P \cdot Vol + \frac{P \cdot [Q(t,G) - G] \cdot (1+r)^{-t}}{1 - (1+r)^{-t}} - P \cdot G \\
&= P \cdot (Vol - G) + \frac{P \cdot Q(t,G) - P \cdot G}{(1+r)^{t} - 1}
\end{aligned}
$$

(9.13)

where $USV_{t,G}$ denotes the present value of the uneven-aged stand for a given cutting cycle t and reserve growing stock G; P denotes stumpage price; Vol denotes the initial volume at the time of harvest; r the real interest rate; and $Q(t,G)$ denotes the volume growth of the uneven-aged stand as a function of the cutting cycle, t, and the structure of the reserve growing stock, G.[26] The entrepreneur's management decision is to determine the optimal cutting cycle, t, and structure of reserve growing stock, G, that maximizes $USV_{t,G}$ given the initial stand condition as described by S.[27]

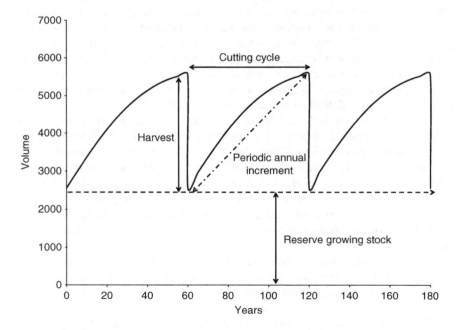

Figure 9.10 The production system for an uneven-aged stand.

Marking guides to regenerate an uneven-aged stand are based on either the concept of a single-tree or group selection (Nyland 1996). The silvicultural objective would be to maintain the classical Arbogast reverse-J structure across diameter classes within the uneven-aged stand (Nyland 1996). Grisez and Mendel (1972), Mendel *et al.* (1973), Godman and Mendel (1978), Herrick and Gansner (1985), Dennis (1987), Davies (1991), Klemperer (1996), Hseu and Buongiorno (1997), Heiligmann (2008), Bettingar *et al.* (2009) and Webster *et al.* (2009) have used an internal rate of return (*irr*) approach to describe the annual value growth rate of a representative individual or single tree of a given diameter to add a financial component to these regeneration marking guides.[28] This is illustrated by equation (9.14)

$$irr = \left(\frac{P \cdot q(T, g)}{P \cdot g}\right)^{\frac{1}{T}} - 1 \qquad (9.14)$$

where P denotes the stumpage price, $q(T, g)$ denotes the volume of the representative individual or single tree of a given diameter as a function of the cutting cycle, T, and its volume at the beginning of the cutting cycle, g, and T denotes the length of the cutting cycle. The general rule is that if the tree's *irr* is greater (less) than an appropriate real interest rate, the tree is (is not) financially mature. Using this general rule implies the annual value growth rate as given by the *irr* is equated with a real interest rate. Unfortunately, this common interpretation of the *irr* is not accurate. The *irr* given in equation (9.14) describes the expected annual value growth of the tree *ex ante*. In addition, *irr* may only be interpreted as an annual growth in value if and only if reinvestment takes place (Price 1993); that is, baring any fires, insect infestations, drought, or damage due to weather, etc. *Ex post*, if $P \cdot q(T, g)$ does occur given $P \cdot g$, then *irr* does describe the annual value growth of the tree. That said, the only accurate statement that can be made concerning the *irr* is that if it is greater than the appropriate real interest rate, there is a greater likelihood that an entrepreneur's wealth will increase. Whether an entrepreneur's wealth increases (or not) is tied directly to the real interest rate not the *irr*. Finally, using equation (9.14) to help develop a marking guide's take-or-leave decision does not include (1) the implications of one or more trees occupying the same growing space and (2) the current tree occupying the site and not allowing the next tree to take its place (i.e. the opportunity cost of the land or land rent as described in equation (9.8b) (Johansson and Löfgren 1985; Hseu and Buongiorno 1997). This is important given a selection system. Consequently, caution is advised when using equation (9.14) to define the concept of financial maturity for a representative individual or single tree of a given diameter using a selection system.

Non-timber forest values and ecosystem goods and services

The focus thus far has been on producing timber or biomass from the forest. Forests produce many other values besides timber or biomass; some may produce income directly and some may not. Familiar examples are maple syrup, mushrooms, pine straw, wildlife, and water flows. In some cases they can be converted into outputs that can be sold in established markets easily; for example, maple syrup, mushrooms, and pine straw. Others can be marketed and sold as the right to participate in an experience; for example, recreation leases for hunting, fishing, hiking, and wildlife viewing. Or, instead of being sold they may be enjoyed directly by the entrepreneur/landowner as a component of the net benefit from owning forestland (e.g. Butler 2008 and 2010). The economic analysis of producing these outputs has more often than not been described as either timber or these non-timber outputs or as a joint production process of timber plus these non-timber outputs; that is, how do these non-timber outputs impact the timber only financially optimal multiple rotation age?

The following authors examine various themes based on the joint production of timber and non-timber outputs or non timber outputs instead of timber: Hartman (1976), Calish *et al.* (1978), Nguyen (1979), Bowes and Krutilla (1985), Peters *et al.* (1989), Stednick (1996), Strange *et al.* (1999), Alexander *et al.* (2002), Haynes and Monserud (2002), and Amacher *et al.* (2009). Students who are interested in these types of production processes should start their review of this literature with these.

Two classic articles that examined this issue are Hartman (1976) and Calish *et al.* (1978). Amacher *et al.* (2009) provide an excellent review of the mathematical economic models and literature resulting from Calish *et al.* and Hartman's work that I will summarize briefly. If the non-timber and ecosystem goods and services values can be described as dependent on the age of the forest, then these values can be included into the Faustmann Model, equation (9.7), given in equation (9.15).

$$
\begin{aligned}
W_T &= \frac{P \cdot Q(T) \cdot (1+r)^{-T} + \sum_{j=0}^{T} B_j \cdot (1+r)^{-t} - C_0}{1 - (1+r)^{-T}} \\
&= \frac{P \cdot Q(T) \cdot (1+r)^{-T} - C_0}{1 - (1+r)^{-T}} + \frac{\sum_{j=0}^{T} B_j \cdot (1+r)^{-t}}{1 - (1+r)^{-T}} \\
&= LEV_T + E_T
\end{aligned}
\tag{9.15}
$$

where W_T denotes the entrepreneur's wealth, B_j denotes the non-timber or ecosystem good or service value that depend on the forest's age and accrues annually, and all other variables are defined as before. The first term on the right-hand side of equation (9.15) defines the land expectation value (LEV_T) or the Faustmann Formula given in equation (9.7). The second term defines the present value of the non-timber or ecosystem good or service value that accumulate until the wealth maximizing optimal rotation age, T, given an infinite sequence of rotations, E_T.

The entrepreneur's decision is to determine the value for T that maximizes their wealth described in equation (9.15). The opportunity costs in terms of revenue and cost for determining the financially optimal multiple rotation age were given in equation (9.8a) without non-timber or ecosystem good or service values. Amacher *et al.* (2009) define an analogous wealth searching rule given in equations (9.16a) to (9.16c)

$$MB_T = P \cdot MP_T + B_T = r \cdot [P \cdot Q(T)] + r \cdot (LEV_T + E_T) = MC_T \qquad (9.16a)$$

$$MB_T = P \cdot MP_T + B_T \qquad (9.16b)$$

$$MC_T = r \cdot [P \cdot Q(T)] + r \cdot (LEV_T + E_T) \qquad (9.16c)$$

where B_T denotes the non-timber or ecosystem good or service value at the wealth maximizing optimal multiple rotation age, and all other variables are defined as before. Equation (9.16b) highlights the marginal benefit, MB_T, component. This includes the marginal revenue product from allowing the timber to grow an additional year, $P \cdot MP_T$, plus the benefit from the non-timber or ecosystem good or service. Equation (9.16c) highlights the marginal cost, MC_T, component. This can still be divided into a stand rent term, $r \cdot [P \cdot Q(T)]$, and a land rent term, $r \cdot (LEV_T + E_T)$. As can be seen, stand rent is defined as before, but land rent now includes the present value of net benefits of the non-timber or ecosystem good or service values for an infinite sequence of rotations, E_T. These opportunity costs can be interpreted in the same manner as described for equations (9.8a) and (9.8b).

Amacher *et al.* (2009) describes the effects of the non-timber or ecosystem good or service value, B_j, on the financially optimal multiple rotation age (as given by equation (9.16a)) depends on the value of $B_T - r \cdot E_T$; namely, the ecosystem good or service value at the wealth maximizing optimal multiple rotation age and the opportunity cost of these values if the harvest is delayed. These authors showed that if the non-timber or ecosystem good or service values increase as forest grows older, $\frac{\Delta B_t}{\Delta t} > 0$ for all relevant values of t, including these values into the rotation age decision will lengthen the financially optimal multiple rotation age; for example, Calish *et al.* (1978) illustrated this with Douglas fir and cutthroat trout, Roosevelt elk, esthetics, or nongame wildlife in the Pacific Northwestern United States. And if the non-timber or ecosystem good or service values decrease as forest grows older, $\frac{\Delta B_t}{\Delta t} < 0$ for all relevant values of t, including these values into the rotation age decision will shorten the financially optimal multiple rotation age; for example, Calish *et al.* (1978) illustrated this with Douglas fir and Columbian black-tailed deer and water yield in the Pacific Northwestern United States.[29] Amacher *et al.* (2009) summarizes these implications:

> The landowner [entrepreneur] should harvest a stand of trees when the sum of the marginal harvest revenue and amenity benefit from delaying harvest for one period equals the opportunity cost of delay harvest, where the opportunity cost is defined as rent on the value of the stand plus the rent on the value of the land (timber plus amenities).
>
> Amacher *et al.* (2009: 51)

Implicit in the above analysis was that the forest was an independent entity; that is, nothing else within the landscape impacted the forest's timber and non-timber or amenity values. This implicit assumption can be relaxed to examine the effects of spatial and temporal interdependence among stands. Amacher *et al.* (2009) also provides an excellent review of the mathematical economic models and literature that examine this issue. The development of these models is beyond the scope of this book. However, I would encourage interested reader to examine the work of Amacher *et al.* (2009).

Recently, the focus has shifted from producing multiple outputs in a sustainable manner to sustaining the forest-based ecosystem that produces the outputs (e.g. Patterson and Coelho 2009; Wunder and Wertz-Kanounnikoff 2009). This evolutionary shift in focus has required re-thinking and describing the production system with respect to forest-based ecosystem processes, goods, and services.[30] The description of the production process begins with defining forest-based ecosystem processes. Brown *et al.* (2007: 332) define ecosystem processes as "the complex physical and biological cycles and interactions that underlie what we observe as the natural world." Ecosystem processes are the cycles and interactions among the abiotic and biotic structures of the ecosystem that produce ecosystem goods and services (Brown *et al.* 2007). These ecosystem processes exist whether humans are present or not and whether humans have preferences for them or not and can be identified using various descriptors; for example nutrient cycling. Brown *et al.* (2007: 332) defines ecosystem services as "the specific results of those [ecosystem] processes that either directly sustain or enhance human life (as does natural protection from the sun's harmful UV rays) or maintain the quality of ecosystem goods (as water purification maintains the quality of stream flow)." Brown *et al.* (2007: 330–331) defines ecosystem goods as "the tangible, material products that result from ecosystem processes." So while water and air purification are forest-based ecosystem services, water and air are ecosystem goods as is biomass in the form of timber and fiber (Brown *et al.* 2007). Boyd and Banzhaf (2005 and 2007) and Kroeger and Casey (2007) narrow the definition of ecosystem goods and services to end products of nature that are directly enjoyed, consumed, or used to produce human well-being. This definition is the most practical because it is well suited to measuring environmental quality and estimating the value of other ecosystem goods and services.[31] This end-product definition avoids the double counting that can result from adding the value of intermediate products to determine the value of a final good. As Boyd and Banzhaf (2007) describe, this does not mean that intermediate ecosystem goods and services are not valuable, rather their value is embodied in the value measurement of the end-product ecosystem good and service. In addition, end-product ecosystem goods and services may not necessarily be the final product consumed. The final product consumed may include end-product ecosystem goods and services and conventional goods and services. For example, fish populations, surroundings, and water body are the end-product ecosystem goods and services used by anglers directly to produce recreational benefits. In this case, fish populations are both end-product ecosystem goods as well as a final economic good (Boyd and Banzhaf 2005 and 2007).

Some forest-based ecosystem goods have markets that are well established, such as those for stumpage and timber (for example http://www.srs.fs.usda.gov/econ/data/prices/index.htm accessed on 21 May 2010). Other markets for forest-based ecosystem goods and services are emerging; for example, the markets for carbon credits – the American Carbon Registry (http://www.americancarbonregistry.org/ accessed on 15 April 2011), the Climate Action Reserve (http://www.climateactionreserve.org/ accessed on 15 April 2011), the Intercontinental exchange (https://www.theice.com/homepage.jhtml accessed on 15 April 2011), and the European Climate Exchange (http://www.ecx.eu/ accessed on 21 May 2010), and wetlands mitigation banking (http://www.epa.gov/wetlandmitigation/ accessed on 21 May 2010). In the case of carbon credits, developing these markets was possible by transforming a process that trees provided into a tradable good that has the following characteristics: (1) additional, (2) real, (3) quantifiable and verifiable, (4) permanent, and (5) exclusive with transferable ownership and legal title (e.g. Ruddell *et al.* 2007). This allows entrepreneurs who own forestland to capture the benefits of producing carbon sequestration. Developing markets for other forest-based ecosystem goods and services will depend on: (1) a perception that the provision of the forest-based ecosystem good or service is scarce; (2) developing tradable commodities with characteristics similar to those for carbon credits; (3) developing institutions (e.g. banks, brokers, etc.) to aid in transactions among consumers and producers; and (4) the ending market structures (e.g. Kemkes *et al.* 2009). Whether any of these forest-based ecosystem goods and services will have any impact on an entrepreneur's management activities will depend on whether the market value set by society is large enough for entrepreneurs to modify their management behavior. For example, Hancock Timberland Investor (2000) showed that when the alternative uses of timberland are commercial or residential development, very high carbon prices would be required to have any meaningful impact on land conversion as development values are just too high relative to forestland values. Time will tell if and how the development of markets for forest-based ecosystem goods and services will affect forestland management (Kline *et al.* 2009).

How to use economic information – *the forest rotation problem* – to make better business decisions?

One of the primary tools a landowner can use to actively manipulate forested ecosystems in a sustainable manner to produce timber and other forest-based ecosystem goods and services (e.g. water quality and quantity, recreation, wildlife, carbon storage, and mushrooms, etc.) consistent with their ownership goals and objectives, is to cut trees according to a sustainable forest management plan. In this chapter, I examined variations on the profit searching rule described in Chapter 5 to address the seemingly simple question of when to cut the trees. For even-aged forests, the entrepreneur's management decision was focused on a single decision variable: the rotation age. I examined four choices: (1) the optimal biological rotation age; (2) the financially optimal single rotation age; (3) the financially optimal multiple rotation age, and (4) forest rent. With uneven-aged forests, the entrepreneur's management decision is to determine the appropriate diameter distribution and species composition of the reserve growing stock and the length of the cutting cycle for each stand that maximized the entrepreneur's wealth.

The mechanics of calculating and the economic implications of the financially optimal single and multiple rotation ages and cutting cycles can be complex and an entrepreneur can get lost in the mechanics of estimating rotation ages and forget if the economic interpretation of what is being done makes sense. If inaccurate or incomplete information is used to build the Pillars of Price, Value, and Cost then Profits – reflected in an optimal rotation age or cutting cycle – will not convey what the entrepreneur believes. After all, the reason you do all these calculations is to generate accurate economic information to help make better business decisions.

As shown by the Architectural Plan for Profit (Figure 9.1), the Pillars of Price, Value, and Cost then Profits are built on the foundation of the Production System. As was illustrated, the foundation of the Architectural Plan for Profit must be built so that it will help provide accurate answers to future questions dealing with cost, value, price, and profit. The importance of the production system in even-aged, uneven-aged, and non-timber values (forest-based ecosystem goods and services) should not be overlooked.

10 Capital theory
Risk

In Chapters 8 and 9, we included the concept of time as a variable input in the Architectural Plan for Profit (Figure 10.1) and examined how including the opportunity cost of time – the interest rate – affected the analysis.

The conclusion was that while all investment analysis techniques had shortcomings, net present value (NPV) was, overall, the best approach because it is consistent with the assumption that humans are maximizers. In addition, the same economic information used to build the Architectural Plan for Profit is used in calculating NPV. In terms of determining the financially optimal rotation age, the Faustmann Model (Chapter 9) was the preferred analysis approach for the same reasons given for NPV.

In Chapter 8, I examined the interest rate and introduced the topic of risk and uncertainty. The definition of risk and uncertainty was the exposure to the chance of loss. If the chance can be described in statistical terms, this is commonly described as risk. If the chance cannot be described in statistical terms, this is commonly described as uncertainty. Risk and uncertainty were categorized as market, biological, or political. While I will not repeat the descriptions of these categories, I would advise the reader to review them. The consequence is that the Production System and the Pillars of Price, Value, and Cost that constitute the Architectural Plan for Profit may now contain risk and uncertainty (valuable economic information) due to market, biological, or political factors. The purpose of this chapter is to describe eight techniques that can be used to include risk and uncertainty in building the Architectural Plan for Profit.

Case study

The case study used in this chapter will be the Loblolly Pine Plantation (Amateis and Burkhart 1985; Burkhart *et al.* 1985). This case study can be found on the Routledge website for this book and its summary will not be repeated here.

Techniques for analyzing risk and uncertainty

I will examine eight techniques for incorporating risk and uncertainty (hereafter just referred to as risk) as economic information into the business decision making

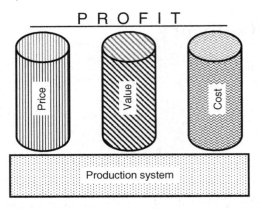

Figure 10.1 The Architectural Plan for Profit.

process. The techniques are: (1) risk as a component of the discount rate; (2) sensitivity analysis; (3) decision trees; (4) equally likely; (5) Bayes decision rule; (6) maximax; (7) maximin; and (8) minimax regret.

The discount rate

The nominal and real interest or discount rates include a risk component as illustrated in equations (8.3a) and (8.3b). There are two implications of including risk as part of the discount rate. First, as the perceived risk increases so would the discount rate. Second, the degree of riskiness [the discounting factor $(1 + r)^{-t}$] then depends solely on when the event occurs. This is illustrated by Table 10.1.

As Table 10.1 illustrates, the weight $(1 + r)^{-t}$ placed on risk is only a function of time and decreases geometrically with time. Consequently, including risk as part of the discount rate does not allow for components of a project to have differing degrees of risk due to presence or scale of costs and benefits unrelated to time. For example, managing forested ecosystem's risk may be a function of stand age (time) but also of species, biological conditions, and weather, etc.

Sensitivity analysis

Sensitivity analysis is defined as a procedure of changing model parameters (e.g. output prices, input wages, discount rate, or production system) systematically to identify those that cause most significant – not necessarily the largest – changes in the final result. The parameters that cause these changes should be determined with a greater level of accuracy than those that do not. In addition, sensitivity analysis will help with understanding the dynamics of the system being modeled. The procedure is simple to describe: to examine the impact of an individual parameter on the system, change one parameter at a time of the model. To examine the impacts of interactions among parameters simultaneously change two

Table 10.1 Interest rate example, $r = 6\%$

Year	$(1 + r)^{-t}$
5	0.74726
10	0.55839
15	0.41727
20	0.31180
25	0.23300
30	0.17411

parameters of the model. As defined by Smith *et al.* (2008), the impacts can be measured using relative-sensitivity elasticities, $RSE_{t,v,}$ given in equation (10.1)[1]

$$RSE_{t,v} = \frac{\dfrac{V_{t,v} - V_{t,0}}{V_{t,0}}}{\dfrac{v - v_0}{v_0}}; \quad t = 1, 2, 3, \dots \tag{10.1}$$

where
   $V_{t,v}$ denotes the value of the final result at time t given the new parameter value β;
   $V_{t,0}$ denote the value of the final result at time t given no parameter change;
   v denotes the new parameter value; and
   v_0 denotes the original parameter value.

The numerator and denominator denote percentage changes. The advantage of using equation (10.1) is that impacts can be compared regardless of units on the parameters; for example, stumpage price is measured as dollars per unit volume – dollars per cubic feet – while regeneration costs are measured as dollars per acre and the real interest rate is a percentage.

   To illustrate sensitivity analysis, I will focus my discussion on the NPV and land expectation value (LEV; i.e. the Faustmann formula) models described in Chapter 9 and given in equations (10.2) and (10.3) respectively[2]

$$NPV_t = PQ(t) \cdot (1 + r)^{-t} - C_0; \quad t = 0, 1, 2, \dots \tag{10.2}$$

$$LEV_t = \frac{PQ(t) \cdot (1 + r)^{-t} - C_0}{1 - (1 + r)^{-t}}; \quad t = 0, 1, 2, \dots \tag{10.3}$$

where
   P = stumpage price of $1.81 per cubic feet (cu.ft) for sawtimber;
   $Q(t)$ = the Loblolly Pine Plantation production system given in equations (2.6) to (2.8);
   r = the 8 percent real interest rate; and
   C_0 = $305.54 per acre regeneration costs.[3]

   Tables 10.2a and 10.2b illustrate the impacts on NPV and LEV if the stumpage price increased or decreased by 25 percent to $2.72 per cu.ft and $1.36 per cu.ft, respectively.

Tables 10.3a and 10.3b illustrate the impacts on NPV and LEV if the real interest rate increased or decreased by 25 percent to 10 percent and 6 percent, respectively.

These sensitivity analyses are presented graphically in Figures 10.2a and 10.2b for NPV and LEV, respectively.

These Tables and Figures are to be read holding plantation age constant. For example, at a plantation age of 15, the magnitudes of NPV and LEV are more

Table 10.2a Sensitivity analysis of a 25% change in stumpage price on NPV and LEV[‡]

Plantation age	NPV ($/acre)	LEV ($/acre)	Price = $2.72 per cu.ft		Price = $1.36 per cu.ft	
			NPV ($/acre)	LEV ($/acre)	NPV ($/acre)	LEV ($/acre)
10	772.00	1,438.13	1,043.56	1,944.02	503.92	938.74
11	897.19	1,570.94	1,200.31	2,101.69	597.97	1,047.02
12	1,005.25	1,667.39	1,335.60	2,215.34	679.14	1,126.49
13	1,095.23	1,732.12	1,448.25	2,290.44	746.73	1,180.98
14	1,167.03	1,769.46	1,538.15	2,332.16	800.67	1,213.99
15	1,221.20	**1,783.40**	1,605.97	**2,345.31**	841.37	**1,228.70**
16	1,258.72	1,777.57	1,652.95	2,334.31	869.55	1,227.99
17	1,280.88	1,755.28	1,680.69	2,303.17	886.20	1,214.42
18	**1,289.16**	1,719.45	**1,691.06**	2,255.49	**892.42**	1,190.29
19	1,285.11	1,672.70	1,686.00	2,194.48	889.38	1,157.61
20	1,270.33	1,617.33	1,667.49	2,122.97	878.28	1,118.18

[‡]NPV denotes net present value measured as dollars per acre ($/acre). LEV denotes land expectation value measured as dollars per acre ($/acre). Price is dollars per cubic feet ($/cu.ft).

Table 10.2b Relative-sensitivity analysis of a 25% change in stumpage price on NPV and LEV[‡]

Plantation age	Price = $2.72 per cu.ft		Price = $1.36 per cu.ft	
	NPV	LEV	NPV	LEV
	Relative-sensitivity elasticities			
10	1.41	1.41	1.39	1.39
11	1.35	1.35	1.33	1.33
12	1.31	1.31	1.30	1.30
13	1.29	1.29	1.27	1.27
14	1.27	1.27	1.26	1.26
15	1.26	1.26	1.24	1.24
16	1.25	1.25	1.24	1.24
17	1.25	1.25	1.23	1.23
18	1.25	1.25	1.23	1.23
19	1.25	1.25	1.23	1.23
20	1.25	1.25	1.23	1.23

[‡]NPV denotes net present value. LEV denotes land expectation value. Price is dollars per cubic feet (cu.ft).

Table 10.3a Sensitivity analysis of a 25% change in the real interest rate on NPV and LEV[‡]

Plantation age	NPV ($/acre)	LEV ($/acre)	r = 0.06		r = 0.10	
			NPV ($/acre)	LEV ($/acre)	NPV ($/acre)	LEV ($/acre)
10	772.00	1,438.13	993.47	2,249.67	591.36	962.41
11	897.19	1,570.94	1,171.75	2,476.16	677.36	1,042.89
12	1,005.25	1,667.39	1,334.85	2,653.62	746.19	1,095.14
13	1,095.23	1,732.12	1,480.53	2,787.34	797.95	1,123.35
14	1,167.03	1,769.46	1,607.51	2,882.39	833.42	**1,131.34**
15	1,221.20	**1,783.40**	1,715.30	2,943.54	853.85	1,122.59
16	1,258.72	1,777.57	1,804.03	2,975.21	**860.75**	1,100.18
17	1,280.88	1,755.28	1,874.29	**2,981.52**	855.76	1,066.83
18	**1,289.16**	1,719.45	1,927.00	2,966.19	840.60	1,024.94
19	1,285.11	1,672.70	1,963.36	2,932.63	816.91	976.59
20	1,270.33	1,617.33	**1,984.69**	2,883.90	786.26	923.54

[‡]NPV denotes net present value measured as dollars per acre ($/acre). LEV denotes land expectation value measured as dollars per acre ($/acre).

Table 10.3b Relative-sensitivity analysis of a 25% change in the real interest rate on NPV and LEV[‡]

Plantation age	r = 0.06		r = 0.10	
	NPV	LEV	NPV	LEV
	Relativity-sensitivity elasticities			
10	−1.15	−2.26	−0.94	−1.32
11	−1.22	−2.30	−0.98	−1.34
12	−1.31	−2.37	−1.03	−1.37
13	−1.41	−2.44	−1.09	−1.41
14	−1.51	−2.52	−1.14	−1.44
15	−1.62	−2.60	−1.20	−1.48
16	−1.73	−2.69	−1.26	−1.52
17	−1.85	−2.79	−1.33	−1.57
18	−1.98	−2.90	−1.39	−1.62
19	−2.11	−3.01	−1.46	−1.66
20	−2.25	−3.13	−1.52	−1.72

[‡]NPV denotes net present value. LEV denotes land expectation value.

sensitive to the 25 percent increase than decrease in stumpage price. In addition, the positive sign of the relative-sensitivity elasticity for stumpage price implies that an increase (decrease) in stumpage price will increase (decrease) the magnitude of NPV and LEV. At a plantation age of 15, the magnitudes of NPV and LEV are more sensitive to the 25 percent decrease than increase in the real interest rate. The negative sign of the relative-sensitivity elasticity for stumpage price implies that an increase (decrease) in the real interest rate will decrease (increase) the

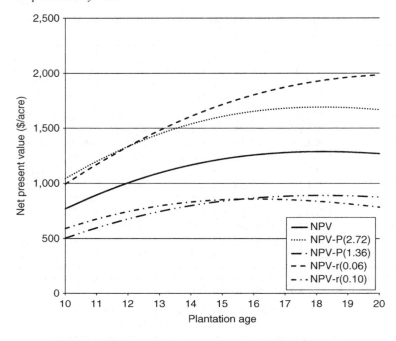

Figure 10.2a Sensitivity analysis: net present value (NPV).[‡]

[‡]P(2.72) denotes a stumpage price increase to 2.72 dollars per cubic feet. P(1.36) denotes a stumpage price decrease to 1.36 dollars per cubic feet. r(0.06) denotes a real interest rate decrease to 6 percent. r(0.10) denotes a real interest rate increase to 10 percent.

magnitude of NPV and LEV. It is left to the reader to conduct a similar sensitivity analysis for a 25 percent increase and decrease in regeneration costs.

Sensitivity analysis can be a very useful tool. However, there are five factors that require careful thought. First, sensitivity analysis is conducted within the range of normal operations. For example, in the Loblolly Pine Plantation sensitivity I chose plantation ages between 10 and 20 as these covered the optimal rotation ages calculated in Chapter 9. The entrepreneur would have to determine this range based on the problem's context. Second, I chose the 25 percent change in stumpage price and real interest rate arbitrarily to illustrate sensitivity analysis. The entrepreneur would have to determine the appropriate ranges for the parameters of the economic model for the given context. Third, time or plantation age impacts the sensitivity analysis in different ways. For stumpage price, the relative-sensitivity elasticities decrease as plantation age increases while for the real interest rate they increased as plantation age increased. This is due to time (or plantation age) having a direct exponential impact on the real interest rate (i.e. $(1 + r)^{-t}$) but not on stumpage price. Smith *et al.* (2008) noted this problem. Fourth, the economic model developed by the entrepreneur is for a specific context. The sensitivity analysis is for that given context. If the context changes (e.g. loblolly pine to select red oak), then the entrepreneur should conduct a new sensitivity analysis.

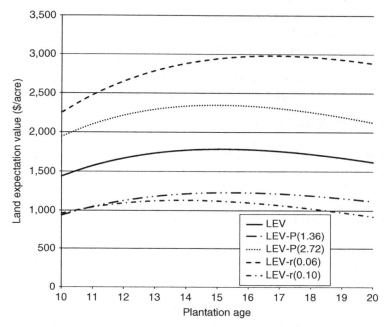

Figure 10.2b Sensitivity analysis: land expectation value (LEV).[‡]

[‡]*P*(2.72) denotes a stumpage price increase to 2.72 dollars per cubic feet. *P*(1.36) denotes a stumpage price decrease to 1.36 dollars per cubic feet. *r*(0.06) denotes a real interest rate decrease to 6 percent. *r*(0.10) denotes a real interest rate increase to 10 percent.

Finally, changing parameters may not only have an effect on the magnitudes of NPV and LEV, but they may also change optimal rotation age. For example, changing stumpage price and regeneration costs in equation (10.2) will not change the financially optimal single rotation age, but increasing (decreasing) the real interest rate will decrease (increase) the rotation age.[4] Johansson and Löfgren (1985) using a methodology called comparative statics showed that for equation (10.3) there is an inverse relationship between changes in stumpage price and changes in the financially optimal multiple rotation age – this inverse relationship also holds for changes in the interest rate – and there is a direct relationship between changes in regeneration costs and changes in the financially optimal multiple rotation age.[5]

Decision trees

A cash flow diagram (see Figure 8.2b for example) describes a set of sequential decisions. There is an implied assumption that the benefits and costs depicted in a cash flow diagram occur with 100 percent certainty. However, some of the benefits and costs may be dependent on circumstances that are out of your direct control. For example, the growth of your trees may be affected by the probability of a wet or dry season, a fire, or an insect/disease infestation, etc. The probability of a dry season may increase the probability of a fire or insect/disease infestation. Because we often want to consider more complex, sequential decisions, a decision tree provides a method to analyze decision making under risk. A decision tree shows the action, states of nature, and outcomes as nodes and branches of a dendritic or tree-like cash flow diagram. Figure 10.3 illustrates a decision tree.

A decision tree is composed of actions you as an entrepreneur can take, states of nature that are beyond your control, and terminal branch that denotes the end of a branch. These are illustrated in Figure 10.3.

> **Actions/decision fork** (denoted by a square) – Represents a point in time where the entrepreneur takes an action or makes a decision.
> **State of nature/event fork** (denoted by a circle) – When an outside force determines which of several random events will occur.
> **Terminal branch** – If no fork (either an event or decision fork) emanates from the branch.

A unique feature of a decision tree is the probabilities that describe if a random event will occur. The probabilities of a specified state of nature must sum to 100 percent.

As illustrated in Figure 10.3, the entrepreneur can choose either action I or II. Given this decision, an event will occur that is beyond their control. This state of nature will happen no matter which action (i.e. I or II) is chosen. The probabilities of all possible events within any state of nature must be defined and described with a probability of occurring. In Figure 10.3, the state of nature, A, has two events that occur with the probabilities α_1 and α_2. These probabilities sum to 100 percent thus defining the state of nature completely.

Once a decision tree has been determined, the entrepreneur can determine the expected value associated with each decision. This is done starting at the terminal branches and working backward toward the initial decision. Just like a cash

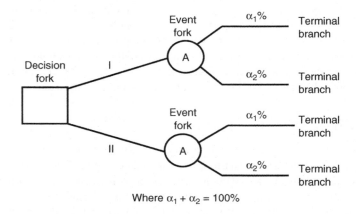

Where $\alpha_1 + \alpha_2 = 100\%$

Figure 10.3 Example decision tree.

flow diagram, the preferred investment analysis tool is net present value. However, in this case we will calculate an expected net present value due to the probabilities associated with each state of nature. To illustrate this I will develop a decision tree for three possible management actions using the Loblolly Pine Plantation case study:

Management Action I – The entrepreneur can precommercially thin the stand at plantation age 15 at a cost of $89.22 per acre, then clearcut the plantation at plantation age 25. An 8 percent real discount rate will be used on all cash flows.

Management Action II – The entrepreneur can use a herbicide spray release at plantation age 10 at a cost of $68.22 per acre, then clearcut the plantation at plantation age 25. An 8 percent real discount rate will be used on all cash flows.

Management Action III – The entrepreneur can use a herbicide spray release at plantation age 10 at a cost of $68.22 per acre plus the precommercial thin at plantation age 15 at a cost of $89.22 per acre, then clearcut the plantation at plantation age 25. An 8 percent real discount rate will be used on all cash flows.

No matter which management action is chosen, the growing season will be described as good, average, or poor. The probabilities associated with each type of growing season are based on historical observations by the entrepreneur and are given as: a good growing season occurs 20 percent of the time; an average growing season occurs 60 percent of the time; and a poor growing season occurs 20 percent of the time. Finally, no matter which management action is chosen, stumpage prices will remain the same, increase, or decrease. Again the probabilities of changing stumpage prices are based on historical observations by the entrepreneur who has determined a 25 percent chance of stumpage price increasing, 50 percent change of stumpage prices remaining the same, and a 25 percent chance of stumpage prices decreasing. These data are given in Table 10.4 and the corresponding decision tree is given in Figure 10.4.

To describe how to read the decision tree, I will examine the branch I → A(1) → B(1). The management action is a precommercial thin, I; the growing season is good with a probability of 20 percent and will result in 6,367 cubic feet per acre (cu.ft/ac) at plantation age 25, A(1); the stumpage price will be $2.27 per cu.ft with a probability of 25 percent, B(1).

Examining Table 10.4, it might seem that Management Action III should be the preferred alternative as this would give the largest expected net present value. However, this branch has the largest cost relative to the other management choices. Is the difference between the present value of the costs and the present value of the expected revenues going to be greater than the other management options?

To determine the expected net present value of each management alternative, you start at the terminal branches and work backward. I will examine each management action separately.

Table 10.4 Decision tree data for Loblolly Pine Plantation

Action	Growing season (A)			Stumpage srice (B)		
	Good (1)	Average (2)	Poor (3)	Increase (1)	No change (2)	Decline (3)
	(%)			(%)		
	20	60	20	25	50	25
	Cubic feet per acre			Dollars per acre		
I	6,367	4,975	3,245	2.27	1.81	1.30
II	5,905	4,799	2,971	2.27	1.81	1.30
III	6,509	5,323	3,359	2.27	1.81	1.30

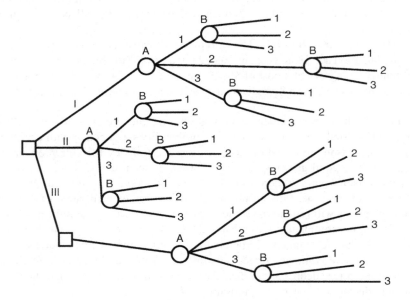

Figure 10.4 Decision tree for the Loblolly Pine Plantation.

Management Action I

Expected stumpage price ($/cu.ft), B

$1.80 = 0.25 \times 2.27 + 0.50 \times 1.81 + 0.25 \times 1.30$

Expected revenue at plantation age 25 ($/ac), A

$8821.05 = 1.80 \times [0.20 \times 6367 + 0.60 \times 4975 + 0.20 \times 3245]$

Expected net present value ($/ac)

$1259.91 = 8821.05 \times (1.08)^{-25} - 89.22 \times (1.08)^{-15}$

Management Action II

Expected stumpage price ($/cu.ft)

$1.80 = 0.25 \times 2.27 + 0.50 \times 1.81 + 0.25 \times 1.30$

Expected revenue at plantation age 25 ($/ac)

$$8366.64 = 1.80 \times [0.20 \times 5905 + 0.60 \times 4799 + 0.20 \times 2971]$$

Expected net present value ($/ac)

$$1259.91 = 8366.64 \times (1.08)^{-25} - 68.00 \times (1.08)^{-10}$$

Management Action III

Expected stumpage price ($/cu.ft)

$$1.80 = 0.25 \times 2.27 + 0.50 \times 1.81 + 0.25 \times 1.30$$

Expected revenue at plantation age 25 ($/ac)

$$9288.40 = 1.80 \times [0.20 \times 6509 + 0.60 \times 5323 + 0.20 \times 3359]$$

Expected net present value ($/ac)

$$1259.65 = 9288.40 \times (1.08)^{-25} - 89.22 \times (1.08)^{-15} - 68.00 \times (1.08)^{-10}$$

Based on the decision tree analysis, the entrepreneur should choose Management Action III.

As illustrated by the above example, decision trees allow the flexibility to assign probabilities to states of nature that are beyond your direct control. In addition, the probability of an event occurring is not solely a function of time. The entrepreneur can use this information to determine the expected net present value of alternative decisions. This flexibility comes at a cost. First, the real world of project evaluation contains many complexities and uncertainties, so building a decision tree that accurately reflects these aspects is a complicated and time-consuming undertaking.

Second, determining the probabilities may be difficult. Probabilities may or may not be affected by management actions. For example, various management actions, such as thinning and herbicide treatments, do not affect the possible states of nature describing wet versus dry years, good or average, or poor growing seasons, etc. However, various management decisions may affect the possible outcomes. For example, softwood tree growth and survival can be affected by various pathogens such as fusiform rust in slash pine (*Pinus elliotii*). The planting of rust-resistant trees can significantly affect growth, survival, and economic value of slash pine (Wagner and Holmes 1999) and thus to reduce the probability of loss, an entrepreneur could plant slash pine which is resistant to fusiform rust. Third, the decision tree described above does not have the probabilities tied to a time period explicitly. For example, it may be possible to determine the probability of a fire, but more difficult to predict the year in which a damaging fire will occur.

Equally likely

If we do not know how likely any of the states of nature are, there is no basis for presuming one state is more likely than any other. Therefore, we should consider all states equally likely and pick the decision with the highest expected average net

Table 10.5 Decision matrix for equally likely decision rule

Decision[‡]	Net present value ($/acre)			Equally likely
	Good	Average	Poor	
I	1,654.62	1,286.73	829.50	1,256.95
II	1,529.15	1,236.84	753.72	1,173.24
III	1,660.66	1,347.21	828.14	**1,278.67**
Probability	$\frac{1}{3}$	$\frac{1}{3}$	$\frac{1}{3}$	

[‡]I denotes precommercial thin at plantation age 15 at a cost of $89.22 per acre, then clearcut the plantation at plantation age 25. II denotes a herbicide spray release at plantation age 10 at a cost of $68.22 per acre, then clearcut the plantation at plantation age 25. III denotes a herbicide spray release at plantation age 10 at a cost of $68.22 per acre plus the precommercial thin at plantation age 15 at a cost of $89.22 per acre, then clearcut the plantation at plantation age 25.

return. To develop this decision rule, I will first introduce a payoff or decision matrix using the three management actions described above as illustrated by Table 10.5.

The cells in the decision matrix represent the payoff of a given action with respect to a state of nature. In Table 10.5, the cells represent the net present value of a management action given a state of nature using a stumpage price of $1.81 cu.ft and a real interest rate of 8 percent. The expected net present values of each management action given equally likely probabilities are given in the last column and the calculations are given as

Management Action I

Expected net present value ($/ac)

$1256.95 = (1654.62 + 1286.73 + 829.50) \times \frac{1}{3}$

Management Action II

Expected net present value ($/ac)

$1173.24 = (1529.15 + 1236.84 + 753.72) \times \frac{1}{3}$

Management Action III

Expected net present value ($/ac)

$1278.67 = (1660.66 + 1347.21 + 828.14) \times \frac{1}{3}$

Based on the equally likely decision rule, Management Action III should be chosen.

Bayes decision rule

The Bayes decision rule uses prior probability to determine the action with the maximum expected payoff. The decision matrix reflecting the Bayes decision rule is given in Table 10.6.

Table 10.6 Decision matrix for the Bayes decision rule

Decision[‡]	Net present value ($/acre)			Bayes
	Good	*Average*	*Poor*	
I	1,654.62	1,286.73	829.50	1,268.86
II	1,529.15	1,236.84	753.72	1,198.68
III	1,660.66	1,347.21	828.14	**1,306.08**
Probability	0.2	0.6	0.2	

[‡]See notes for Table 10.5.

The probabilities in Table 10.6 are based on those in Table 10.4. The expected net present values of each management action given prior probabilities are shown in the last column and the calculations are given as:

Management Action I

Expected net present value ($/ac)

$$1268.86 = 0.2 \times 1654.62 + 0.6 \times 1286.73 + 0.2 \times 829.50$$

Management Action II

Expected net present value ($/ac)

$$1198.68 = 0.2 \times 1529.15 + 0.6 \times 1236.84 + 0.2 \times 753.72$$

Management Action III

Expected net present value ($/ac)

$$1278.67 = 0.2 \times 1660.66 + 0.6 \times 1347.21 + 0.2 \times 828.14$$

Based on the Bayes decision rule, Management Action III should be chosen.

Maximax

The maximax decision rule selects the action that contains the best of the best possible payoffs. This rule is described as the optimistic approach as it chooses the best of the best. The decision matrix reflecting the maximax decision rule is given in Table 10.7.

The steps to determine the maximax are: (1) calculate the maximum possible payoff given each decision or action (this is illustrated by the last column in Table 10.7); and (2) determine the action associated with the maximum payoff from the above set. As illustrated by the last column in Table 10.7, Management Action III should be chosen as it gives the most optimistic outcome of $1,660.66 per acre.

Maximin

While the maximax decision rule is considered the optimistic approach, maximin is considered the pessimistic approach. The pessimistic decision maker wants to

252   Capital theory: risk

Table 10.7 Decision matrix for the maximax decision rule

Decision[‡]	Net present value ($/acre)			Maximum outcome
	Good	Average	Poor	
I	1,654.62	1,286.73	829.50	1,654.62
II	1,529.15	1,236.84	753.72	1,529.15
III	1,660.66	1,347.21	828.14	**1,660.66**

[‡]See notes for Table 10.5.

choose the action by considering the worst consequences of each possible action and picking the action with the least worst among them. The decision matrix reflecting the maximin decision rule is given in Table 10.8.

The steps to determine the maximin are: (1) calculate the minimum possible payoff given each decision or action (this is illustrated by the last column in Table 10.8); and (2) determine the action associated with the maximum payoff from the above set. As illustrated by the last column in Table 10.8, Management Action I should be chosen as it gives the most conservative outcome of $829.50 per acre.

Minimax regret

The minimax regret decision rule selects the action that minimizes the maximum regret. The decision maker may be worried most about recriminations or "I told you so" remarks after an action has been chosen and nature reveals itself. The difference between the return from the best possible decision for a given state of nature and the return from the action taken is a measure of potential "regret" or the opportunity lost and is to be minimized.

The best possible outcome for each state of nature is calculated to determine the minimax regret. This is given by the last row of the decision matrix of Table 10.9a.

The regret matrix is calculated by differences between the best possible outcome for a given state of nature and all possible decisions given that state of nature. The regret matrix is illustrated by Table 10.9b.

The last column of Table 10.9b shows that maximum possible regret of each decision. It is this regret that is minimized. As illustrated by the last column in Table 10.9b, Management Action III should be chosen because it minimizes the regret of the three management actions.

How to use economic information – *capital theory: risk* – to make better business decisions?

Because not all costs and revenues may be known with 100 percent certainty due to events beyond the control of an entrepreneur, I have introduced eight techniques that can be used to analyze decisions that include risk. The first two techniques examined (i.e. including risk in the interest rate and sensitivity analysis)

Table 10.8 Decision matrix for the maximin decision rule

Decision[‡]	Net present value ($/acre)			Minimum outcome
	Good	*Average*	*Poor*	
I	1,654.62	1,286.73	829.50	**829.50**
II	1,529.15	1,236.84	753.72	753.72
III	1,660.66	1,347.21	828.14	828.14

[‡]See notes for Table 10.5.

Table 10.9a Decision matrix for maximin regret decision rule[‡]

Decision[‡]	Net present value ($/acre)		
	Good	*Average*	*Poor*
I	1,654.62	1,286.73	829.50
II	1,529.15	1,236.84	753.72
III	1,660.66	1,347.21	828.14
Maximum	1,660.66	1,347.21	829.50

[‡]See notes for Table 10.5.

Table 10.9b Regret matrix for maximin regret decision rule[‡]

Decision[‡]	Net present value ($/acre)			Maximum
	Good	*Average*	*Poor*	
I	6.03	60.48	0.00	60.48
II	131.51	110.36	75.79	131.51
III	0.00	0.00	1.37	**1.37**

$6.03 = 1660.66 - 1654.62$

[‡]See notes for Table 10.5.

do not result in a decision rule per se, but allow the entrepreneur to understand the dynamics of the system being modeled.[6] Decision trees, equally likely, Bayes decision rule, maximax, maximin, and minimax regret all share the common traits of (1) defining a decision rule to identify the optimal course of action for mutually exclusive decisions and (2) combining actions that are under the direct control of the entrepreneur with states of nature that are not. Decision trees, equally likely, and Bayes decision rule are described as decision problems under risk as probabilities associated with the various states of nature that are defined explicitly. Maximax, maximin, and minimax regret are described as decision problems under uncertainty as there are no probabilities associated with the various states of nature.

The techniques described are relatively straightforward to use by entrepreneurs. The analysis of these types of problems falls under the rubric of Decision Theory and Operations Research. If a reader is interested in further information on this topic, I would suggest starting your search with:

Winston, W. (2004) *Operations Research: Applications and Algorithms*, 4th ed., Thomson Brooks/Cole Publishing.
Hillier, F.S. and Lieberman, G.J. (2005) *Introduction to Operations Research*, 8th ed., New York, NY: McGraw-Hill.
Levin, R.I., Rubin, D.S., Stinson, J.P., and Gardner, Jr. E.S. (1992) *Quantitative Approaches to Management*, 8th ed., New York, NY: McGraw-Hill.

Risk and uncertainty can affect every component of the Architectural Plan for Profit and are more often than not relevant to the decision making process. Which technique is used depends on the questions that the entrepreneur asks and the level and detail of the information available. That said, I feel that the sensitivity analysis approach is often a very useful technique to use because it:

- requires the entrepreneur to identify appropriate ranges for the production system, cost, value, and price of the Architectural Plan for Profit given normal operating conditions;
- results in a better understanding of the system being modeled by the Architectural Plan for Profit; and
- results in describing a decision analysis space of likely solution alternatives.

McIntyre *et al.* (2010) provide an example of this type of approach.

11 Forest taxes

Benjamin Franklin's famous quote in his letter to Jean Baptiste Le Roy in 1789 – "But in this world nothing can be said to be certain, except death and taxes" – seems just as relevant today as then. What are taxes and why do governments collect taxes? Simply, taxes are compulsory levies on private individuals and organizations by governments for two general purposes: (1) to raise revenue to finance expenditures on maintaining a satisfactory economic growth rate, promoting economic stability (e.g. legal, infrastructure, education, and defense, etc.), and distributing income to conform to society's currently held standards of equity; and (2) to discourage the consumption of goods and services with high social costs (e.g. alcohol and tobacco). Taxes can be levied on many aspects of economic activity; for example, income, wage, property, value added, sales, and inheritance, etc. Focusing on the Architectural Plan for Profit, taxes can affect all of its various components in one way or another.

For example, a tax on the use of a specific input in the production system may cause the entrepreneur to search for a lower cost substitute. A tax on the waste generated as the result of a production system may cause the entrepreneur to modify the product system to abate the waste. A tax on profit may cause the entrepreneur to change the amount of output produced.

There are many different types of taxes at the local, state, and national level that may affect an entrepreneur's management decisions. The primary focus will be on three common state taxes that impact managing forest and natural resources. These are property, productivity, and yield taxes. I will, however, briefly introduce two federal taxes – the income and inheritance tax. To provide readers with a general context in which to think about taxes, I will introduce six principles of taxation: neutrality, certainty and understandability, equity, efficiency, adequacy, and time bias. To analyze the potential impacts on entrepreneur's management decisions, I will only focus on two of them: neutrality and time bias.

Case study

The case study used in this chapter will be the Loblolly Pine Plantation (Amateis and Burkhart 1985; Burkhart et al. 1985). This case study can be found on the Routledge website for this book and its summary will not be repeated here.

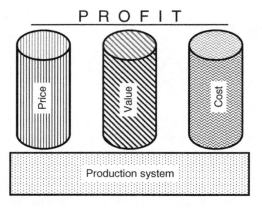

PROFIT

Figure 11.1 The Architectural Plan for Profit.

Economic principles of taxation

Neutrality

A tax should interfere as little as possible with attaining the pre-tax optimum allocation and use of resources. A caveat to this principle is when a government's policy goal is to modify individual decision behavior away from consuming goods and services with high social costs (e.g. alcohol and tobacco).

Certainty and understandability

The tax payment should be certain and non-arbitrary. The assessment methodology should be agreed upon and used by all. The time of payment, the manner of payment, and the quantity to be paid should be clear and plain to all.

Equity

Equity, as described in Chapter 1, denotes the concept of a just or fair (not necessarily equal) distribution of goods, services, output, income, etc., among all consumers. The classical descriptions of equity are horizontal and vertical.

Horizontal equity – Fairness of justice in the treatment of individuals in similar circumstances.
Vertical equity (ability to pay) – Fairness or justice in the treatment of individual in different circumstances. Vertical equity is further divided into three categories:

- Progressive: The net benefits represent a higher proportion of the lower-income person's income than of the rich person's income. Or the tax burden – the tax rate or tax payment as a proportion of income – increases as the level of income increases; the higher an individual's income, the larger the tax burden.

•   Regressive: The net benefits represent a higher proportional net benefit to high-income people than to low-income people. Or, the tax burden – the tax rate or tax payment as a proportion of income – increases as the level of income decreases; the lower an individual's income, the larger the tax burden.

•   Proportional: The net benefits affect individuals of each income level in the same proportion – the tax rate or tax payment as a proportion of income.

There are three additional concepts of equity when considering taxes based on a property's value:

Assessment equity – Comparable properties are assessed comparably within an assessment jurisdiction.
Tax equity – Comparable properties pay comparable taxes within a taxing jurisdiction or across jurisdictions.
Everybody pays – There is 100 percent compliance and everyone bears their share of the burden.

Efficiency

Tax collection costs should not be an inordinately high percentage of tax revenues. The real costs of collecting the tax should be a minimum inconvenience to taxpayers.

Adequacy

Taxes should raise enough net revenue such that government projects can be funded adequately without burdening taxpayers too much.

Time bias

Taxes based on a property's value are an annual expense; for example, state property and productivity taxes. However, forestlands often do not produce annual revenues that can be used to cover these annual taxes; revenues occur only periodically. Thus, there is a disconnect between the annual taxes paid on the value of the property and the cash generated by the property that can be used to pay the tax. For example, an entrepreneur cannot harvest the annual growth of the forest to pay the annual tax without harvesting the trees that generate the growth.

Federal income and inheritance taxes

An income tax is a tax on the flow of net revenue that an entrepreneur receives in a given year. For federal income, taxable income generated as a result of a capital gain is distinguished from ordinary income because income defined as a capital gain is taxed at a lower rate than income defined as ordinary income. For example, revenue from a commercial timber harvest may be classified as a capital gain depending on the context, while ordinary income includes wages, salaries, interest, dividends, rents and recreational leases, royalties, and business profits (Haney *et al.* 2001). These differences can affect specific management actions an entrepreneur may choose because a tax is an additional cost that must be taken into account. More information about how federal income taxes can impact management can be found in *The Forest Landowners' Guide to the Federal Income Tax*, 2001, United States Forest Service Agricultural Handbook No. 718 (http://www.timbertax.org/publications/aghandbook/ or http://www. fs.fed.us/publications/ 2001/01jun19-Forest_Tax_Guide31201.pdf accessed on 5 August 2010).

Estate taxes or the inheritance tax is of concern to most family farm landowners as one of every five acres of family forest land is owned by someone who is at least 75 years old (Butler 2008). These owners will be looking to pass their land to their heirs. Improper estate planning, which includes management of their forest proprieties, may force the heirs to liquidate the timber or land assets to pay inheritance taxes. More information about estate taxes can be found in *Estate Planning for Forest Landowners: What Will Become of Your Timberland?*, 2009, United States Forest Service Southern Research Station General Technical Report SRS-112 (Siegel *et al.* 2009, http://www.srs.fs.fed.us/pubs/31987 accessed on 5 August 2010).

Property tax

Property tax is an *ad valorem* or "according to value" tax based on the assessed value of the real property. Real property is defined as land, buildings (warehouses, offices factories, and mills), and anything affixed to the land. In the case of forestland, real property includes standing timber. Assessed value is generally based on the highest and best use of the property which may or may not be its current use. Highest and best use valuation is based on four conditions: (1) what is physically possible; (2) what is financially feasible; (3) what is legally permissible; and (4) what is most productive from a financial perspective. Often market transactions of comparable properties are used to determine highest and best use. Due to the expenses involved, real property is generally reassessed approximately every three to ten years depending on how active the markets are.

In the case of forestland, while standing timber is part of real property, it may not be included as part of the assessed value for a variety of reasons. A principal one is that many states combine a modified property tax with a yield or severance tax as part of tax incentive programs to promote forest management (http://www. timbertax.org/statetaxes/quickreference/ accessed on 8 December 2010).[1] However, to examine the potential impacts of a property tax on an entrepreneur's management decisions (i.e. the financially optimal multiple rotation age), I will examine two general cases: (1) the assessment includes both land and timber[2] and (2) the assessment includes only land.

The annual property tax payment is based on the assessed value of the real property multiplied by the property tax rate. This tax rate is often called the millage rate which is $1.00 per $1,000.00 of assessed value. The millage rate, τ, for a taxing district (e.g. town or county) is based on the district's projected budget, B, over the assessed property values in the district, V, as shown in equation (11.1):

$$\tau = \frac{(B/V)}{1000} \tag{11.1}$$

I will assume a tax rate of 8 mils or $\tau = 0.008$ to illustrate calculating a property tax.

Table 11.1 illustrates the annual property tax payment for the Loblolly Pine Plantation case study assuming the assessed value of the land is the land expectation value of $1,783.40 per acre at a plantation age of 15 and the assessed value of the timber is the liquidation value of the timber which will be reassessed annually.

Table 11.1 shows that the land component is constant while the timber component increases with time. Consequently, the annual tax payment increases with time and can be described as a variable cost. If the timber is only reassessed periodically, then the property tax will still be a variable cost if the periodicity of reassessment is less than the rotation age. If the periodicity of the reassessment is greater than or equal to the rotation age, then the annually property tax will be a fixed cost.

Equation (11.2) will be used to illustrate the impact of a property tax on the financially optimal multiple rotation age if described as a variable cost. The after-tax

Table 11.1 Property tax payments given the Loblolly Pine Plantation case study[‡]

Plantation age	Yield (cu.ft/acre)	Tax on inventory ($/acre/year)	Tax on land ($/acre/year)	Total tax payment ($/acre/year)
8	803.11	11.63	14.27	25.90
9	1,035.13	14.99	14.27	29.26
10	1,285.26	18.61	14.27	32.88
11	1,549.36	22.43	14.27	36.70
12	1,823.64	26.41	14.27	40.67
13	2,104.73	30.48	14.27	44.74
14	2,389.62	34.60	14.27	48.87
15	2,675.73	38.74	14.27	53.01

[‡]The stumpage price is equal to $1.81 per cubic feet (cu.ft) and the tax rate is 8 mils
Volume at plantation age 12 = 1,823.64 cu.ft/acre
Liquidation value = 1823.64 × 1.81 = $3,300.79 per acre
Tax payment on inventory = 3,300.79 × 0.008 = $26.41 per acre per year
Tax payment on land = 1783.40 × 0.008 = $14.27 per acre per year
Total property tax payment = 26.41 + 14.27 = $40.67 per acre per year

land expectation value for the financially optimal rotation age T, $LEV_T(\tau)$, is given by equation (11.2) and based on work by Amacher *et al.* (2009)

$$
\begin{aligned}
LEV_T(\tau) &= LEV_T - \frac{\displaystyle\sum_{t=1}^{T} \tau\cdot(PQ(t) + LEV_T)\cdot(1+r)^{-t}}{1-(1+r)^{-T}} \\[2em]
&= LEV_T - \frac{\displaystyle\sum_{t=1}^{T} \tau\cdot PQ(t)\cdot(1+r)^{-t}}{1-(1+r)^{-T}} - \frac{\tau\cdot LEV_T}{r}
\end{aligned}
\tag{11.2}
$$

where

LEV_T denotes the pre-tax land expectation value calculated using equation (9.7) at plantation T;

τ denotes the property tax rate;

r denotes real interest rate;

P denotes a stumpage price;

$Q(t)$ denotes the production system; and

$PQ(t)$ denotes the liquidation value of the inventory.

The right-hand side of equation (11.2) contains three components. The first is defined as the pre-tax land expectation value at rotation age T, LEV_T. The second is the present value of annual tax payment on the reassessed timber inventory. The third is the present value of the annual tax payment on the land. Table 11.2 illustrates the management implications if the annual property tax is a variable cost, equation (11.2), for the Loblolly Pine Plantation case study using a mil rate of 0.008, a real interest rate of 8 percent, a stumpage price of $1.81 per cubic feet, and the production system given by equations (2.6) to (2.8).

Table 11.2 After-tax land expectation value given the Loblolly Pine Plantation case study – variable cost

Plantation age	Yield (cu.ft/acre)	Present value of the tax on inventory ($/acre)	Present value of the tax on land ($/acre)	Pre-tax land expectation value ($/acre)	After-tax land expectation value ($/acre)
8	803.11	44.19	104.37	1,043.68	895.12
9	1,035.13	55.66	126.41	1,264.07	1,082.00
10	1,285.26	67.87	143.81	1,438.13	1,226.44
11	1,549.36	80.64	157.09	1,570.94	1,333.20
12	1,823.64	93.79	166.74	1,667.39	1,406.87
13	2,104.73	107.15	173.21	1,732.12	1,451.76
14	2,389.62	120.58	176.95	1,769.46	**1,471.93**
15	2,675.73	133.98	178.34	**1,783.40**	1,471.08

Table 11.2 shows that if the property tax is described as a variable cost it is non-neutral. That is, the financially optimal multiple rotation age decreases as a result of the annual property tax on land and timber.[3] In addition, Klemperer (1996) showed that if the property tax is defined as a variable cost it is also biased against high establishment costs.

If reassessment occurs only once during any potential rotation, then the annual property tax is a fixed cost. Equation (11.3a) illustrates the after-tax land expectation value $LEV_T(\tau)$, assuming the assessed value of the timber inventory and land is based on a defined rotation age T:

$$LEV_T(\tau) = LEV_T - \frac{\tau \cdot (LEV_T + PQ(T))}{r} \tag{11.3a}$$

Equation (11.3b) illustrates the after-tax land expectation value $LEV_T(\tau)$, assuming the assessment is based only on the land value defined rotation age T:[4]

$$LEV_T(\tau) = LEV_T - \frac{\tau \cdot LEV_T}{r} \tag{11.3b}$$

Table 11.3 illustrates the management implications if the annual property tax is a fixed cost, equations (11.3a) and (11.3b), for the Loblolly Pine Plantation case study using a plantation age of 15 to define the land and timber inventory's assessed values.

As shown by Table 11.3, the after-tax financially optimal multiple rotation age is the same as the pre-tax rotation age. In this case, the annual property tax is neutral with respect to management.[5] A property tax can cause you to stop management if the annual property tax payment is larger than the annualized pre-tax land expectation value. A property tax can also cause land use to change if the tax payment is larger than the opportunity cost of alternative land uses. For example, the fixed cost annual property tax payment is $53.01 or $14.27 per acre per year based

Table 11.3 After-tax land expectation value given the Loblolly Pine Plantation case study – fixed cost

Plantation age	Present value of the tax on inventory plus land ($/acre)	Pre-tax land expectation value ($/acre)	After-tax land expectation value inventory plus land ($/acre)	After-tax land expectation value land only ($/acre)
9	662.65	1,264.07	601.42	1,085.73
10	662.65	1,438.13	775.48	1,259.79
11	662.65	1,570.94	908.30	1,392.60
12	662.65	1,667.39	1,004.75	1,489.06
13	662.65	1,732.12	1,069.48	1,553.78
14	662.65	1,769.46	1,106.81	1,591.12
15	662.65	**1,783.40**	**1,120.75**	**1,605.06**
16	662.65	1,777.57	1,114.93	1,599.23

on the assessed value of land and timber or land only, respectively (Table 11.1). The pre-tax land value can be annualized using equation (11.4):

$$LEV_A = LEV_T \times r \qquad (11.4)$$

For the Loblolly Pine Plantation case study the annualized pre-tax land expectation value, LEV_A, is \$142.67 (= 1,783.40 × 0.08) per acre per year.

The property tax can be neutral or non-neutral depending on if they are realized by the entrepreneur as a fixed or variable cost, respectively. Property taxes are also considered to have a time bias, as forest properties generally do not provide annual positive net revenues to pay the tax. As a result, the burden of the property tax increases as the income from the property is moved into the future, or the periodic incomes from the property become less frequent.

Productivity tax

A productivity tax is an annual tax based on the "ability" of the land to grow timber.[6] Forest productivity, FP_t, is defined as monetary value of the annual net yield minus average costs (Klemperer 1996) and is given by equation (11.5)

$$
\begin{aligned}
FP_t &= \frac{\dfrac{P \cdot Q(t) - C}{t}}{r} \\
&= \frac{P \cdot \dfrac{Q(t)}{t} - \dfrac{C}{t}}{r} \\
&= \frac{P \cdot MAI_t - AC}{r}
\end{aligned}
\qquad (11.5)
$$

where
   $Q(t)$ denote the estimated volume per acre for a given site quality or index at time t,
   P denotes stumpage price,
   C denotes the per acre costs,
   r denotes the real interest rate,
   MAI_t denotes the mean annual increment at stand age t or the average annual growth, equations (2.9) and (2.10), and
   AC denotes the annual average costs or the annual per acre costs.[7]

The numerator gives the average annual net revenue for that stand based on the yield potential of a given site quality or index irrespective of the actual harvest or standing timber (Klemperer 1996; Koskela and Ollikainen 2003; Mutanen and Toppinen 2005).[8] The denominator is used to capitalize the expected ability of the stand to produce annual net revenues. The annual per acre tax payment is given by equation (11.6)

$$FP_t(\tau) = \tau \times FP_t \qquad (11.6)$$

where the tax rate, τ, is given by equation (11.1). Using the Loblolly Pine Plantation case study as an example, the annual forest productivity tax is $28.23 per acre per year

Forest productivity at plantation age 15

$$FP_{15} = \frac{P \cdot \dfrac{Q(t)}{t} - \dfrac{C}{t}}{r} = \frac{1.81 \times 167.23 - \dfrac{305.54}{15}}{0.08} = 3,529.03 \ \$/acre$$

Annual forest productivity tax payment

$$FT_{15}(\tau) = FT_{15} \times \tau = 3,529.03 \times 0.008 = 28.23 \ \$/acre/year$$

The capitalization of forest productivity as given in equation (11.5) assumes that a net revenue of $P \times MAI_t - AC$ is received every year. In the case of an even-aged forest, this would require a fully regulated forest structure.[9] In the case of an uneven-aged forest, this would require sufficient stocking such that an annual harvest could take place (Klemperer 1996). If the landowner does not have either of these forest structures, the forest would not produce the annual net revenue needed to cover the productivity tax. Thus, the productivity tax is generally considered to have a time bias.

As defined, the productivity tax is an annual fixed cost. As illustrated in Chapter 5, fixed costs do not change the profit searching rule. Thus, the productivity tax is considered neutral. A productivity tax can cause you to stop management if the annual tax payment is larger than the annualized pre-tax land expectation value. A productivity tax can also cause land use to change if the tax payment is larger than the opportunity cost of alternative land uses. For the Loblolly Pine Plantation case study the annualized pre-tax land expectation value, LEV_A, is $142.67 (= 1,783.40 $\times$ 0.08) per acre per year.

Yield tax

A yield tax is normally levied at the time of harvest – every time there is a *positive revenue* or merchantable timber is removed – and is a percent of the value of the forest yield or harvest.[10] A yield tax payment, $Y_t(\tau)$, is given by equation (11.7)

$$Y_t(\tau) = P \times Q(\tau) \times \tau \tag{11.7}$$

where τ denotes the yield tax rate and all other variables are defined as before. Using the Loblolly Pine Plantation case study as an example, the yield tax is $290.58 per acre using a yield tax rate of 6 percent.

Yield tax payment

$$Y_{15}(\tau = 6\%) = P \times Q(15) \times \tau = 1.81 \times 2675.73 \times 0.06 = 290.58 \ \$/acre$$

This tax eliminates the time bias of the property and productivity taxes; the entrepreneur only pays the tax when they have the revenue from a merchantable harvest.

A yield tax effectively reduces the stumpage price, and the revenue that the entrepreneur receives for any harvest. This is illustrated in equation (11.8):

$$\begin{aligned} \text{Revenue} &= P \times Q(t) - (P \times Q(t) \times \tau) \\ &= (P - P \times \tau) \times Q(t) \\ &= P \times (1 - \tau) \times Q(t) \end{aligned} \tag{11.8}$$

The pre-yield tax revenue for the Loblolly Pine Plantation case study is $4,843.07 (= 2,675.73 × 1.81) per acre and the post-yield tax revenue is $4,552.49 (= 2,675.73 × 1.70). Chang (1982) and Johansson and Löfgren (1985) show that the impact of a reduction in stumpage price is to lengthen the financially optimal multiple rotation age. If the entrepreneur does not harvest, then they do not have to pay the yield tax. The yield tax is considered non-neutral.

A yield tax is often used to replace the general property tax and is generally applied to renewable resources. A yield tax would seem to offer a logical means for eliminating some of the difficulties imposed on timber production by the property and productivity tax. However, a yield tax by its nature cannot provide a predictable, dependable source of income to the taxing entity (e.g. town, municipality, or state).

Forestry tax incentive programs

Many states in the United States have a policy of promoting forestry and forest products by private landowners. Unfortunately, neither the annual property nor productivity tax helps meet this policy goal. These programs take two general forms: (1) a modified annual property or productivity tax, or (2) a modified annual property or productivity taxed coupled with a yield tax. A modified property or productivity tax is often developed using a reduced tax rate. The land is assessed in its current use rather than highest and best use, or a simple flat per acre rate is charged. Landowners must meet certain restrictions and there are penalties for withdrawing from the program. The National Timber Tax Website provides a summary of the forest tax structures of all 50 states (http://www.timbertax.org/statetaxes/quickreference/ accessed on 16 August 2010).

In New York State, for example, Real Property Tax Law 480a provides forest landowners with a lower annual property tax coupled with a 6 percent yield tax on any commercial harvest (http://www.orps.state.ny.us/assessor/manuals/vol4/part2/section4.08/sec480-a.htm accessed on 16 August 2010 and http://www.dec.ny.gov/regulations/2422.html accessed on 16 August 2010). To be eligible for this program, the following conditions must be met:

- A management plan prepared by a qualified forester *must be approved* by New York State's Department of Environmental Conservation listing all harvests, stocking levels, etc.
- Ten-year commitment and annual re-commitment.
- Sufficient stocking of trees to allow a merchantable harvest of timber in 30 years.

- The forest land cannot have been cut within the previous three years (with some exceptions).
- At least 50 contiguous acres are used exclusively for forest crops.
- The owner can use ten standard cords (4 foot × 4 foot × 8 foot) or equivalent without paying a yield tax. Any noncommercial cutting as defined by the approved management plan is not subject to the yield tax.

How to use economic information – *forest taxes* – to make better business decisions?

Paraphrasing Benjamin Franklin, two things in life are certain – death and taxes. Taxes are a cost that the entrepreneur must take into account. The cash flows that you analyze for your clients will probably have tax implications at the local, state, or federal level. Taxes can be described as a loss in revenue, as with a yield tax, that affects the Pillar of Price, or a cost, as with the property and productivity tax, that affects the Pillar of Cost. Ignoring the tax implications of any management actions could add unexpected costs and penalties that you must cover and any legal actions brought by the taxing entity against you.

From the entrepreneur's point of view, the concern is the impact taxes will have on management decisions given the existing forest condition (e.g. species composition, age, and density), financial parameters (e.g. prices, costs, and discount rates), and planning horizon and within the context of their ownership goals and objectives. As Johansson and Löfgren have stated "All kinds of taxes will decrease the value of the land" (Johansson and Löfgren 1985: 100). Besides the effect on land values, to what extent are management decisions, such as rotation age, thinning timing and intensity, and regeneration method, affected by taxes? Klemperer (1996) proposed using a spreadsheet to simulate the degree to which a given tax will effect management actions. This approach is similar in nature to sensitivity analysis as discussed in Chapter 10 in Tables 10.2a, 10.2b, 10.3a, and 10.3b. The argument presented in Chapter 10 for using sensitivity analysis also holds in this case.

12 Estimating nonmarket values

Forested ecosystem functions and processes provide a variety of ecosystem goods and services through passive and active management.[1] For example, tree biomass, an ecosystem good used to produce timber – a direct benefit – is derived from tree productivity, an ecosystem process. Carbon sequestration is also an ecosystem service derived from tree productivity (e.g. Hines *et al.* 2010). Fish populations, surroundings, and water body are ecosystem goods and services, used by anglers directly to produce recreational benefits, are the result of ecosystem processes. Air quality is an ecosystem service that contributes to human health – a direct benefit – that results from atmospheric deposition, an ecosystem process. Water quality is an ecosystem good – a direct benefit – that results from nutrient and hydrologic cycling, an ecosystem process. Many of these ecosystem goods and services are traded in formal markets (e.g. stumpage) and emerging markets (e.g. carbon sequestration). Many, however, are not traded in formal markets. Consequently, with these ecosystem goods and services there is no observable market equilibrium to determine the optimal level to produce (Chapter 7). There is no demand curve to help distribute them among various potential users (Chapter 6). Finally, there is no pricing mechanism (a measure of relative scarcity) to help allocate the resources used in their production (Chapter 7).

That said, individuals value these nonmarket ecosystem goods and services by observing their choices. Individuals are observed recreating (hiking, hunting, fishing, viewing wildlife, birding, etc.) and the primary reasons for families owning forestland are for esthetics, privacy, recreation, and protection of nature (Butler 2008 and 2010). At a state, regional, and national scale, people prefer clean water and air and wilderness, etc. Consequently given the lack of economic information generally available from a well-functioning market, the entrepreneur, landowner, or policy maker needs to develop economic information analogous to that obtained from markets to make informed decisions concerning how much of these ecosystem goods and services to provide.[2] This can be illustrated by referring to the Architectural Plan for Profit in Figure 12.1.

The Production System is describing the ecosystem processes and functions that provide the ecosystem goods and services. In prior chapters, the boundaries of the Pillars of Cost, Value, and Price were all solid. The economic information used to build these pillars was available from the market. The methods discussed

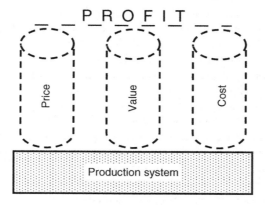

Figure 12.1 The Architectural Plan for Profit.

in this chapter will allow the developing of the Pillars of Cost, Value, and Price. Finally, the concept of profit should be revisited briefly. I have used the term Profit in the Architectural Plan for Profit as a proxy for individuals comparing what they gave up to obtain a good or service (its opportunity cost) with the benefits received from obtaining it. Simply, individuals search to find the greatest positive difference between benefits and costs. With nonmarket ecosystem goods and services the benefits, as measured by the Pillars of Price and Value, and net benefits or profit are inferred from observable choices individuals have made or choices individuals state they would make in a given context.

As the focus of estimating nonmarket benefits is measuring an economic value, the economic concept of value will be revisited. The two methods I will describe for estimating nonmarket benefits are revealed preference and stated preference.

Case studies

The case studies used so far have all had well-developed markets. Thus, they are not useful in this chapter. However, I will illustrate the concepts with published studies.

Economic value – revisited

According to a seminal work on the economic concept of value by Brown (1984), an assigned value is the expressed relative importance or worth of an object to an individual in a given context. Therefore, an assigned value is observable based on the choices individuals make. Assigned values are relative; it is not the intrinsic nature of the object but the object relative to all other objects that gives rise to an assigned value (Brown 1984). Assigned values are relative, not absolute, and depend on the context surrounding an individual's choice. If the context changes, the value an individual assigns to an object relative to all other objects will also change. Price is related to assigned value and is defined as a per-unit measure of assigned value and a measure of relative scarcity. As the price of a good or service increases relative to other goods and service, this particular good or service is scarcer relative to the other goods and services.

In well-functioning markets, measures of assigned value derived from the intersection of the supply and demand curves for a particular good (see Figure 7.3c) are identified as consumer surplus (the net benefit to the consumer for purchasing the good or service; see Figure 6.15b) and producer surplus (the net benefit to the producer for selling the good or service; see Figure 6.10), respectively. When an observable exchange takes place between consumer and producer, these net benefits are maximized and are captured by the consumer and producer. If well-functioning markets do not exist for a particular ecosystem good or service, then these measures of economic value are not observable directly.

The objective of the Architectural Plan for Profit is maximizing profits or net benefits. As was described in Chapter 5, the greatest use of the Architectural Plan for Profit is not to estimate total profits or net benefits, but to develop a systematic rule to compare the incremental (i.e. marginal) net benefits of one choice relative to another. As illustrated by Figures 6.15a and 6.15b, it is summing the incremental or marginal net benefits from individual choices that leads to estimates of total net benefit. These individual choices are represented by the various price–output combinations (Figures 6.15a and 6.15b). If well-functioning markets are absent, revealed and stated preference can be used to estimate these measures of economic value.

Assigned measures of value

Revealed and stated preference methods focus on estimating an assigned value for a nonmarket ecosystem good or service. There are two measures of assigned value that are relevant: use and nonuse values.

Use value

Use values are composed of three types of assigned values. First is a direct use value. These values include consumptive use values observed through market exchanges (e.g. stumpage, mushrooms, and maple syrup) and end-products of nature (i.e. ecosystem goods and services) that are consumed directly to produce human well-being, and nonconsumptive use values that are end-products of nature that are enjoyed or used directly to produce human well-being. For example, fish populations (consumptive use) and surroundings and water body (nonconsumptive use) are the end-product ecosystem goods and services used by anglers directly to produce recreational benefits.

Second is an indirect use value. These values are derived from ecosystem services, such as water filtration, soil and water conservation, nutrient cycling, and a genetic library, etc. (Pagiola *et al.* 2004).[3] Third is option value. This is an individual's expressed relative importance to retain the opportunity to consume a resource in the future; that is, the preservation of the stock is valuable merely to keep our options open. For example, people may be willing to pay for preserving biodiversity or genetic materials to ensure the option of having these goods in the future. Option value may also reflect your willingness to pay to preserve the Alaska wilderness as you might want to visit it. This value component is a controversial element of the total value.[4]

Nonuse values

Nonuse values are composed of two types of assigned values. First is existence value; that is, the expressed relative importance to simply preserve the existence of some resource. Second is bequest value; that is, the expressed relative importance to leave resources for future generations or their heirs. The difference between existence and bequest values is subtle and they are often combined.[5]

Total economic value

In Chapter 7, market equilibrium was determined by maximizing the sum of producer and consumer surplus or the area under the demand curve bounded by a given quantity minus the relevant opportunity costs (Chapter 6). This is illustrated in Figure 7.3c. Market equilibrium price and quantity provide economic information used in consumption and production decisions by consumers and producers, respectively, and defines a market value for a good or service. The market equilibrium illustrated in Figure 7.3c also describes the total economic value (TEV) of a good or service and reflects the assigned value of a good or service. Figure 12.2 illustrates the decomposition of TEV.

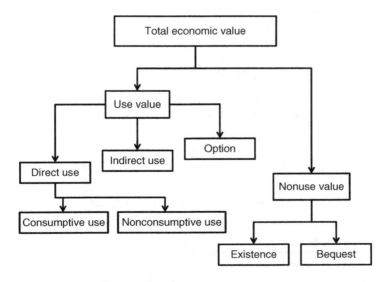

Figure 12.2 Total economic value.

Figure 12.2 shows that TEV is comprised of use plus nonuse values or direct plus indirect use plus option plus existence plus bequest values. For a good such as maple syrup, TEV reflects a consumptive direct use value. For an ecosystem good such as trout, the TEV would reflect a consumptive direct use value if the angler keeps it or a nonconsumptive direct use value if the angler practices catch and release or possibly a nonuse existence value for an individual.

Many ecosystem goods and services are not traded in formal markets and, thus, there is no observable demand curve and no estimate of TEV. Consequently, the focus on estimating nonmarket values is deriving an individual's expressed willingness and ability to pay for a given quantity at a given place and time; namely, the demand curve.[6] The terminology "expressed relative importance or worth" was used deliberately as the use and nonuse values being estimated are assigned values. It bears repeating; assigned values are relative, not absolute, and depend on the context surrounding an individual's choice. If the context changes, the value an individual assigns to an object relative to all other objects will also change.

Revealed preference or indirect methods

The core of the revealed preference techniques is that the analyst observes choices people have made but does not ask them directly how much they value the ecosystem good or service. The logic of revealed preference methodology is that because people are consuming these nonmarket goods and services, we can observe the choices they are making in consuming them. When producing any output (tennis shoes, clean air, reduction in noise pollution, purchase of a house, or recreational experiences, etc.), inputs must be used. Based on the behavioral assertion that people are maximizers (Chapter 1), if we observe the purchases of

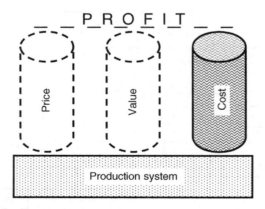

Figure 12.3 Revealed preference methodology –
the Architectural Plan for Profit.

inputs, we can infer that the value of the output must be at least as great as the cost of the inputs. Figure 12.3 illustrates this logic using the Architectural Plan for Profit.

Based on the description of the production system used to produce a defined set of nonmarket goods and services in a given context, the set of inputs are enumerated. Observing the expenditures individuals make on these inputs, the Pillar of Cost is estimated. As a result, a monetary value for the other components of the Architectural Plan for Profit can be inferred.

Revealed preference methods are used to estimate an individual's willingness to pay for nonmarket goods or services based on their existing choices.[7] In this respect the analysis is from the point of view of the consumer. In previous chapters, the analysis has been from the point of view of the entrepreneur who produces or provides a good or service. The student should keep in mind this change in perspective when studying this section.

The greatest disadvantage of revealed preference methods is that they cannot be used without observable individual choice behavior that can be used to infer nonmarket values. In the case of nonuse values, there are no observable market interactions between the individual and the resource in question; thus, in the case of measuring nonuse values these methods are inappropriate.[8]

I will describe two common revealed preference techniques: hedonic model and travel cost model.[9]

Hedonic model

Rosen (1974) formalized the theory supporting hedonic models and laid the foundation upon which they are built. Rosen's hedonic theory states that the market

Market price $300,000   Market price $200,000

Figure 12.4a Hedonic model: What is the value of the lake?

price of a good is a function of its characteristics or attributes. For example, what is the "use value" of a lake?[10] Or, what is a potential homeowner's willingness to pay for lake-front property? Figure 12.4a illustrates this problem.

The lake is an attribute or characteristic associated with one house but not the other. The houses are otherwise identical. The use value of the lake is not explicit; it is included in the price of the lake house. In the example shown in Figure 12.4a, the use value of the lake is $100,000. Unfortunately, not all comparisons are going to be as straightforward as that illustrated by Figure 12.4a. In Figure 12.4b, again the comparison is between two houses, one of which still has the lake attribute.

However, now the houses are not identical. The lake house also has a garage. Now the price difference of $175,000 is for the combined net benefit of the lake and garage plus any other indirect use values. The hedonic model allows for estimating the combined use value or the implicit price of each attribute.

Rosen's theory that the market price of a good is a function of observable and measurable attributes can be generalized using the hedonic price function illustrated by equation (12.1):

$$\text{Market Price} = f(\text{positive attributes, negative attributes}) \qquad (12.1)$$

Using the hedonic model starts with creating a list of attributes that are proposed to be related to the good. Each of these attributes will affect the use value of the good positively or negatively. Implicit price functions for these attributes can be quantified through statistical estimation techniques such as a regression model with the good's market price, the dependent variable, and its various attributes, the independent variables.

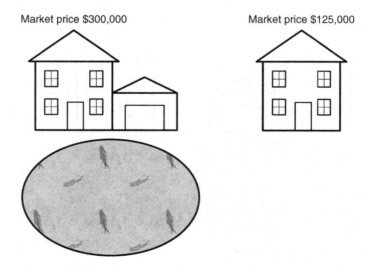

Figure 12.4b Hedonic model: What is the value of the lake?

Once the observable and measurable market prices and attributes have been identified and recorded, implicit price functions for these attributes can be quantified through statistical estimation techniques such as a regression model. Equation (12.2) illustrates a linear–linear hedonic price functional form for equation (12.1)

$$P_j = \beta_0 + \beta_1 X_{j1} + \beta_2 X_{j2} + \beta_3 X_{j3} + \cdots + \beta_K X_{jK} + \varepsilon_j$$

$$= \beta_0 + \sum_{k=1}^{K} \beta_k X_{jk} + \varepsilon_j \qquad (12.2)$$

where the βs are the coefficients to be estimated statistically and the Xs denote the attributes.[11] In addition, the βs denote the implicit price or value the consumer places on a specific attribute.[12] The implicit price is defined as dollars per unit of the attribute holding the level of all other attributes constant; for example, if the attribute is the number of feet of lake frontage, then the units on the implicit price are dollars per foot lake frontage. If the sign of a coefficient for a given attribute is positive (negative), then an additional unit of that attribute will increase (decrease) the consumer's willingness to pay for the good. In this manner, an implicit price function could be estimated for both the lake and garage shown in Figure 12.4b. Finally, workable competition is a key assumption to using a hedonic model: the resulting implicit price will represent a buyer's willingness to pay for that attribute accurately (Rosen 1974).

For the purpose of discussing hedonic models, I will re-write equation (12.2) as equation (12.3)

$$P = f\left(\frac{X_1}{[\pm]}, \frac{X_2}{[\pm]}, \frac{X_3}{[\pm]}, \cdots, \frac{X_k}{[\pm]} \right) \qquad (12.3)$$

where the signs in the brackets indicate whether the attribute has a positive or negative effect on market price. For example, in the lake-front property illustrated in Figure 12.4b, assuming the attributes (number of bedrooms, number of bathrooms, and size, etc.) are the same, the hedonic model will be shown as

$$P = f\left(\frac{Lake}{[+]}, \frac{Garage}{[+]} \right)$$

where the attributes of interest are the lake and garage. In this example, a tenable proposition is that the implicit prices of the attributes are positive. Thus, having lake frontage and a garage would both increase the market price of a house. The magnitudes of the implicit prices would indicate the per unit value the buyer places on each attribute (e.g. dollars per foot of lake frontage or dollars per garage square feet). Each attribute that is included in the hedonic model will have an estimated implicit price. Caution should be used with estimating hedonic pricing models. For example, the statistically estimated implicit prices are dependent on the market being in equilibrium and defined as workably competitive and if the hedonic model is underspecified (that is important attributes are not included) or multicollinearity among attributes will cause bias in the estimated implicit prices. A more detailed discussion of the statistical analysis of hedonic models is beyond the scope of this book. If a reader is interested in these topics, I would recommend they review the topics of regression analysis and econometrics.

I will use two examples to illustrate the hedonic model: timber sale prices and amenity values.

Timber sale prices

From the point of view of a harvester (the buyer): What is their willingness to pay for standing trees (stumpage price)? Their willingness to pay would be affected by the various attributes associated with the trees from a timber sale; for example, species (Sp), quality or grade (G), hauling distance (HD), diameter at breast height (DBH) and volume (Vol). Using the market price of a timber sale, the hedonic price function is given in equation (12.4):

$$Timber\ Sale\ Price = f \left(\frac{Sp}{[?]}, \frac{G}{[+]}, \frac{HD}{[-]}, \frac{DBH}{[+]}, \frac{Vol}{[+]} \right) \tag{12.4}$$

The proposed effects the various attributes have on the harvester's willingness to pay are given in the brackets below each attribute holding the level of all other attributes constant (*ceteris paribus*). For example, as the volume (Vol) of the sale increases holding the level of all other attributes constant, the tenable proposition is that the amount a harvester is willing to pay would also increase. As the DBH of the trees within a sale increases *ceteris paribus* so should a harvester's willingness to pay for the sale. An increase in hauling distance (HD) would increase a harvester's cost of transporting the stumpage from where it is being harvested to a sawmill, *ceteris paribus*. A harvester would be expected to cover this increase in cost by lowering their willingness to pay for the sale. Tree grade (G) is a measure of quality (Chapter 2). As tree quality increases, higher quality and valuable products can be produced from the tree. A harvester would be willing to pay more for higher than grade trees *ceteris paribus*. Finally, it seems obvious that different tree species (Sp) would have different market prices; for example, a black cherry (*Prunus serotina*) has a higher stumpage price per unit volume than a white pine (*Pinus strobes*). The reason for the question mark in equation (12.4) is that the relationship between species and stumpage price is dependent on the functional form used to model it. For example, the relationship would be positive if a sale with predominately White Pine (with lower stumpage price) is compared to a sale with predominately Black Cherry (with a higher stumpage price) and negative if a sale with predominately Black Cherry is compared to a sale with predominately White Pine.

Puttock *et al.* (1990)[13] examined 344 timber sales throughout Southwestern Ontario from 1982 to 1987. They developed a hedonic price function illustrated by equation (12.5)

$$P_j = f\left(\begin{array}{ccccccc} TVol_j & PctSpVol_{1j} & PctSpVol_{2j} & AVol_j & G_j & HD_j & PI_j \\ [+] & [+] & [+] & [+] & [-] & [-] & [+] \end{array} \right)$$

(12.5)

where

P_j denotes the lump-sum sale price of the jth timber sale,[14]

$TVol_j$ denotes the total volume of the jth timber sale;

$PctSpVol_{1j}$ denotes the proportion of the total sales volume from highest value species group present in the jth timber sale (species are divided into three groups according to value and species);

$PctSpVol_{2j}$ denotes the proportion of the total sales volume from medium value species group present in the jth timber sale;

$AVol_j$ denotes the average volume per tree on the jth timber sale;

G_j denotes a quality index for the timber on the jth timber sale (this index is based on the percent of timber in grade 1, grade 2, and so on, with 1 denoting excellent quality and 4 denoting poor quality);

HD_j denotes hauling distance from the jth timber sale to the purchasing mill; and

PI_j denotes an annual industry price index (this variable will account for inflation differences for the period 1982 to 1987).

The terms in the brackets are the hypothesized signs for each attribute. I would advise the reader to take a moment and compare the hypothesized signs for each attribute given by Puttock *et al.* (1990) in equation (12.5) and the hypothetical hedonic model in equation (12.4). Are the hypothesized signs in equation (12.5) tenable?

Puttock *et al.* (1990) examined three different functional forms for the hedonic model. I will only examine the results of linear–linear form which are given in Table 12.1.

Where the first column lists the variables from equation (12.5), the second column defines the statistically estimated β coefficients for a liner–linear regression equation, and the third column define the units associated with the statistically estimated β coefficients.

The terms in the second column of Table 12.1 also define the implicit prices. For example, if the total volume of a timber sale increased by 1 m^3, this would add 42.91 Canadian dollars (CDs) to the timber sale value. A 1 percent increase in volume from the highest value species group would add 4,528.80 CDs to the timber sale value. A 1 kilometer increase in hauling distance would decrease the timber sales value by 14.05 CDs.

Amenity values

Urban trees and open space provide many different types of benefits to citizens. These benefits include end-product ecosystem goods and services such as air quality improvement and shading by urban trees. Social benefits from urban trees and open space include improved quality of life. A study by Sander *et al.* (2010)

Table 12.1 Results of the hedonic model of Southwestern
Ontario timber sales from 1982 to 1987
(Puttock *et al.* 1990)[‡]

Variable	Estimated coefficient value	Units
$TVol_j$	42.91	$\dfrac{CD}{m^3}$
$PctSpVol_{1j}$	4,528.8	$\dfrac{CD}{PctSpVol_{1j}}$
$PctSpVol_{2j}$	2,433.2	$\dfrac{CD}{PctSpVol_{2j}}$
$AVol_j$	2,699.5	$\dfrac{CD}{m^3}$
G_j	−1,170.7	$\dfrac{CD}{G_j}$
HD_j	−14.05	$\dfrac{CD}{km}$
PI_j	254.0	$\dfrac{CD}{CD}$

[‡]CD denoting Canadian dollars, m^3 denoting cubic meters, *km* denoting
kilometers, and all other variables defined as before.

examined the values of urban trees to single family home property values as well
as how values vary with different levels of tree cover. In addition, they examined
whether tree cover affects home prices beyond the local parcel. The proposed
log–linear hedonic price function is given in equation (12.6)

$$\ln(P) = f(S, N, E) \tag{12.6}$$

where $\ln(P)$ represents natural log of the property's sale price, S represents parcel
and structural attributes for the property (e.g. finished square feet, home age, lot
acreage), N represents neighborhood attributes for the property (e.g. distance to
shopping centers, school quality), and E represents environmental attributes of the
property (e.g. proximity to lakes, percentage of tree cover).

A study by Poudyal *et al.* (2009) examined the quality of open space and how
it affects the property value in a neighborhood. The proposed linear–linear hedonic
price function is given in equation (12.7)

$$P = f(S, N, O) \tag{12.7}$$

where P represents the property's sales price, S represents parcel and structural
attributes of the property, N represents neighborhood attributes of the property,
and O represents land use amenities in the neighborhood.

While the hedonic pricing models presented in equations (12.6) and (12.7)
appear straightforward, the list of variables that define and describe each set of
attributes is too extensive to provide here. In addition, a discussion of the methods

used to statistically estimate the hedonic prices is beyond the scope of this book. However, I will discuss the implications of the resulting implicit prices for each study starting with Sander *et al.* (2010). While past studies have shown that urban trees increase the sales price of homes, the study by Sander *et al.* (2010) is unique in adding a spatial component to the analysis. That is, to what extent do trees on adjacent and neighborhood properties affect property values? Referring to Figure 12.2, Sander *et al.* are using the statistically estimated implicit prices to provide economic information about use and indirect use values. Their results indicate that increasing tree cover increases home sale value up to a limit. For example, increasing tree cover to 44 percent within a 328-feet buffer increases home sale value but increasing tree cover beyond this amount will decrease home sale value. Basically, homeowners value trees in their local neighborhoods, at distances that roughly correspond to the length of a city block. This value may reflect a preference for tree-lined streets and the shading and esthetic environment they offer. Homeowners appear to place less value on tree cover beyond their immediate local neighborhood and on tree cover over 40 percent in their immediate local neighborhood. Urban trees provide positive externalities (Sander *et al.* 2010).

Poudyal *et al.* (2009) examined how urban residents value variety, spatial configuration, and patterns of open space in their neighborhoods. They hypothesized that residents prefer heterogeneity (i.e. more diversity) within open space to homogeneity and fewer but larger plots of open space to many smaller plots. Again the magnitudes and the signs associated with the statistically estimated implicit prices were used to test these hypotheses. Their results indicated that urban residents prefer a neighborhood with a variety of land uses within open spaces types (e.g. pine forest, hardwood forest, wetland, and pasture) to one with less diverse and homogenously composed open spaces (e.g. pine forest or hardwood alone), prefer open spaces that are more even and square/rectangular shape to those in crooked or convoluted shapes, and did not prefer neighborhoods with residential areas mixed with industrial or commercial land uses (Poudyal *et al.* 2009).

The values estimated by both of these studies are use values. The utility of these studies goes beyond determining housing prices. These studies provide urban planners, residents, and decision makers with economic information concerning broader issues of land use planning and urban tree planting goals.

Travel cost model

The travel cost model (TCM) begins with the realization that a major cost of outdoor recreation is the travel and time costs incurred to get to the recreation site (Clawson and Knetsch 1966). Because individuals reside at varying distances from the recreation site, the variation in distance, number of trips, and travel costs can be used to determine the willingness to pay for a recreation site (Whitehead *et al.* 2008). There are two general categories of TCMs: zonal travel cost model and individual travel cost model. After introducing the zonal and individual travel cost models, I will use a simple zonal travel cost model to illustrate developing a demand curve based on a study by Wagner and Choi (1999).

Zonal travel cost model

Zonal travel cost models (ZTCM) are based on defining travel zones from an identified recreational site. For example, Figure 12.5 for the High Peaks of the Adirondack Park in New York State.

Within each travel zones, travels costs to the recreation site are assumed to be similar and individuals within the zones are assumed to have similar socio-economic characteristics. Referring to Figure 12.5, within a 50-mile radius of the High Peaks travel costs are assumed to be zero, between 50 and 150 miles the travel costs are positive and constant for all people who live in that zone and travel to the High Peaks, and between 150 and 250 miles the travel costs are greater than the previous travel zone and constant for all people who live in that zone and travel to the High Peaks. This same pattern holds for all subsequent defined travel zones. The trips per capita from each zone are determined as well as the travel costs from each zone. The ZTCM is useful when using secondary data sources such as recreation permits or fee receipts that contain limited data (Loomis *et al.* 2009). These data can be used to develop a demand curve relating price to the number of visitors. If sufficient data are collected, a statistical relationship between price and number of visitors can be developed similar to the hedonic models presented above.[15] If not, numerical methods can be used to develop a demand curve. A discussion of the statistical analysis of ZTCMs is beyond the scope of this book. If a reader is interested in these topics, I would recommend they review the topics of regression analysis and econometrics.

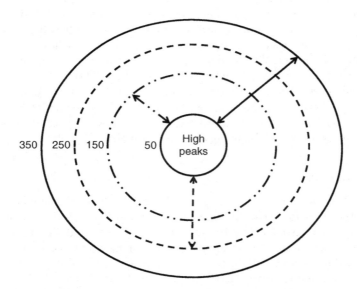

Figure 12.5 Zonal travel cost model: What is the willingness to pay to recreate in the High Peaks of the Adirondack Park in New York State?

There are five concerns with using a ZTCM. First, in ZTCMs average travel cost and zone average travel time cannot be separately included because they are nearly perfectly correlated (Brown and Nawas 1973); that is, separating an individual's opportunity cost of time spent in-transit and the cost of travel (e.g. the cost of gas) due to the potential co-relationship between these two variables; for example, the longer the trip the greater the cost of gas and the opportunity cost of time spent in-transit. In addition, accounting for an individual's opportunity cost of time spent in-transit is also problematic. For example, assuming the opportunity cost of time spent in-transit is a function of the individual's wage or salary would require collecting socio-economic information. If just in-transit time is collected, then the opportunity cost of time spent in-transit is the same for all individuals within a given zone and each individual within a given zone has the same socio-economic status. Second, the ZTCM does not address the problem of accounting for travel costs associated with visiting multiple sites in a single trip or substitute sites. Third, the ZTCM also does not account for users with zero travel costs as illustrated by Figure 12.5. Fourth, the ZTCM does not include the cost of equipment (e.g. tents, boots, fishing poles, and cameras, etc.) as they are not travel costs.[16] Finally, the ZTCM estimates a direct use value (see Figure 12.2).

Individual travel cost model

The individual travel cost model (ITCM) requires collecting data directly from visitors to a recreation site using a survey. Data collected by the survey would include, for example, location of the visitor's home, how far they traveled to the site, how many times they have visited the site in a given time period, length of trip, travel expenses (e.g. gas, food, lodging, etc.), demographic information (income, age, education, etc.), how many different recreation sites or destinations were part of this trip, and recreation site quality metrics, etc. Based on these data, an implicit ITCM can be defined by equation (12.8)

$$N_{ij} = f(C_{ij}, SED_{ij}, \theta_{ij}) \tag{12.8}$$

where N_{ij} are the number of visits by the ith individual during a defined time period to site j, C_{ij} denotes travel costs of the ith individual to site j, SED_i denotes a set of social, economic, and demographic variables of the ith individual at site j, and θ_{ij} denotes a set of recreation site quality metrics of site j identified by the ith individual.

As with hedonic models, the ITCM may take many different forms. Again the most common are linear–linear, log–linear, linear–log, and log–log. Readers may note the similarities of the ITCM and the hedonic models presented above. The ITCM has been used to address the concerns raised by the ZTCM. In addition, Vaughan and Russell (1982) pioneered an extension of equation (12.8) called varying parameters model that can be used to account for site quality differences (e.g. water quality). In addition Bockstael *et al.* (1989) and Kaoru *et al.* (1995) illustrate the use of a random utility model that can account for site quality differences. The ITCM can estimate direct and indirect use value (see Figure 12.2).

However, a discussion of the statistical analysis of ITCMs is beyond the scope of this book. If a reader is interested in these topics, I would recommend they review the following articles Bockstael *et al.* (1989), Willis and Garrod (1991), Dobbs (1993), Hesseln *et al.* (2003 and 2004), and Loomis *et al.* (2009) and regression analysis and econometrics.

Zonal travel cost model example

A large landowner was interested in developing a stewardship plan to help manage the approximately 3,612 acres on and near the Moose River that flows through Fowlersville, New York. A study by Wagner and Choi (1999) examined the recreational use value for fishing access on the Moose River. The data used to develop the ZTCM were from the New York State Department of Environmental Conservation's 1996 Angler Survey (NYS DEC 1997). The five steps used were based on those described by Rideout and Hesseln (2001).

First, determine contiguous zones around the study area. New York Department of Environmental Conservation's regional administrative structure was used to define the travel zones based on the data collected by the 1996 Angler Survey.[17] The estimated one-way driving distance from the most populated city from each region to Fowlersville, New York rounded to the nearest 40-mile increment defined the radii for each zone. A travel distance of approximately 300 miles or less defined the limit of the ZTCM based on research by Loomis and Walsh (1997). Second, determine the population in each zone. The population of each region is obtained from the US Census Bureau. Third, determine the number of anglers visiting Region 6, containing the Moose River in Fowlersville, New York from each DEC administrative regions, using the 1996 Angler Survey. These data are given in Table 12.2a.

The Region 6 Fisheries Unit in Watertown New York estimated that approximately 221 anglers visit the study area annually. Using the last column from Table 12.2a the number of anglers visiting Fowlersville was determined. Fourth, estimate the per capita participation rates from each zone. These data are given in Table 12.2b.

Constructing the demand curve is the last step. Due to the limited data, a numerical rather than a statistical approach was used. The travel cost per angler was given by equation (12.9):

$$Travel\ Cost\ (\$/angler) = \frac{0.31\ (\$/mile) \times [Distance\ (mile/vehicle) \times 2]}{1.6\ (anglers/vehicle)} \quad (12.9)$$

In 1999, the United States Internal Revenue Service used $0.31 per mile as a cost of traveling by cars. In addition, the Region 6 Fisheries Unit in Watertown New York estimated there are approximately 1.6 anglers per vehicle. Table 12.3 gives the travel cost per angler assume no change in travel costs.

The demand curve was developed using the information in Table 12.3 and four assumptions. First, the sole purpose of the travel was to go fishing. Fortunately for most anglers, single purpose and destination are the rule rather than the exception.

Table 12.2a Basic zonal travel costs data for anglers fishing in Region 6 containing the Moose River in Fowlersville, New York[‡]

DEC administrative region	Center of the DEC administrative region	Number of anglers	One-way travel (Miles)	Anglers visiting region 6[§] (%)
6	Utica	65,290	40	34.6
7	Syracuse	39,570	80	21.0
5	Saratoga Springs	4,700	120	2.5
4	Albany	5,880	120	3.1
8	Rochester	31,470	160	16.7
9	Buffalo	16,450	240	19.2
Other	New York City	5,220	280	2.8
Out of state		20,240	320	10.7
Total		188,820		100

[‡]These data are based on New York State Department of Environmental Conservation (DEC) 1996 Angler Survey (NYS DEC 1997). The estimated driving distance from the most populated city from each region to Fowlersville, New York were calculated by using the Yahoo's travel map web site (http://maps.yahoo.com) and rounded to the nearest 40-mile increment.

[§]The percent anglers visiting the ith Region is determined as the number anglers visiting the ith Region divided by the total number of anglers visiting Region 6; for example, $\frac{65,290}{188,820} \times 100 = 34.6$.

Table 12.2b The estimated number of anglers by region of residence for anglers fishing in on the Moose River in Fowlersville, New York[‡]

DEC region	Center of the DEC administrative region	Anglers visiting region 6 (%)	Zone population	Anglers visiting the study site	Participation rate per 100,000 population
6	Utica	34.6	566,346	76.4	13.49
7	Syracuse	21.0	1,201,502	46.3	3.85
5	Saratoga Springs	2.5	528,946	5.5	1.04
4	Albany	3.1	895,611	6.9	0.77
8	Rochester	16.7	1,331,908	36.8	2.77
9	Buffalo	19.2	1,508,394	19.3	1.28
Other	New York City	2.8	11,957,748	6.1	0.05
Out of state		10.7	31,590,846[§]	23.7	0.07
Total		100		221.0	

[‡]The New York State Department of Environmental Conservation (DEC) Region 6 Fisheries Unit in Watertown New York, estimated that approximately 221 anglers visit the Moose River in Fowlersville, New York annually.

[§]Out of state population includes parts of Connecticut, Massachusetts, New Hampshire, New Jersey, Pennsylvania, Rhode Island, and Vermont.

Table 12.3 Initial travel costs per angler for fishing on the Moose River in
Fowlersville, New York

DEC region	Center of the DEC administrative region	One-way travel (miles)	Travel cost per angler	Anglers visiting the study site	Participation rate per 100,000 population
6	Utica	40	15.50	76.4	13.49
7	Syracuse	80	31.00	46.3	3.85
5	Saratoga Springs	120	46.50	5.5	1.04
4	Albany	120	46.50	6.9	0.77
8	Rochester	160	62.00	36.8	2.77
9	Buffalo	240	93.00	19.3	1.28
Other	New York City	280	108.00	6.1	0.05
Out of state		320	124.00	23.7	0.07
Total				221.0	

Second, there were no benefits from the time spent traveling. Third, within each zone, anglers were assumed to face the same travel distance, prices, incomes, tastes, and substitute recreation opportunities. Therefore, within each zone anglers faced roughly the same travel cost, so that only travel costs by car would explain the difference in the decision of whether or not to visit the site. Consequently, per capita participation rates were assumed to be only a function of travel costs. Increasing the travel costs faced by each zone and multiplying the participation rate associated with the increased travel costs with total population in each zone, total number of visits to the site can be calculated. This is illustrated by Table 12.4.

Finally, the approach assumes the individuals respond in the same way to an entrance fee as they would to an increase in travel costs. The corresponding total quantities of visitors at varying travel costs (i.e. prices) define the demand curve and are given in Table 12.5 and shown graphically in Figure 12.6.

Given the zonal travel cost demand curve, the consumer surplus at each different price level can be estimated. This is given in the last column of Table 12.5 and shown graphically in Figure 12.7.

Stated preference or direct method

As was stated earlier, the core of the revealed preference techniques is that the analyst observes choices people have made but does not ask directly how much they value the resource. The core of the state preference techniques is simply to ask people directly how much they value the ecosystem good or service. There are two approaches that are used to elicit how much an individual values a resource directly based on a hypothetical situation: willingness to pay (WTP) and willingness to accept (WTA). I will first introduce the concepts of WTP and WTA and then describe contingent valuation – a common stated preference technique. I will

Table 12.4 Travel costs per angler assuming an increase in travel costs of $15.50 for anglers fishing on the Moose River in Fowlersville, New York

DEC region	Center of the DEC administrative region	One-way travel (miles)	Travel cost per angler	Anglers visiting the study site	Participation rate per 100,000 population
6	Utica	40	31.00	21.8	3.85
7	Syracuse	80	46.50	10.9	0.90[‡]
5	Saratoga Springs	120	62.00	14.6	2.77
4	Albany	120	62.00	24.8	2.77
8	Rochester	160	77.50[§]	26.9	2.02[§]
9	Buffalo	240	108.50	0.8	0.05
Other	New York City	280	124.00	9.0	0.07
Out of state		320	139.50	0.0	0.00
Total				108.7	

[‡]Regions 4 and 5 were defined within the same travel cost zone. Their participation rates from Table 12.3 were averaged; thus, the participation rate for Syracuse, NY given a $15.50 increase in travel costs was $0.90 = \dfrac{1.04 + 0.77}{2}$.

[§]A travel cost of $77.50 is not given in Table 12.3. However, $77.50 is the average of Regions 8 and 9s travel costs $77.50 = \dfrac{62 + 93}{2}$. The participation rate was calculated in a similar manner $2.02 = \dfrac{2.77 + 1.28}{2}$.

Table 12.5 Zonal travel cost demand schedule and consumer surplus for anglers fishing on the Moose River in Fowlersville, NY

Travel cost increase ($)	Number of anglers	Consumer surplus ($)
0.00	221.00	6,236.55
15.50	108.74	3,681.07
31.00	85.27	2,177.51
46.50	58.81	1,060.91
62.00	28.52	384.16
77.50	8.91	94.11
93.00	1.19	15.81
108.50	0.42	3.29
124.00	0.00	0.00

conclude with a description of a contingent valuation study concerning deer management in an urban area.

Willingness to pay versus willingness to accept

The basic format for WTP is to ask the individual how much they would be willing to pay to enter a wilderness area, to canoe, backpack, fish, hike, or have clean

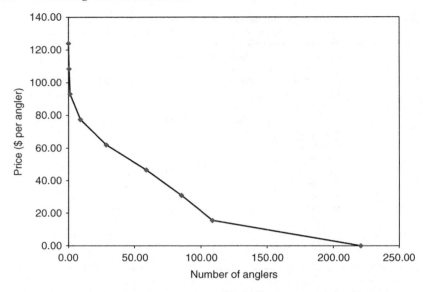

Figure 12.6 The zonal travel cost demand curve for anglers fishing in Fowlersville, NY.

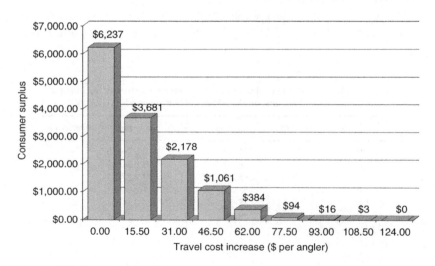

Figure 12.7 Consumer surplus estimated from the zonal travel cost demand curve.

water or air, etc.[18] There are three provisos associated with WTP. First, there is an implicit condition of no existing endowment.[19] In any market transactions, the buyer does not have the good or service, the seller does. Second, there is a potential for a gain. If the buyer obtains the good or service from the seller, then the buyer's net benefit increases as a result of the transactions. Third, WTP is constrained by the individual's income.

The basic format for WTA is to ask the individual how much they would be willing to accept in compensation for *not* being able to enter a wilderness area, to canoe, backpack, fish, hike, or have clean water or air, etc.[20] There are two provisos associated with WTA. First, there is an implicit condition of existing endowment. Basically, the individual is giving up something they have. Second, WTA is not constrained by the individual's income level.

Theoretically, WTA and WTP should converge; however, a review of WTA and WTP studies by Horowitz and McConnell (2002) and Venkatachalam (2004) concluded that WTA is generally greater than WTP. The debate as to why this occurs is wide ranging and there is no agreement as to the interpretation of this result.[21] This debate centers on the following areas: (1) Income effect as WTP is constrained by income while WTA is not. (2) Endowment effect as what exists is seen as a reference point (i.e. part of your endowment) and attitudes to surrendering some of what is already owned or experienced are quite different from those that come into play when there is the prospect of gain. Thus, the potential loss of something known or you have an existing property right to is greater than the potential gain of something unknown or you do not have an existing property right to. (3) Substitution effect as a lack of substitutes would make it difficult to compensate individuals for the removal of a good leading to large WTA values; the less the good is like an ordinary market good (i.e. the dichotomy between a public versus private good), the greater the difference.[22] (4) Transactions cost effect as WTP is not impacted by transactions costs while WTA is. (5) Experimental design effect as poorly designed surveys could lead to bias in WTP versus WTA estimations.[23]

Given these concerns, a legitimate question is which approach to chose: WTP or WTA would elicit assigned values that would correspond closest to an individual's preferences. To address this concern it is useful to review briefly the circumstance surrounding WTP and WTA assessments; in a word, the circumstance is the lack of markets. A market structure contains the economic information that individuals use to express their preferences by making observable choices. This structure is based on an individual willingness to pay (demand) for a good or service. Thus, to elicit assigned values it would seem most appropriate to develop a hypothetical structure that closely resembles a market and an individual's behavior within that market. That behavior is best described as willingness to pay. Consequently, WTP is generally preferred to WTA (Carson *et al.* 2003; Venkatachalam 2004).[24]

Contingent valuation model

The contingent valuation model (CVM) is often attributed to Ciriacy-Wantrup (1947) and the earliest use is attributed to Davis (1963) who used it to estimate WTP for big game hunting in Maine. CVM as a tool for nonmarket evaluation came to prominence with the grounding of the oil tanker Exxon Valdez in Prince William Sound of the northern part of the Gulf of Alaska on 24 March 1989. At the time, this was the largest oil spill from a tanker in United States history spilling over 11 million gallons of crude oil (http://www.epa.gov/oem/content/learning/exxon.htm accessed on 16 September 2010). As part of the damage settlement, not only was Exxon liable for the direct use values associated with Prince William Sound but also nonuse values in the form of existence values (Figure 12.2). These nonuse values were estimated by using CVM. Carson *et al.* (2003) describes the CVM analysis that was used to estimate these values.

The focus of using CVM is to have individuals state their values for a nonmarket good such as a forest-based ecosystem goods and services. Basically, the analyst is trying to get individuals to state their assigned value. An assigned value is the expressed relative importance or worth of an object to an individual in a given context (Brown 1984). Therefore, a CVM must define the context in which an individual will make an observable choice. The Architectural Plan for Profit in terms of the CVM model is given in Figure 12.8.

For example, the context will define a cause and effect relationship between different forest management actions and different levels of water quality (the Production System). The individual will be given a range of choices that describe how much it will cost or how much compensation will be given for the different management actions associated with the different levels of water quality (the Pillar of Cost). The choice made will define the Pillars of Price, and Value and the Profit. The dashed lines are used to represent contextual or situational nature in which individuals state their choices. That is, a CVM sets up a hypothetical market and an individual's assigned values are contingent upon the set up of this hypothetical market and the information provided. The tool used to elicit these choices is a survey.[25]

The preceding paragraph provides a broad outline of the steps of implementing a CVM. The first step is to define the problem and the context and identify the target audience. That is, what are the cause and effect relationships that individuals will be asked to state preferences about? Are these relationships described in

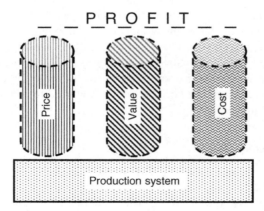

Figure 12.8 Contingent valuation model (stated revealed preference methodology) – the Architectural Plan for Profit.

an understandable manner and provide enough information so that respondents within the target audience can make reliable and valid choices?[26] The second step is to develop a survey instrument and sampling methodology. As WTP is constrained by income, respondents stated WTP in this hypothetical market would imply they have less income to spend on other goods. The survey instrument would also include relevant social, demographic, and economic questions to help describe the target audience. Face-to-face surveys, while very expensive, are preferred to telephone surveys which are preferred to the least expensive, mail surveys (Arrow *et al.* 1993). Sampling methodology should recognize the possibility of non-response and self-selection bias (i.e. the difference between a survey estimate and the actual population value due to a low response rate and a strong difference between respondents and non-respondents) and develop strategies to address this possibility. Developing an appropriate survey instrument and sampling methodology are not trivial exercises as the value measurements derived depend directly on the soundness of the survey instrument and sampling methodology. The third step is to administer the survey instrument based on the sampling methodology. The final step is to analyze the surveys using appropriate statistical techniques and report the results.

The CVM can provide measures of use and nonuse values (see Figure 12.2). This ability provides an advantage over revealed preference approaches if the values that are to be estimated include nonuse components. However, care should be taken when implementing a contingent valuation study. Poor experimental design and survey can introduce bias into the analysis. Venkatachalam (2004) lists some of these. First, hypothetical bias – as previously stated, CVM sets up a hypothetical market and an individual's assigned values are contingent upon the set up of this hypothetical market and the information provided. In a real situation, if wrong choices are made, individuals suffer real consequences. The use and nonuse values estimated are based on what individuals say they would do in a given context, not what they actually do. Consequently, the hypothetical bias can lead to, in some cases significant, overestimates of economic value (Arrow *et al.* 1993). Second, embedding or scope bias – different values for the same good depending on whether the good is valued on its own or valued as a part of a more inclusive package. Third, sequencing bias – the value for a particular good differs depending on the order of the good in a sequence in multi-good valuation studies. Fourth, information bias – the estimated values depend directly on the level and nature of information provided to the respondents. Fifth, elicitation bias – the approach (bidding, auctions, payment cards, or take-it-or-leave-it choices, etc.) to get individuals to state their preferences. Sixth, perception bias – the WTP for a potential gain only makes sense if the individual can observe that a relevant change will take place. Finally, strategic bias – individuals may underestimate their WTP (i.e. free riding, individuals expect others would pay enough such that the good would be provided) or overestimating their WTP (i.e. individuals think their stated WTP value would influence if the good is provided, but that their stated WTP would not be used in any future pricing policy). A variation is accounting for protest and zero bids (Halstead *et al.* 1992; Jorgensen and Syme 2000; Bowker *et al.* 2003; Meyerhoff and Liebe 2006). Protest could be positive bid outliers resulting from individuals who are opposed to the situation described in the survey for whatever reason. A zero bid could also be a protest bid or an individual who places no economic value on the good or service. These potential biases highlight the caution that should be used when developing an appropriate survey instrument and sampling and statistical analysis procedures.

Contingent valuation study – deer management

Wildlife–human interactions can produce positive and negative effects. Bird watching or viewing deer in an open field could be characterized as producing a positive net benefit. However, collisions between deer and cars or destruction created by deer browsing on gardens and other landscape plants could be characterized as a negative net benefit. A study conducted by Bowker *et al.* (2003) examined the willingness of residents of Sea Pines Plantation on Hilton Head Island, South Carolina, to pay to reduce the damage to landscaping by deer. The deer population was approximately one deer per hectare compared to an estimated carrying capacity of around one deer per five to ten hectares. A CVM study was conducted to estimate the residents' WTP to reduce damage caused by the deer by 25 or 50 percent using lethal means versus nonlethal contraceptive. This approach was used because while there is a correlation between deer number and damage, determining an exact relationship is difficult due to environmental factors beyond the control of the researchers. The lethal means had a lower cost and an estimated greater degree of confidence in the outcome.

A mail survey was sent to 100 percent of the residents of this community. The survey contained questions about residents' perceptions of present and future damage to their landscape caused by deer; general social, economic, and demographic backgrounds; and WTP for either the lethal or nonlethal means for decreasing damage caused by deer. Due to the emotions associated with the proposed control methods, accounting for a zero bid accurately was critical and the survey designed and sampling and statistical analysis procedures were devised to address this concern. Examples of the WTP question for the lethal and nonlethal means are:

> **Lethal means** – This program would involve the killing of deer by trained wildlife professionals. The program would be conducted during the winter months to maximize efficiency and safety and to minimize public conflict and inconvenience. All meat would be donated to food banks or other charitable organizations. Lethal removal is relatively cost efficient, and is an option currently available for deer on Sea Pines. Assume it might be possible to implement a lethal removal program that would remove enough deer each year to reduce by 25 percent and/or 50 percent your economic losses resulting from deer damage to landscape plantings in your yard. How much would you be willing to pay each year for this benefit? _____$/year.
> **Nonlethal means** – Scientists are currently developing this form of nonlethal control. If successfully developed and approved, this program would involve

the treatment of deer by trained wildlife professionals during late summer prior to the breeding season. Presently, contraception is experimental, relatively expensive to apply, and not currently available for management of deer on Sea Pines. However, assume that a contraceptive program could be developed, and that this program would treat enough deer each year to reduce by 25 percent and/or 50 percent your economic losses resulting from deer damage to landscape plantings in your yard. How much would you be willing to pay each year for this benefit? _____$/year.

The results of the study are given in Table 12.6.

Residents were willing to pay more for lethal versus nonlethal control means at both the 25 and 50 percent damage reduction levels; that is, residents were not willing to spend more for the nonlethal alternative. The lethal zero bidders were opposed to the lethal means and nonlethal zero bidders did not think the nonlethal alternative would be effective. Zero bidders comprised a large enough component that if put to a referendum either method would have been voted down.

Market versus nonmarket evaluation

Care should be taken when comparing estimates of value derived from nonmarket methods with those observed from market. This caution was pointed out by Rosenthal and Brown (1985) and Canham (1986) and is illustrated in Figures 12.9a and 12.9b.

Figure 12.9a illustrates a demand curve estimated for a recreational activity using a nonmarket method (e.g. Figure 12.6) and the consumer surplus derived from the demand curve (e.g. Figure 12.7).[27] Figure 12.9b illustrates the observed market transactions from a timber sale, for example. Comparing Figures 12.9a and 12.9b reveals two important observations.

Table 12.6 Results of the willingness to pay for lethal versus nonlethal control methods for deer in the community of Sea Pines Plantation on Hilton Head Island in South Carolina

Without zero bids	Average willingness to pay	
	Lethal control	Nonlethal control
25% damage reduction	$59.28	$39.05
50% damage reduction	$88.23	$52.02

With zero bids	Average willingness to pay	
	Lethal control	Nonlethal control
25% damage reduction	$43.64	$27.37
50% damage reduction	$56.34	$45.75

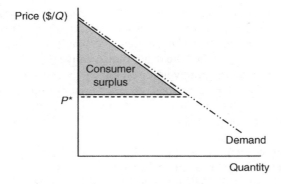

Figure 12.9a Estimated value (consumer surplus) based on a nonmarket valuation method.

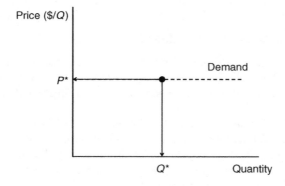

Figure 12.9b Observed market value.

First, the estimated market demand curve in Figure 12.9a has a negative slope and the demand curve faced by the entrepreneur of this recreational activity has a negative slope. The implication is that there are few substitutes for this recreational activity due to the possible uniqueness of the site and spatial monopoly due to distances between potential recreations sites (Rosenthal and Brown 1985; Canham 1986). The entrepreneur in this case has some market power with respect to pricing (Chapter 7). The observed market transaction from a timber sale (the intersection of market supply and demand) illustrated in Figure 12.9b is described by a point, not a line or curve. An incorrect interpretation of this observation would be that the market demand curve is identical to the price line and the market demand curve faced by the entrepreneur selling the timber would be perfectly elastic. That is, there are many substitutes, and the seller of the timber has no market power with respect to pricing (Chapter 7). This may not be the case, as the timber being sold may be a species that is relatively scarce and of high quality to create a spatial monopoly. If a market demand curve would have been developed, it would also have had a negative slope. Second, due to a negatively sloped demand curve shown in Figure 12.9a, estimates of total economic value include measures of consumer surplus. However, the measure of total economic value derived from Figure 12.9b includes no measure of consumer surplus. To make a reasonable comparison between the total economic value derived from a market versus nonmarket good or service, a consistent methodology should be used in both cases. Consequently, care should be taken when comparing measures of total economic value between market and nonmarket ecosystem goods and services.

The above example implies a choice of either timber or recreation. However, most often the decisions are not characterized as mutually exclusive or an all-or-nothing choice as Heyne *et al.* (2010) describe. Projects often produce multiple outputs in terms of market and nonmarket ecosystem goods and services. Decisions involve production levels for the different outputs. That is, should additional units of a market or nonmarket ecosystem good or service be produced or provided? This question describes an incremental or marginal analysis similar to the profit searching rule described in Chapter 5; namely, weighing the additional, incremental, or marginal benefits against the additional, incremental, or marginal costs. Various approaches have been described to examine these tradeoffs. However, I will describe two approaches briefly. Both focus on estimating the costs and if the decision is to add incremental units of the market or nonmarket ecosystem good or service, then the incremental benefits were expected to be greater than the incremental costs. These approaches have relied on simple to complex cost allocation procedures.

The first approach determines those cost that are specific to the provision of a given output, separable to the provision of a given output, and those that are common or joint with respect to the project. Specific costs are attributed to a single output; for example, the cost of recreational docks around a multipurpose reservoir. Separable costs are costs resulting from an output being included in a multipurpose project and are often described as the incremental cost of adding the output as the last output in a multipurpose project. The output's cost allocation comprises its specific and separable costs and a share of the common or joint costs. A variety of procedures have been used to allocate the common or joint costs to each output.[28] The classic example of this cost allocation approach is called separable cost–remaining benefits and was first described in 1958 by the Federal Inter-Agency River Basin Committee. The outputs allocated costs can then be weighed against the expected benefits. If the outputs benefit equals or exceeds (less then) the smallest incremental cost, the output is included (not included) in the project at its current production level. The literature on separable cost–remaining benefit is too vast to list here. I would recommend interested students to search peer-reviewed literature databases to find articles on this topic.

The second approach was described by Montgomery *et al.* (1994). These authors note explicitly that preserving an endangered species like the northern spotted owl is not an all-or-nothing decision. The appropriate unit of analysis for benefits and costs is the likelihood of survival and how certain society wants to be of species survival. Montgomery *et al.* develop a relationship between the probability of survival and the amount of habitat associated with that probability. As using this habitat–owl survival probability relationship, they develop a "marginal physical cost curve" relating estimated thousand board feet of timber from the potential habitat areas to a survival rate. While this marginal physical cost curve does not have the same dimensions as a traditional marginal cost curve, it allowed the authors to develop a marginal cost curve describing the opportunity cost of the lost timber harvest with respect to survival rate. This was done using the Timber Assessment Market Model developed by Adams and Haynes (1980). The resulting marginal cost curve allows decision makers to examine the incremental costs in terms of increasing the probability of survival by 1 percent with estimated incremental benefits. A detailed discussion of the methods they used to develop this marginal cost curve is beyond the scope of this book. However, I would encourage interested students to read Montgomery *et al.* (1994) and subsequent articles using this approach.

How to use economic information – *estimating nonmarket values* – to make better business decisions?

Markets provide economic information that entrepreneurs can use to determine what to produce, how much to produce, and for whom to produce. However, not all of the ecosystem goods and services individuals use are exchanged in markets. Consequently, there is no observable market equilibrium to determine the optimal level to produce, no observable demand curve to help distribute them among various potential users, and no observable pricing mechanism to help allocate the resources used to produce them. That said, individuals value these nonmarket ecosystem goods and services as can be seen by observing their choices. Given the lack of economic information generally available from a well-functioning market, the entrepreneur, landowner, or policy maker needs to develop economic information – the opportunity cost faced by an entrepreneur, landowner, or policy maker – analogous to that obtained from markets in order to make informed decisions concerning how much of these ecosystem goods and services to provide.

The two approaches to develop this economic information discussed were revealed and stated preference methods. What is the nature of the economic information generated from these methods? In a nutshell, the economic information generated is a measure of an individual's willingness to pay. If you are an entrepreneur, landowner, or policy maker, this basic economic information will help you make informed decisions concerning providing these ecosystem goods and service. The concept of willingness to pay is different depending on which method is chosen. Revealed preference methods estimate an individual's willingness to pay based on their existing choices. The individual is not asked directly their willingness to pay for an ecosystem good or service. Consequently, revealed preference methods require observable choice behavior and are used to estimate use values but not nonuse values as these are unobservable. The estimated value measures inferred from individual's choices using a hedonic pricing model are direct and indirect use values while the travel cost model can be used to estimate direct use values (Figure 12.2). The stated preference method examined was the contingent valuation model. This model is based on asking an individual directly their willingness to pay for an ecosystem good or service. Thus, it can be used to estimate use and nonuse values (Figure 12.2).

The nonmarket valuation methods discussed develop different measures of value. If the questions being asked by an entrepreneur, landowner, or policy maker involve only direct use values, then any one of the methods discussed is appropriate. If the questions being asked involve direct and indirect use values, then either the hedonic or contingent valuation model is appropriate. If the questions being asked involve nonuse values, then only the contingent valuation model is appropriate. Finally, caution should be used when comparing the measure of value estimated using one of these methods from those observed from well-functioning markets. Using economic information inappropriately – whether it is derived using a nonmarket valuation method or observed from a market directly – will probably lead to poor business decisions.

Appendix 1

Mathematical appendix

Marginal product (MP) is defined as the change in output per unit change of the identified input. In mathematical terms, MP defines a rate of change (or slope) at a specific point on the total product curve or production function. The slope of a point on a curve is calculated by taking the first derivative of the function with respect to the identified input. Let $Q(x_1, x_2,..., x_j)$ or $Q(\bullet)$ denotes the system of producing or providing a good or service, Q, using inputs that the manager has direct control over, $x_1, x_2,..., x_j$. MP for the ith input is define as

$$MP = \frac{\partial Q(x_1, x_2,..., x_j)}{\partial x_i} = \frac{\partial Q(\bullet)}{\partial x_i}; \quad i = 1, 2, 3,..., j \tag{A1.1}$$

Mathematically, MP can only be calculated for production processes that are continuous; for example, the Select Red Oak and Loblolly Pine Plantation case studies. MP is calculated for a range of outputs, not for just the last unit produced as shown in Tables 2.7 through 2.9.

Figures 2.10 and 2.11 show that when MP is greater than Average Product (AP), AP is increasing and when MP is less than MP, AP is decreasing. Finally, these figures show that when MP equals AP, AP is at its maximum. These relationships can be illustrated mathematically by using the definition of AP and its first derivative with respect to an identified input. Equation (A1.2) defines AP

$$AP = \frac{Q(x_1, x_2, x_3,..., x_j)}{x_i} = \frac{Q(\bullet)}{x_i}; \quad i = 1, 2, 3,..., j \tag{A1.2}$$

Taking the first derivative of AP with respect to x_i using the quotient rule and simplifying gives:

$$\frac{\partial AP}{\partial x_i} = \frac{1}{x_i} \frac{\partial Q(\bullet)}{\partial x_i} - \frac{Q(\bullet)}{x_i^2}$$

$$= \frac{1}{x_i}\left(\frac{\partial Q(\bullet)}{\partial x_i} - \frac{Q(\bullet)}{x_i} \right)$$

$$= \frac{1}{x_i}(MP - AP) \tag{A1.3}$$

Analyzing the last expression of equation (A1.3) shows that if $MP > AP$ then $\partial AP/\partial x_i > 0$ or AP is increasing. If $MP < AP$ then $AP/\partial x_i < 0$ or AP is decreasing. Finally if $MP = AP$ then $\partial AP/\partial x_i = 0$ and AP is at its maximum.

Appendix 2

Technical efficiency versus production cost efficiency

Technical efficiency, as described in Chapter 2, are combinations of inputs and outputs such that it is not possible to increase output without increasing input or when the most amount of output is produced with the least amount of input. Therefore, all the descriptors of the production system; i.e. total product, average product, and marginal product, must reflect technical efficiency. Thus, every point on a graph of Total Product for which Marginal Product is greater than or equal to zero must be technically efficient. The least cost model, equation (A2.1), includes the production system of producing or providing Q as part of the constraint:

$$Min\ TVC = \sum_j w_j \cdot x_j$$

s.t.

$$Q(x_1, x_2,..., x_j) = Q^0 \tag{A2.1}$$

By definition any descriptors of the production system, $Q(x_1, x_2,..., x_j) = Q(\bullet) = Q^0$, must be technically efficient.

Equation (A2.1) can be formulated and solved using a Lagrangian equation:[1]

$$Min\ L = \sum_j w_j \cdot x_j - \lambda[Q(\bullet) - Q^0] \tag{A2.2}$$

where λ is the Lagrangian multiplier. The first order conditions require that:

$$\frac{\partial L}{\partial x_1} = w_1 - \lambda \frac{\partial Q(\bullet)}{\partial x_1} = 0$$

$$\frac{\partial L}{\partial x_2} = w_2 - \lambda \frac{\partial Q(\bullet)}{\partial x_2} = 0$$

$$\vdots \qquad \vdots \qquad \vdots$$

$$\frac{\partial L}{\partial x_j} = w_j - \lambda \frac{\partial Q(\bullet)}{\partial x_j} = 0$$

$$\frac{\partial L}{\partial \lambda} = -[Q(x_1, x_2,..., x_j) - Q^0] = 0 \tag{A2.3}$$

I will assume that the production function, $Q(\bullet)$, is twice differentiable with $\partial^2 Q(\bullet)/\partial x_j^2 < 0$ for all j (in other words the production function is consistent with the Law of Diminishing Returns) and the Karush (1939) and Kuhn and Tucker (1951) conditions are satisfied for optimal values of x_j; $j = 1, 2,\dots$. Solving the first j equations for λ gives:

$$\frac{\partial Q(\bullet)/\partial x_1}{w_1} = \frac{\partial Q(\bullet)/\partial x_2}{w_2} = \dots = \frac{\partial Q(\bullet)/\partial x_j}{w_j}$$

$$\frac{MP_1}{w_1} = \frac{MP_2}{w_2} = \dots = \frac{MP_j}{w_j}$$

(A2.4)

To illustrate the economic interpretation of (A2.4), I will use a production system with only two inputs, x_1 and x_2, and one output, Q. Figure A2.1a illustrates graphically the relationship between the technically efficient levels of the two inputs required to produce two different levels of output, Q^0 and Q^1.

These curved lines are called isoquants or lines of equal production levels. Given this production system, equation (A2.4) reduces to

$$\frac{\partial Q(\bullet)/\partial x_1}{w_1} = \frac{\partial Q(\bullet)/\partial x_2}{w_2}$$

$$\frac{MP_1}{w_1} = \frac{MP_2}{w_2}$$

$$\frac{MP_2}{MP_1} = \frac{w_2}{w_1}$$

(A2.5)

The left-hand side of the last expression in equation (A2.5), MP_2/MP_1, defines the marginal rate of technical substitution between inputs x_1 and x_2 in producing a given

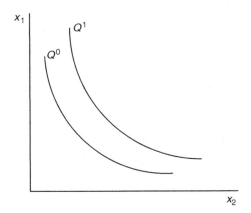

Figure A2.1a Isoquants.

level of output, Q^0; for example, the substitution of labor for capital. The slope of an isoquant at any given point is defined as MP_2/MP_1 (Silberberg and Suen 2001; Chiang 1984; Henderson and Quandt 1980), and thus defines the marginal rate of technical substitution between inputs x_1 and x_2 for a given level of output, Q^0.

The right-hand side of the same expression, w_2/w_1, defines the slope of the objective function given in equation (A2.1)

$$TVC = w_1x_1 + w_xx_2 \qquad (A2.6)$$

Solving equation (A2.6) for x_1 gives

$$x_1 = \frac{TVC}{w_1} - \frac{w_2}{w_1}x_2 \qquad (A2.7)$$

If all the variable costs are used to purchase input x_1 then from equation (A2.6), the total amount of x_1 purchased is TVC/w_1. Similarly, if all the variable costs are used to purchase input x_2 then from equation (A2.6), the total amount of x_2 purchased is TVC/w_2. Given the prices of the inputs, x_1 and x_2, point A defines the least cost way of producing output level Q^0.

Both points A and B in Figure A2.1b are technically efficient. At input prices of w_1 and w_2, point A is production cost efficient and technically efficient while point B is only technically efficient. As the price of x_1 increases relative to the price of x_2, less x_1 and more x_2 will be used to produce Q^0. Given the change in prices of the inputs, point B is now production cost efficient and technically efficient while point A is only technically efficient. Technical efficiency requires that the input–output combinations be on the isoquant. Production cost efficiency requires technical efficiency (input–output combinations on the isoquant) and the level of inputs used depends on the prices paid for the inputs.

The relationship between technical efficiency, production cost efficiency, and economic efficiency is described in Chapter 5.

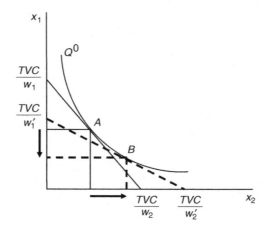

Figure A2.1b Production cost efficiency.

Appendix 3

Average and marginal cost

Marginal cost (MC) is interpreted as the incremental variable production cost per unit of output. In mathematical terms, MC defines a rate of change (or slope) at a specific point on the TC or TVC curves. The slope of a point on a curve is calculated by taking the first derivative of the function with respect to the identified input. As MC is defined with respect to TVC (as TFCs are irrelevant) and output, let $C[Q(x_1, x_2, \ldots, x_j)]$ or $C(\bullet)$ denote the variable cost function. Equation (A3.1) defines MC:

$$MC = \frac{dC[Q(x_1, x_2, \ldots, x_j)]}{dQ} = \frac{dC(\bullet)}{dQ} \tag{A3.1}$$

Equation (A3.1) can only be calculated for variable cost functions that are continuous.

Average variable cost measures the variable production costs per unit of output. The mathematical definition of AVC is

$$AVC = \frac{C(\bullet)}{Q} \tag{A3.2}$$

The mathematical relationship between MC and AVC (or TC) can be illustrated by using the definition of AVC and its first derivative with respect to output. Taking the first derivative of AVC with respect to output using the quotient rules and simplifying gives:

$$\begin{aligned}
\frac{dAVC}{dQ} &= \frac{Q\left(\frac{dC(\bullet)}{dQ}\right) - C(\bullet)}{Q^2} \\
&= \frac{1}{Q}\left(\frac{dC(\bullet)}{dQ} - \frac{C(\bullet)}{Q}\right) \\
&= \frac{1}{Q}(MC - AVC)
\end{aligned} \tag{A3.3}$$

Analyzing the last expression of equation (2.15) shows that if $MC < AVC$ then $dAVC(\bullet)/dQ < 0$ or AVC (and ATC) is decreasing. If $MC > AVC$ then

$dAVC(\cdot)/dQ > 0$ or AVC (and ATC) is increasing. Finally if $MC = AVC$ then $dAVC(\cdot)/dQ = 0$ and AVC (and ATC) is at its minimum.

Equation (A3.3) also identifies two additional important ideas. First, the concepts of average cost (ATC and AVC) and marginal cost are tied to the production system directly. This relationship is based on the definitions of average and marginal cost as illustrated mathematically as $C[Q(x_1, x_2,\ldots, x_j)]$ or $C(\cdot)$ in equations (A3.1), (A3.2), and (A3.3). Thus it is imperative that you as a manager must answer the questions "What are the inputs?", "What are the outputs?", and "How do you define the production process?" to the best of your ability if you want to control your costs. Second, the shapes of the average and marginal cost curves mirror those of AP and MP and equation (A3.3) mirrors equation (A1.3) of Appendix 1. Again this is due to the direct tie between average and marginal costs and the production system.

Appendix 4
Profit and least cost models

Let equation (A4.1) define the profit model of a production system that uses only one input and produces only one output:

$$\Pi = P \cdot Q(x) - wx - TFC \tag{A4.1}$$

where the notation used is defined by equation (5.1). The first order conditions for maximizing equation (A4.1) with respect to input is given in equation (A4.2a)

$$\frac{d\Pi}{dx} = P\frac{dQ(x)}{dx} - w = 0 \tag{A4.2a}$$

given the production system is twice differentiable with $\frac{dQ(x)}{dx} > 0$ and $\frac{d^2Q(x)}{dx^2} < 0$. Equation (A4.2a) can be rewritten as:

$$P\frac{dQ(x)}{dx} = w \tag{A4.2b}$$

The left-hand side of equation (A4.2b) is $P \cdot MP$ or marginal revenue product (MRP; see Chapter 4) and the right-hand side is the marginal factor cost or wage paid for the input. The optimal amount of the input, x^*, is such that the equalities of equations (A.2a) and (A.2b) hold and $Q(x^*)$ defines the optimal amount of output.

Equation (5.2) can be formulated and solved using a Lagrangian equation:[1]

$$Min\ L = wx - \lambda(Q(x) - Q(x^*)) \tag{A4.3}$$

where λ is the Lagrangian multiplier and $Q(x^*)$ identifies the given level of production. The first order conditions require that:

$$\frac{\partial L}{\partial x} = w - \lambda\frac{\partial Q(x)}{\partial x} = 0 \tag{A4.4}$$

$$\frac{\partial L}{\partial \lambda} = Q(x) - Q(x^*) = 0 \tag{A4.5}$$

the production system is twice differentiable as above and the Karush (1939) and Kuhn and Tucker (1951) conditions are satisfied for optimal value of x. Equation

(A4.4) can be rewritten as:

$$\lambda \frac{\partial Q(x)}{\partial x} = w \tag{A4.6a}$$

where λ is the shadow price for the value of the marginal product. Equation (A4.5) requires that $Q(x) = Q(x^*)$. Inserting this requirement into equation (A4.6a) gives

$$\lambda \frac{\partial Q(x^*)}{\partial x^*} = w \tag{A4.6b}$$

In this case, the value of the marginal product, λ, is what you could sell the additional product for in the market. Thus, the shadow price is the output price and equation (A4.6b) is the same as equation (A4.2b). Therefore, if profits are maximized, costs must be minimized.

As equations (5.2) and (5.3) define a primal-dual relationship, the result described above also holds for equation (5.3).

Appendix 5
Calculus of profit maximization

Output approach

Let equation (A5.1) define the profit model:

$$\Pi = P \cdot Q(\bullet) - \sum_j w_j x_j - TFC \tag{A5.1}$$

where the notation used is defined by equation (5.1). In Chapter 3, a relationship was developed between the total variable cost, $\sum_j w_j x_j$, and the various levels of output from the production system, $Q(\bullet)$. Based on this relationship, an implicit functional relationship defining total variable costs could be expressed as $\sum_j w_j x_j = C(Q(\bullet))$. The first order conditions for maximizing equation (A5.1) with respect to output is given in equation (A5.2a):

$$\frac{\partial \Pi}{\partial Q(\bullet)} = P \frac{\partial Q(\bullet)}{\partial Q(\bullet)} - \frac{\partial C(Q(\bullet))}{\partial Q(\bullet)}$$

$$= P \frac{\partial Q(\bullet)}{\partial Q(\bullet)} - \frac{\partial \left(\sum_j w_j x_j \right)}{\partial Q(\bullet)} = 0 \tag{A5.2a}$$

The second order condition for a maximum would require that $\partial^2 \Pi / \partial Q(\bullet)^2 < 0$ or the profit function is concave. Equation (A5.2a) can be rewritten as

$$P = \frac{\partial \left(\sum_j w_j x_j \right)}{\partial Q(\bullet)} \tag{A5.2b}$$

The left-hand side of equation (A5.2b) is marginal revenue (MR) or price, P (see Chapter 4), and the right-hand side is marginal cost (MC, see Chapter 3). The optimal amount of the output, $Q^*(\bullet)$, is such that the equalities of equations (A5.2a) and (A5.2b) hold. Equations (A5.2a) and (A5.2b) also show the profit maximizing searching rule is derived directly from the profit maximization.

Input approach

Again, let equation (A5.1) define the profit model:

$$\Pi = P \cdot Q(\bullet) - \sum_j w_j x_j - TFC \tag{A5.1}$$

The first order conditions for maximizing equation (A5.1) with respect to the inputs are given in equation (A5.3):

$$\frac{\partial \Pi}{\partial x_j} = P \frac{\partial Q(\bullet)}{\partial x_j} - \frac{\partial(w_j x_j)}{\partial x_j} = 0; \, \forall j \tag{A5.3}$$

where $\forall j$ denotes for all j. The second order conditions for maximizing equation (A5.1) are given in equations (A5.4a) and (A5.4b):

$$\frac{\partial^2 \Pi}{\partial x_j^2} < 0; \, \forall j \tag{A5.4a}$$

$$\frac{\partial^2 \Pi}{\partial x_j^2} \cdot \frac{\partial^2 \Pi}{\partial x_k^2} - \frac{\partial^2 \Pi}{\partial x_j \partial x_k} > 0; \, \forall j \neq k \tag{A5.4b}$$

Due to the fact that $\partial^2 \Pi / \partial x_j^2 = P \cdot \partial^2 Q(\bullet) / \partial x_j^2$, Silberberg and Suen (2001) show that equations (A5.4a) and (A5.4b) can be written as:

$$\frac{\partial^2 Q(\bullet)}{\partial x_j^2} < 0; \, \forall j \tag{A5.4c}$$

$$\frac{\partial^2 Q(\bullet)}{\partial x_j^2} \cdot \frac{\partial^2 Q(\bullet)}{\partial x_k^2} - \frac{\partial^2 Q(\bullet)}{\partial x_j \partial x_k} > 0; \, \forall j \neq k \tag{A5.4d}$$

Equations (A5.4a) and (A5.4c) define the Law of Diminishing Returns (see Chapter 2). Equations (A5.4b) and (A5.4d) state that changing the amount used of the jth input will affect its own marginal product and the marginal products of all the other inputs. The overall impact must still be similar to the Law of Diminishing Returns (Silberberg and Suen 2001).

The term $P \partial Q(\bullet)/\partial x_j$ in equation (A5.3) is $P \cdot MP$ or marginal revenue product (MRP; see Chapter 4) of the jth input. The second term in equation (A5.3) defines the derivative of total factor cost of the jth input, $w_j x_j$, with respect to the jth input. This term, $\partial(w_j x_j)/\partial x_j$, defines the marginal factor cost (MFC) of the jth input or its wage, w_j. Equation (A5.3) can be rewritten as:

$$P \frac{\partial Q(\bullet)}{\partial x_j} = \frac{\partial(w_j x_j)}{\partial x_j} = w_j \frac{\partial x_j}{\partial x_j} = w_j; \, \forall j \tag{A5.5}$$

The interpretation of equation (A5.5) parallels equation (A5.2b). First, equation (A5.5) states that the optimal amount of the jth input is defined when it is $MRP_j = MFC_j = w_j$. This condition must hold for all inputs simultaneously. Thus, the profit searching rule with respect to input is similar to that of output: weighing some form of additional or incremental revenue with some form of additional or incremental cost. Second, marginal revenue equals price given the assumption of workable competition in the output market. If the same assumption is made with respect to the input market, the parallel would be that marginal factor cost of the jth input equals its wage or $MFC_j = w_j$. What the profit searching rule described in equation (A5.5) tells me is that for profit to increase the added revenue from the sale of the additional output produced using the jth input (MRP_j) must more than cover the jth input's marginal factor cost (MFC_j) or the price (wage) that I have to pay for the additional input, w_j. This is illustrated by Figures A5.1a and A5.1b.

The reader will notice the parallels between Figures A5.1a and A5.1b and Figures 5.5a and 5.5b.

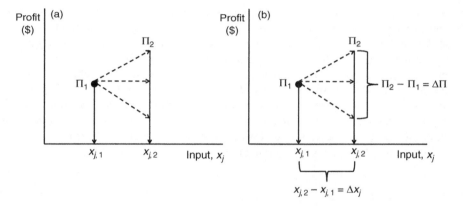

Figure A5.1 (a, b) Profit searching.

Appendix 6
Price searching

Price searching given market power

Reiterating, the goal of your pricing strategy is to (1) set the price to cover explicit and implicit measures of total costs; (2) reflect your consumers' willingness and ability to pay; and (3) recognize any competition. Examining the discussed pricing strategies relative to these three goals shows that they basically met the first but not the second and third goals. In Chapter 6, the concept of elasticity was introduced to examine the sensitivity of consumers' reactions to changes in prices and income. Focusing on own-price elasticity of demand will allow developing of a pricing strategy that will include all three goals but at the cost of additional information; namely, determining own-price demand elasticity (Table 6.7).

Equation (A6.1) defines the pricing rule:

$$P = MC \cdot \left[\frac{1}{1 + \left(\frac{1}{E_P} \right)} \right] \tag{A6.1}$$

where P denotes price, MC denotes marginal cost, and E_p denotes own-price elasticity of demand. This pricing rule is derived from the definition of marginal revenue (MR) in Note 8, the definition of own-price elasticity (Table 6.7), and profit searching rule of $MC = MR$. Demand curves have a negative or zero slope, thus own-price elasticity is negative or $E_p \leq 0$. By definition: (1) if $0 < |Ep| < 1$ then $MR < 0$ and the demand and MR curves are inelastic; (2) if $|Ep| = 1$ then $MR = 0$; (3) if $|Ep| > 1$ then $MR > 0$ and the demand and MR curves are elastic; and (4) if $|Ep|$ = infinity then the demand curve is perfectly elastic, the market structure is defined as workable competition and $MR = P$.

If the entrepreneur has market power in their commodity market, then the demand curve has a negative slope and $E_p < 0$. Given the definition of own-price elasticity from Table 6.7, the entrepreneur can change price and observe the amounts their consumers purchase. This would allow them to, at the minimum, calculate an arc own-price elasticity. If the entrepreneur determines that the arc own-price elasticity is $0 < |Ep| < 1$ this implies that $MR < 0$. The entrepreneur would be able to increase total revenue by decreasing the amount of the commodity sold by increasing price. An entrepreneur with market power in the commodity market would search for the market price given by equation (A6.1) that satisfies the profit searching rule such that $MR = MC$ with $MR > 0$. For the entrepreneur with commodity market power, the economic information in these results is that: (1) their profit maximizing search is focused on the portion of demand curve that is elastic; and (2) the demand curve is the economic model of the opportunity cost of the consumer and their sensitivity to own-price changes. This economic information is summarized by the market solution described in Figure 7.6a.

Profit margins

Although profit margins are not considered a pricing strategy, per se, they can be used as an indicator of pricing policy and a firm's ability to control costs. A profit margin is defined as the after-tax profits per dollar of revenue. They are measured as a percentage (i.e. 0 < profit margin < 1). Equation (A6.2) defines a profit margin mathematically

$$\text{Profit Margin} = \frac{\text{After tax Profit}}{\text{Total Revenue}} = \frac{\text{After tax Profit}}{P \cdot Q} \qquad (\text{A6.2})$$

where P denotes price and Q denotes the amount of output sold. For example, a profit margin of 30 percent means that for every $1.00 of total revenue $0.30 of after tax profits are generated. Consequently, a low profit margin indicates a low margin of safety in terms of controlling cost and a higher risk that a decline in sales will erase profits and result in a loss.

Appendix 7
Financial formulae

The following are common financial formulae. Table A7.1 gives the notation that are used in the formulae.

Future value of a terminating periodic series

$$V_t = a \cdot \frac{(1+r)^t - 1}{(1+r)^w - 1}$$

Future value of a terminating every-period series

$$V_t = a \cdot \frac{(1+r)^t - 1}{r}$$

Sinking fund formula

$$a = V_t \cdot \frac{r}{(1+r)^t - 1}$$

Present value of a terminating periodic series

$$V_0 = a \cdot \frac{(1+r)^t - 1}{[(1+r)^w - 1] \cdot (1+r)^t}$$

Table A7.1 Financial formulae notation

Variable	Description
V_0	The value of a cost or revenue in time 0
V_t	The value of a cost or revenue in time t
r	The annual interest rate expressed as a decimal
t	The number of annual periods over which interest is charged
a	An equal periodic payment
w	The years between periodic payments

Present value of a terminating every-period series

$$V_0 = a \cdot \frac{(1+r)^t - 1}{r \cdot (1+r)^t}$$

Annual installment payment

$$a = V_0 \cdot \frac{r \cdot (1+r)^t}{(1+r)^t - 1}$$

Present value of a perpetual periodic series

$$V_0 = \frac{a}{(1+r)^w - 1} = \frac{a \cdot (1+r)^{-w}}{1 - (1+r)^{-w}}$$

Present value of a perpetual every-period series

$$V_0 = \frac{a}{r}$$

Appendix 8

Sustainability and the interest rate

The weight the interest rate places on determining the present value of any cash flow is illustrated by this simple example in Table A8.1.

As can be seen in Figure A8.1, the weight placed on any revenue or cost in the future decreases exponentially. One of the tenets of sustainability is that the use of resources should not have an impact on future generations. However, based on Table A8.1, the weight placed on any future revenue or cost is very small. This is troublesome when the future is when most of the revenues may occur, which brings to mind the old saying that the future is worth nothing. The implication of Table A8.1 is that the incentive is to use resources in the present going against the tenet of sustainability just described. This is what Chichilnisky (1997) termed the Dictatorship of the Present. Discussions surrounding the choice of the interest rate and its impact on sustainable forest management are not new and draw on the debate about social versus private interest rates (e.g. Mikesell 1977; Baumol 1968; Feldstein 1964a and 1964b). Harou (1985) discusses a methodology to develop a social interest rate which would be less than the private interest rate reducing the Dictatorship of the Present. Kant (2003) continues this discussion, arguing against using an interest rate obtained from the capital markets for valuing the flow of forest-based ecosystem goods and service that are not traditionally traded in markets.

An alternative proposed to the discounting factor $(1 + r)^{-t}$ is called hyperbolic discounting (Hepburn and Koundouri 2007; Gowdy and Erickson 2005; Price 2004; Pearce *et al.* 2003; Casalmir 2000; Li and Löfgren 2000; Cropper and Laibson 1999; Chichilnisky 1997; Laibson 1997; Munasinghe 1993). This concept is drawn from the Weber–Fechner Law establishing that human responses to a change in a stimulus are nonlinear and are inversely proportional to the existing level of the stimulus. The notion of a hyperbolic discount rate is that a "higher value is placed on benefits delivered in the near term, followed by a sharp drop and flattening out in the medium term, so that the value of something stays fairly constant out into the distant future" (Gowdy and Erickson 2005: 214–215).

Equation (A8.1) defines $r(t)$ as the non-constant discount rate at time t:

$$r(t) = \frac{\dfrac{d\Delta(t)}{dt}}{\Delta(t)} = \frac{\dot{\Delta}(t)}{\Delta(t)} = \frac{\kappa}{t} \qquad (A8.1)$$

where κ is a negative constant. The hyperbolic discounting factor, $\Delta(t)$, is given by equation (A8.2):

$$\Delta(t) = e^{\kappa \ln t} = t^{\kappa} \tag{A8.2}$$

where e denotes the exponential function and ln denotes the natural log. The relationship between a hyperbolic discounting factor and a continuous time discounting factor e^{-it} are as follows.[1] The non-constant discount rate, $r(t)$, and the non-hyperbolic discounting factor approach zero as $t \to \infty$. If $\kappa < 0$ and constant, then $r(t) < 0$. Nevertheless, $\kappa < 0$ implies that $0 < \Delta(t) = e^{\kappa \ln t} \leq 1, \forall t \geq 1$. This is

Table A8.1 Interest rate example, $r = 6\%$

Year	$(1 + r)^{-t}$
5	0.74726
10	0.55839
15	0.41727
20	0.31180
25	0.23300
30	0.17411

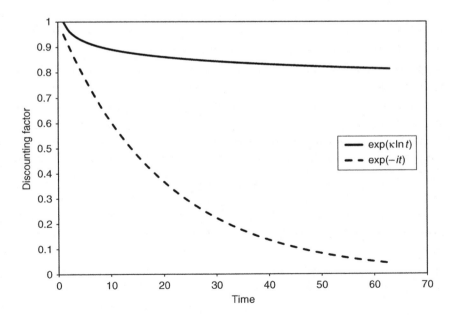

Figure A8.1 Hyperbolic discounting factor and continuous time discounting factor.[‡]

[‡] $\exp(\kappa \cdot \ln t) = e^{(\kappa \cdot \ln t)}$ denotes the hyperbolic discounting factor and $\exp(-it) = e^{(-it)}$ denotes the constant time discounting factor with $|\kappa| = i = 0.05$.

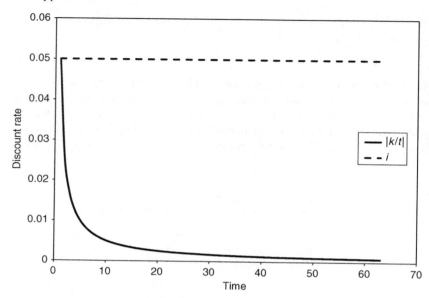

Figure A8.2 Non-constant discount rate and an interest rate.[‡]

[‡] $\left|\dfrac{\kappa}{t}\right|$ denotes the non-constant discount rate and i denotes the interest rate with $|\kappa| = i = 0.05$.

consistent with $0 < x(t) = e^{-it} \leq 1$, $\forall t \geq 0$ where i denotes a market interest rate (e.g., $0 < i < 1$). In addition, given the definition of $r(t)$ and that $\left[\dfrac{dx(t)}{dt}\right] \cdot \left[\dfrac{1}{x(t)}\right] = -i, \left|\dfrac{\kappa}{t}\right|$ would imply a non-constant time preference concept, at the minimum. If $t \geq 1$, $\kappa < 0$, and $|\kappa| = i$, then $0 < e^{-it} < e^{\kappa \ln t} \leq 1$.[2] This is illustrated by Figure A8.1.

With a non-constant hyperbolic discount rate, $\left|\dfrac{\kappa}{t}\right|$, the preferences between periods (e.g. t and $t+1$) are dynamically inconsistent (Laibson 1997). With a capital market determined interest rate, i, the preferences between periods (e.g. t and $t+1$) are dynamically consistent and equal to the discount rate. This is illustrated by Figure A8.2.

As illustrated by Figures A8.1 and A8.2, hyperbolic discounting does address the problem of the future being worth nothing given the discounting factor $(1 + r)^{-t}$. However, hyperbolic discounting factors are not accepted as the rule in financial markets, so they can be described as an academic exercise.

Appendix 9
Misinterpreting the internal rate of return in forest management planning and economic analysis

John E. Wagner

SUNY – College of Environmental Sciences and Forestry, Syracuse, NY

Sustainable forest management planning includes accounting for revenues and costs that accrue throughout time. While debate continues as to the how to account for these cash flows, the most used techniques are Net Present Value, Benefit Cost Ratios, and Internal Rate of Return (irr). Managing forests sustainably depends critically on interpreting the results and management implications of these techniques accurately. It is appealing to equate the irr with a market-derived rate of return given its definition and accept/reject criterion. Unfortunately, its mathematical derivation does not support this interpretation and past use of irr often illustrates this confusion and misinterpretation. The irr only reflects the amount and timing of the net cash flows for a given venture and does not include any social, economic, or other external factors found in market-derived discount rates. Therefore the irr does not reflect an appropriate rate of return or opportunity cost of capital for sustainable forest management. My purpose is to provide a theoretical argument that can be used to help correct this misinterpretation and stimulate discussions on the economics of sustainable forest management.

KEY WORDS: Internal Rate of Return, Sustainable Forest Management

Introduction

There is general agreement that sustainable forest management includes positive weights given to the concepts of ecological, social, and economic sustainability (Floyd, 2002). Economic sustainability often deals with accounting for revenues and costs that accrue throughout time. While debate continues as to the appropriate methods to account for these cash flows, the Internal Rate of Return (*irr*) – as well as other economic analysis tools, e.g., Net Present Value (NPV) and Benefit-Cost Ratios (BCR) – has been used in forest products analyses and forest management planning for many years. The following is not meant to be an all encompassing review of *irr* as used in forestry but to illustrate the history of its use: Bentley and Teeguarden (1965), Webster (1965), Gansner and Larson (1969), Marty (1970), Samuelson (1976), Fortson and Field (1979), Schallau and Wirth (1980), Kurtz et al. (1981), Mills and Dixon (1982), Clutter et al. (1983), Dennis (1983),

Bengston (1984), Harpole (1984), Leuschner (1984), Harou (1985), Johansson and Löfgren (1985), Baily (1986), Hansen (1986), Gregory (1987), Shaffer et al. (1987), Hu and Burns (1988), Pearce (1990), Uys (1990), Dangerfield and Edwards (1991), Fight et al. (1993), Hatcher et al. (1993), Buongiorno et al. (1995), Klemperer (1996), Hseu and Buongiorno (1997), Mehta and Leuschner (1997), Terreaux and Peyron (1997), Kadam et al. (2000), Brukas et al. (2001), Davis et al. (2001) [and earlier editions], Rideout and Hesseln (2001), Battaglia et al. (2002), Buongiorno and Gilles (2003), Duerr (2003), van Gardingen et al. (2003), Jagger and Pender (2003), Zinkhan and Cubbage (2003), Siry et al. (2004), Venn (2005), Fox et al. (2007), Legault et al. (2007), Heiligmann (2008), Bettinger et al. (2009), and Webster et al. (2009).

The unique feature of the *irr* as compared to NPV or BCR is that while the inputs of NPV and BCR are: 1) the amount and timing of all cash flows and 2) an appropriate opportunity cost of capital given as a discount or interest rate, and the result is a number that represent the NPV or the BCR; the input of *irr* consists only of the amount and timing of all cash flows and the result is the "internal rate of return" that sets the present value of the net cash flow equal to zero. The *irr*, as with the NPV or BCR, is used in project evaluation (Hartman and Schafrick, 2004; Copeland et al., 2005; Brealey et al., 2008). The general rule is if the *irr* is greater than (less than) an appropriate opportunity cost of capital the investment is acceptable (not acceptable).[1] As Hazen (2003) describes, and is evident by the literature review, the appeal of using the *irr* is interpreting it as a rate of return. Unfortunately, representing the *irr* as a rate of return similar to the interest rate used to calculate NPV or BCR is, as the title of the article states, misinterpreting the *irr*. The choice of a discount rate in sustainable forest management is important as capital is often invested for long periods of time (Bullard et al., 2002; Hepburn and Koundouri, 2007). The level of sustainable forest management is tied to the discount rate; for example, the direct impacts on rotation age calculations (e.g., Johansson and Löfgren, 1985). Indirect impacts could include the social–political implications of dramatic changes in harvest levels and resulting changes in forest structure (Brukas et al., 2001; Sandelescu et al., 2007). In addition, the indirect impacts on the provisions of forest-based ecosystem goods and services other than timber (Kant 2003). This misinterpretation appears to be common among practitioners (e.g., Heiligmann, 2008; Webster et al., 2009; Wagner, 2009; and Pickens et al., 2009) and maybe inadvertently advanced by academicians.

The purpose of this article is to provide a theoretical basis that can be used in helping correct this enduring misinterpretation and stimulate discussions among all those involved in sustainable forest management (e.g., academics, practitioners, and stakeholders). Addressing the issues surrounding the use of the *irr* as an

---

1 While no investment analysis criterion is criticism free, the following authors discuss the caveats of using *irr* as an investment criterion especially when choosing among mutually exclusive projects (e.g., Alchian, 1955; Dudley, 1972; Beaves, 1988; Klemperer, 1996; Cary and Dunn, 1997; Hajdasinski, 1996 and 1997; Luenberger, 1998; Hazen, 2003; Hartman and Schafrick, 2004; Copeland et al., 2005; Brealey et al., 2008; Johnstone, 2008; and Kierulff, 2008).

investment evaluation tool is **not** the purpose of this article. This has been done, is being done, and will continue being done in the literature; for example, Just et al. (2004), Copeland et al. (2005), and Brealey et al. (2008) examine the consistency of *irr* as an investment evaluation tool relative to NPV. Nor, is the purpose to address issues of optimal rotation age; for example, Samuelson (1976), Johansson and Löfgren (1985), and Terreaux and Peyron (1997) have shown that using *irr* to determine economically optimal rotation ages provides a different answer than using the "Faustmann" model (Faustmann, 1849; Pressler, 1860; Ohlin, 1921). Thus, I will divide the article into four sections. First is a discussion of what an appropriately defined discount or interest rate represents. Second is a discussion of the mathematical derivation of the *irr* and the resulting feasible economic interpretation of the *irr* as a rate of return. Using these sections as a theoretical basis, the third is a discussion of interpreting the *irr* in the context of even-age to uneven age sustainable forest management. Finally, I will conclude with a few brief remarks.

What is the Interest or Discount Rate?

The interest or discount rate describes the opportunity cost of capital or the fact that there are alternative investment opportunities. The opportunity cost of capital reflects not only the amount and timing of the net cash flows (internal factors), but also the purchasing power of the net cash flow in the different periods, risk, and other relevant business information such as patents, research and development, and entrepreneurial history, etc. (external factors) (Fisher, 1930; Henderson and Quandt, 1980; Luenberger, 1998; Silberberg and Suen, 2001; Copeland et al., 2005; Brealey et al., 2008). For any given risk level, the investment chosen should increase an investor's wealth no less than the next best alternative investment. Thus, the opportunity cost of capital is set by the capital markets (Brealey et al., 2008; Johnstone, 2008).[2]

Interest rates are most often defined as real and nominal. Fisher (1930) developed the relationship between the nominal and real interest rate as:

$$i = (1 + r) \cdot (1 + \omega) - 1 = r + \omega + r \cdot \omega \tag{A9.1a}$$

where *i* denotes the nominal interest rate, *r* denotes the real interest rate, and ω denotes the anticipated inflation rate ($\omega \geq 0$). If the anticipated inflation rate is close to zero, equation (A9.1a) is often written as

$$i = r + \omega \tag{A9.1b}$$

In addition, Silberberg and Suen (2001) and Brealey et al. (2008) note that historically market interest rates have included a risk premium to account for exter-

---

2 It should be noted that not all investor's face perfect capital markets (Johansson and Löfgren, 1985; Price, 1993; Klemperer, 1996; Luenberger, 1998; Silberberg and Suen, 2001). Often there is a difference between their borrowing and lending rates, with the rate charged for borrowing capital being usually greater than they receive for lending or investing.

nal factors other than inflation. Taking some liberty so as to be brief, let the risk adjusted real interest rate be given by equation (A9.2a)

$$\tilde{r} = r_f + \eta \tag{A9.2a}$$

where r_f denotes a risk-free real interest rate (e.g., the return on 1-year U.S. Treasury bills) and η denotes a risk premium ($\eta \geq 0$). Thus equation (A9.2b) can be expanded to include a risk premium:

$$\tilde{i} = \tilde{r} + \omega = r_f + \eta + \omega \tag{A9.2b}$$

Equations (A9.2a) and (A9.2b) can be used as models for the real and nominal opportunity costs of capital as defined by the capital markets.[3] As Klemperer (1996), Davis et al. (2001), and Silberberg and Suen (2001), Brealey et al. (2008) point out there is no universal or unique risk premium. Thus there is no unique real or nominal interest rate which could be used to examine all investments.

A market-derived appropriate opportunity cost of capital is often described as the hurdle rate of return, the guiding rate of return, or the minimum acceptable rate of return. Bullard et al. (2002) illustrates this by examining different interest rates required by nonindustrial private forest landowners in Mississippi. In business the weighted average cost of capital (see Copeland et al., 2005; Brealey et al., 2008) is used to approximate a market-derived opportunity cost of capital. Consequentially, an appropriate opportunity cost of capital reflects internal and external factors associated with alternative investment opportunities.

While my primary focus is on market-derived opportunity costs of capital, there are additional bodies of literature that discuss a market-derived versus socially-derived versus hyperbolic interest rates. I will briefly discuss if the *irr* is consistent with the concepts of social and hyperbolic discount rates in the conclusions.

The economic Interpretation of The Internal Rate of Return as a Rate of Return

The *irr* is generally attributed to Fisher (1930). However, Boulding (1935) and Keynes (1936) also published manuscripts discussing the *irr* at roughly the same time. I will use Boulding (1935) as my primary reference as he most clearly lays out the foundation of the *irr* as used in forestry. Based on Boulding (1935), the *irr* is commonly defined as the "interest rate" or "discount rate" that sets the NPV equal to zero as illustrated by equation (A9.3)

$$\sum_{t=0}^{T} \frac{R_t}{(1 + irr)^t} - \sum_{t=0}^{T} \frac{C_t}{(1 + irr)^t} = 0 \tag{A9.3}$$

3 Breley et al. (2008) provide a very detailed discussion of this topic that is much too long to be repeated here.

where R_t and C_t denote revenue and cost flows at time t, respectively, R_t and C_t are real numbers with R_t, $C_t \geq 0$, $(R_0 - C_0) < 0$, and $(R_t - C_t) > 0$ for at least one value of $1 \leq t \leq T$, T is a positive finite whole real number, and *irr* is the internal rate of return.

Interpreting the *irr* as a rate of return is tied directly to its appearance as a discount or interest rate in equation (A9.3). However, this interpretation does not follow from its mathematical derivation. While the issues surrounding determining the *irr* have been discussed in the literature, I will provide brief overview of calculating the *irr* to setup its interpretation as a rate of return and interpretation within the context of even-age to uneven-aged sustainable forest management.

Calculating the Internal Rate of Return

As Boulding (1930) and others (e.g., Lorie and Savage, 1955; Gansner and Larson, 1969; Norstrøm, 1972; Clutter et al., 1983; Klemperer, 1996; Luenberger, 1998; Bidard, 1999; Hazen, 2003; Hartman and Schafrick, 2004; Copeland et al., 2005; Brealey et al., 2008; Johnstone, 2008) have pointed out, calculating *irr* can be complicated and result in unique or multiple solutions. Equation (A9.3) can be rewritten as a polynomial equation of degree 'T'

$$P(N_T) = \sum_{t=0}^{T} N_t X^t = 0 \tag{A9.4}$$

where $N_t = (R_t - C_t)$ and

$$X = \frac{1}{1 + irr} \tag{A9.5}$$

While the following mathematical statements may seem obvious they are important to restate as interpreting the *irr* as a rate of return and in the context of sustainable forest management depends on them directly. The X's are the zeros or roots of this polynomial which depend only on the magnitudes of N_t and t. In addition, N_t is a constant real number for all t. By definition there are 'T' possible roots or zeros for this polynomial which may be positive or negative real numbers or complex numbers. Equation (A9.5) shows that *irr* is a function of the root of the polynomial described in equation (A9.4).

Before a searching technique is used to determine the root(s) of equation (A9.4), one can determine the number of positive real roots. The general approach to determine the number of positive and negative roots (both real and complex numbers) is Descartes' Rule of Signs (Aufmann et al., 1993; Newnan, 1983; Levin, 2002) which states that the maximum number of positive real roots of $P(N_T)$ is equal to the number of sign changes between consecutive nonzero coefficients or less by an even number. If we assume that the roots of the polynomial in equation (A9.4) are positive, real, and rational, Descartes' Rule of Signs does not guarantee however, that *irr*(s) will be positive. Solving equation (A9.5) for *irr* gives

$$irr = \frac{1}{X} - 1 \tag{A9.6}$$

if $1/X < 1$ then *irr* is negative. Various other approaches have been employed to determine if a cash flow will provide a unique positive real root (e.g., Norstrøm, 1972; de Faro, 1973; Luenberger, 1998).[4]

The *irr*(s) obtained from solving equations (A9.4), (A9.5), and (A9.6) can be expressed as either a decimal or percent and are a function of the roots of $P(N_T)$. The interest or discount rate can also be expressed as either a decimal or percent. They, however, reflect the market's assessment of the internal and external factors surrounding an investment.

Unique Internal Rate of Return

As illustrated above and defined by Hazen (2003, p. 32) interpreting *irr*s as rates of return is "not part of their mathematical definition". Thus, interpreting any *irr*, calculated using equation (A9.3), as a rate of return is based on two assumptions. First, all revenues and costs are known with certainty (Boulding, 1935; Luenberger, 1998; Hazen, 2003). Second, *any* net cash flow invested and *any* net cash not withdrawn are reinvested at the *irr* (Bentley and Teeguarden, 1965; Marty, 1970; Dudley, 1972; Beaves, 1988; Price, 1993; Copeland et al., 2005; Johnstone, 2008; Kierrulff, 2008). These assumptions imply that if the initial cost of the project (e.g., C_0 in equation (A9.3)) was deposited in a "bank" and compounded annually (i.e., reinvested annually) using the *irr* for T years, at the end of T years you would be able to withdraw the required amount (e.g., R_T in equation (A9.3)) zeroing out the account (Foster and Brooks, 1983; Price, 1993; Luenberger, 1998; Copeland et al., 2005; Johnstone, 2008).[5] In this bank, the capital markets do not define the rates of return, transactions costs are zero, all revenues and cost are known with certainty, and there is no risk (Luenberger, 1998; Johnstone, 2008). Interpreting the *irr* as a rate of return is illustrated by Example 1 (Table A9.1 and Figure A9.1) and Example 2 (Table A9.2 and Figure A9.2).

In both cases the projects last for 15-years. In Example 1 there is an initial cost and three unequal withdrawals in years 5, 10, and 15. The *irr* is 10.26%. As can be seen any net cash not withdrawn is reinvested at the *irr*. At the end of 15 years the final withdrawal of $1,900.00 zero's out the account. In Example 2 there is an initial $2,000.00 cost and an additional $2,000.00 deposit in year 10. This cash

---

4 Lin (1976) and Hazen (2003) also provide a review of this literature.
5 It should be noted that NPV and BCR general assume revenues and costs are also known with certainty. They also have an implied reinvestment assumption. The difference is that with NPV and BCR the net cash flows are reinvested annually using a market-derived discount or interest rate while with *irr* the net cash flows are reinvested annually at the *irr* which is not a market-derived discount or interest rate (Copeland et al. 2005).

Table A9.1 Example 1 of a unique real internal rate of return (irr)‡

Year	Revenue	Cost	N	PV(N)	
0		2000.00	−2000.00	−2000.00	irr = 0.10258
5	1500.00		1500.00	920.56	
10	1700.00		1700.00	640.28	
15	1900.00		1900.00	439.17	
Sum			3100.00	0.00	

	Years			
	0 to 5	6 to 10	11 to 15	
2000.00	1758.90	1166.03	Beginning Balance
1258.90	1107.14	733.96	Interest Income
−1500.00	−1700.00	−1900.00	Withdraw
1758.90	1166.03	−0.01	Ending Balance

‡N denotes Net Cash flow and PV(N) denotes the present value of the net cash flow.

Table A9.2 Example 2 of a unique real internal rate of return (*irr*)‡

Year	Revenue	Cost	N	PV(N)	
0		2000.00	−2000.00	−2000.00	*irr* = 0.0602
5	2500.00		2500.00	1866.41	
10		2000.00	−2000.00	−1114.72	
15	3000.00		3000.00	1248.31	
Sum			1500.00	0.00	

	Years			
	0 to 5	*6 to 10*	*11 to 15*	
	2000.00	178.94	2239.68	Beginning Balance
	678.94	60.74	760.30	Interest Income
	−2500.00	2000.00	−3000.00	Withdraw/Deposit
	178.94	2239.68	−0.01	Ending Balance

‡N denotes Net Cash flow and PV(N) denotes the present value of the net cash flow.

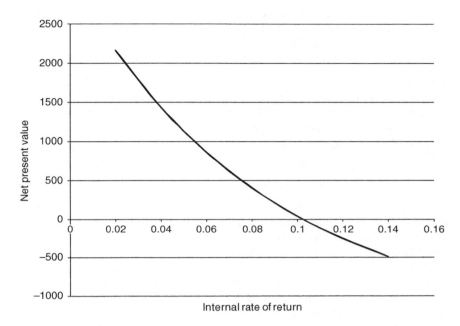

Figure A9.1 Example 1 of a unique real internal rate of return (*irr*).

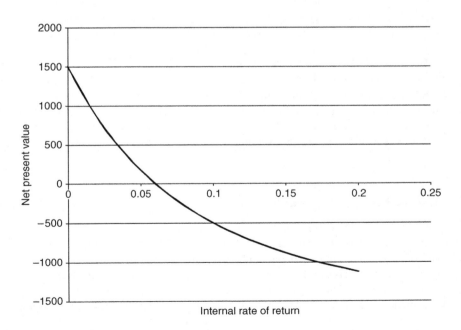

Figure A9.2 Example 2 of a unique real internal rate of return (*irr*).

flow has no negative roots as the coefficients of the polynomial are real and alternate in sign (Levin, 2002). The *irr* is 6%. Again, any net cash flow not withdrawn or deposited is reinvested at the *irr*.

Examples 1 and 2 are defined as pure investments with $\dfrac{\partial P(N)}{\partial \hat{i}} < 0$ where $\hat{i}$ denotes rates of return with $-1 < \hat{i} < \infty$ (Hartman and Schafrick, 2004). A unique real root does not however preclude the possibility of complex roots. I will use an example given by Hartman and Schafrick (2004) to illustrate this point. The net cash flow is given in Example 3 (Table A9.3 and Figure A9.3).

While there is a unique real *irr* of 219% there are two complex roots where $\left(\dfrac{\partial P(N)}{\partial \hat{i}} = 0\right)$. Consequently, the net cash flow cannot be defined as a pure investment. There is a portion of the graph were $\dfrac{\partial P(N)}{\partial \hat{i}} > 0$ defining a loan or money is withdrawn from the project (Hartman and Schafrick, 2004). Examining Example 3 illustrates this point. At the end of year 1 there is an ending balance of −$2.81 as a result of withdrawing $6.00. There is a penalty in year 2 of −$6.15 for withdrawing more than the balance. In year 2 there is also a deposit of $11.00 this creates a positive ending balance of $2.04. Finally, in year 3 there is a withdrawal of $6.50 which is equal to year 2's ending balance plus interest zeroing out the account. In this case the bank reinvests any positive ending balances and charges a penalty on negative ending balances at the *irr*. Again, the capital markets do not define the rates of return, transactions costs are zero, all revenues and cost are known with certainty, and there is no risk.

Consistent with its mathematical derivation, the *irr* defines an annual "rate of return" for a project that depends *exclusively* on the amount and timing of a given project's cash flow (Boulding, 1935; Copeland et al., 2005; Brealey et al., 2008). As Boulding (1935, p. 482) states: "Thus, bound up in the very structure of any net revenue series there is a rate of return which pertains to it, and which can be calculated if we know all the terms of the net revenue series and nothing else." The opportunity cost of capital represents both internal and external factors. There is no information provided in Examples 1, 2 or 3 to determine if any of these investments would increase an investor's wealth no less than an alternative investment of similar risk.

Multiple Internal Rates of Return

Consistent with Descartes' Rule of Signs, Lorie and Savage (1955, p. 237) describe the classic general net cash flow that will result in multiple roots and consequently, multiple *irr*s as one with "initial cash outlays, subsequent net cash inflows, and final cash outlays". Nonetheless, the assumptions used for interpreting multiple *irr*s, calculated using equation (A9.3), as rates of return are the same as defined above. Example 4 (Table A9.4 and Figure A9.4) defines an investment with three *irr*s; namely, 0%, 14.87%, and 24.58%.

The net cash flow is characterized by an initial investment of $100.00, an additional investment of $110.00 in year 10, and withdrawals of $60.00 in years

Table A9.3 Example 3 of a unique real internal rate of return (*irr*) with complex roots (Hartman and Schafrick 2004)[‡]

Year	Revenues	Cost	N	PV(N)		Years		
						0 to 1	1 to 2	2 to 3	
0		1.00	−1.00	−1.00		1.00	−2.81	2.04	Beginning Balance
1	6.00		6.00	1.88	*irr* = 2.19	2.19	−6.15	4.46	Interest Income
2		11.00	−11.00	−1.08		−6.00	11.00	−6.50	Withdraw/Deposit
3	6.50		6.50	0.20		−2.81	2.04	0.00	Ending Balance
Sum			0.50	0.00					

[‡]N denotes Net Cash flow and PV(N) denotes the present value of the net cash flow.

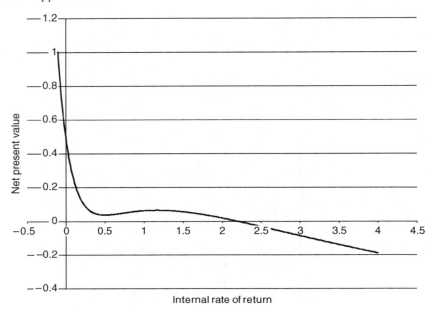

Figure A9.3 Example 3 of a unique real internal rate of return (*irr*) with complex roots (Hartman and Schafrick 2004).

5 and 15. As shown by Table A9.4, for all three *irr*s the ending balance in year 5 is negative requiring the payment of a penalty (for the *irr*s > 0) in year 10. In year 15 there is a withdrawal which is equal to year 10's ending balance plus interest zeroing out the account. Each *irr* defines an annual "rate of return" for the project that depends *exclusively* on the amount and timing of a given project's cash flow. As with Example 2 (Table A9.2 and Figure A9.2) the bank reinvests any positive ending balances and charges a penalty on negative ending balances at the *irr*. As before, the capital markets do not define the rates of return, transactions costs are zero, all revenues and cost are known with certainty, and there is no risk. Finally, as Hazen (2003) and Johnstone (2008) describe, there is nothing in the calculation of the *irr*s that leads to determining which of the three is the "correct" rate of return as each provide sufficient funds to zero out the account at the end of 15-years given the net cash flow.[6] As Hazen (2003, p. 46) points out it is "not meaningful to compare [multiple] internal rates of return within a single project... The magnitude of the internal rate by itself carries no further information."

6 While this statement is technically accurate, there are various methodologies that have been described to determine the "correct" *irr* if using it as a profitability measure (e.g., Beaves, 1988; Hazen, 2003; Hartman and Schafrick, 2004; Brealey et al., 2008; Kierulff, 2008).

Table A9.4 Example 4 of a multiple real internal rate of returns (*irrs*)‡

Year	Revenues	Cost	N	PV(N)			Years		
							0 to 5	6 to 10	11 to 15
0		10.00	-10.00	-10.00	irr = 0	Beginning Balance	10.00	-50.00	60.00
5	60.00		60.00	60.00		Interest Income	0.00	0.00	0.00
10		110.00	-110.00	-110.00		Withdraw/Deposit	-60.00	110.00	-60.00
15	60.00		60.00	60.00		Ending Balance	-50.00	60.00	0.00
Sum			0.00	0.00					

Year	Revenues	Cost	N	PV(N)			Years		
							0 to 5	6 to 10	11 to 15
0		10.00	-10.00	-10.00	irr = 0.1487	Beginning Balance	10.00	-40.00	30.00
5	60.00		60.00	30.00		Interest Income	10.00	-40.00	30.00
10		110.00	-110.00	-27.50		Withdraw/Deposit	-60.00	110.00	-60.00
15	60.00		60.00	7.50		Ending Balance	-40.00	30.00	0.00
Sum			0.00	0.00					

Year	Revenues	Cost	N	PV(N)			Years		
							0 to 5	6 to 10	11 to 15
0		10.00	-10.00	-10.00	irr = 0.24578	Beginning Balance	10.00	-29.99	20.00
5	60.00		60.00	20.00		Interest Income	20.01	-60.01	40.01
10		110.00	-110.00	-12.22		Withdraw/Deposit	-60.00	110.00	-60.00
15	60.00		60.00	2.22		Ending Balance	-29.99	20.00	0.01
Sum			0.00	0.00					

‡N denotes Net Cash flow and PV(N) denotes the present value of the net cash flow.

Revisiting the Reinvestment Assumption

As noted by Bentley and Teeguarden (1965), Marty (1970), Dudley (1972), Beaves (1988), Price (1993), Davis et al. (2001), Copeland et al. (2005), Brealey et al. (2008), Johnstone (2008), and Kierrulff (2008), the assumption of any net cash flow invested and any net cash not withdrawn are reinvested at the *irr*(s) is an implied component of using equation (A9.3). An alternative approach is to calculate a Modified Internal Rate of Return (*mirr*).[7] Lin (1976) and Kierulff (2008) develop a *mirr* that requires defining an appropriate opportunity cost of capital for any net cash flows that are reinvested or withdrawn (i.e., net cash flows that are not used to finance any further investments into the project) and an appropriate opportunity cost of capital for any net cash flows that are financed. This formulation is given in equation (A9.7):

$$
mirr = \left(\frac{\displaystyle\sum_{t=0}^{T} R_t \, (1 + r_{(R)})^{T-t}}{\displaystyle\sum_{t=0}^{T} C_t \, (1 + r_{(C)})^{-t}} \right)^{\frac{1}{T}} - 1
$$

$$
= \left(\frac{FV(R)}{PV(C)} \right)^{\frac{1}{T}} - 1 \tag{A9.7}
$$

where $r_{(R)}$ and $r_{(C)}$ represent the appropriate opportunity costs of capital for reinvesting and financing any net cash flows, respectively. The denominator defines the present value of the costs discounted using the appropriate opportunity cost of capital for any net cash flows that are financed, PV(C). The numerator defines the future value of any net cash flows that are available to be reinvested or withdrawn using the appropriate opportunity cost of capital for reinvesting, FV(R). As described by Beaves (1988), the denominator and numerator convert a net cash flow with positive and negative components into an equivalent one with an indexed negative net cash flow in time zero and an indexed positive net cash flow in time T.

Table A9.5 gives *mirr* calculation for Example 4 assuming $r_{(R)}$ = 7% and 5% and $r_{(C)}$ = 7% and 5%.

As can be seen, equation (A9.7) converts the net cash flow from one with two negative and two positive net cash flows to an equivalent one with an indexed negative net cash flow at time 0 and an indexed positive net cash flow at time 15. By definition the new equivalent cash flow will have a unique real positive root that will result in a unique positive *mirr*. The *mirr* is an annual rate of return

---

7 A Modified Internal Rate of Return is also used to convert a net cash flow with multiple *irr*s into an equivalent cash flow with a unique real positive *irr* (i.e., a *mirr*) that can then be used as a profitability measure. While there are various formulations of the *mirr* (e.g., Hirshleifer, 1970; Marty, 1970; Lin, 1976; Schallau and Wirth, 1980; Harpole, 1984; Beaves, 1988; Liu and Wu, 1990; Hajdasinski, 1996 and 1997; Cary and Dunn, 1997; Davis et al., 2001; Hazen, 2003; Brealey et al., 2008; Kierulff, 2008), I will highlight only one as the interpretation of the *mirr* as a rate of return is the same regardless of the formulation.

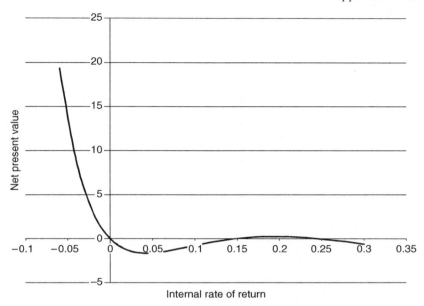

Figure A9.4 Example 4 of a multiple real internal rate of returns (*irr*s).

weighted by the appropriate opportunity costs of capital for financing and reinvesting. Interpreting *mirr* as a rate of return given the equivalent cash flow is similar to that described in Examples 1, 2, or 3.

Equation (A9.7) also shows that the weights (i.e., the appropriate opportunity costs of capital) placed on the positive and negative net cash flows are critical in calculating *mirr*. Equation (A9.8) gives the first order condition of *mirr* with respect to FV(R)/PV(C):

$$\frac{\partial mirr}{\partial \frac{FV(R)}{PV(C)}} = \frac{1}{T}\left[\frac{FV(R)}{PV(C)}\right]^{\frac{1}{T}-1} > 0 \qquad (A9.8)$$

If T is constant and the ratio FV(R)/PV(C) is positive, greater than one, and increasing, then *mirr* will increase but at a decreasing rate. As PV(C) approaches 0 then *mirr* approaches infinity. Table A9.5 shows that if costs are discounted at a larger (smaller) appropriate opportunity cost of capital than revenues are compounded the *mirr* increases (decreases).

Interpretation of the Internal Rate of Return Within the Context of Sustainable Forest Management

Net cash flows of even-aged to uneven-aged forest management activities can be simple to complex. Although in general, these net cash flows often result in a

Table A9.5 Calculation of a Modified Internal Rate of Return (*mirr*) using Example 4‡

Year	Revenues	Cost	N	PV(Cost)	FV(Revenue)		Years 0 to 15	
0		10.00	−10.00	10.00			77.53	Beginning Balance
5	60.00		60.00		118.03		100.50	Interest Income
10		110.00	−110.00	67.53			−178.03	Withdraw
15	60.00		60.00		60.00		0.00	Ending Balance
Sum			0.00	77.53	178.03			

$mirr = 0.05698$
$r_{(C)} = 0.05$
$r_{(R)} = 0.07$

Year	Revenues	Cost	N	PV(Cost)	FV(Revenue)		Years 0 to 15	
0		10.00	−10.00	10.00			65.92	Beginning Balance
5	60.00		60.00		118.03		112.11	Interest Income
10		110.00	−110.00	55.92			−178.03	Withdraw
15	60.00		60.00		60.00		0.00	Ending Balance
Sum			0.00	65.92	178.03			

$mirr = 0.06848$
$r_{(C)} = 0.07$
$r_{(R)} = 0.07$

Year	Revenues	Cost	N	PV(Cost)	FV(Revenue)		Years
							0 to 15
0		10.00	-10.00	10.00		Beginning Balance	65.92
5	60.00		60.00		97.73	Interest Income	91.82
10		110.00	-110.00	55.92		Withdraw	-157.73
15	60.00		60.00		60.00	Ending Balance	0.00
Sum			0.00	65.92	157.73		

$mirr = 0.05989$

$r_{(C)} = 0.07$

$r_{(R)} = 0.05$

[+]$mirr$ denotes the modified internal rate of return.

$r_{(R)}$ denotes the opportunity cost of capital for reinvesting any positive net cash flows.

$r_{(C)}$ denotes the opportunity cost of capital for financing any negative net cash flows.

PV(Cost) denotes present value of the costs.

PV(Revenue) denotes present value of the revenues.

(Table A9.5 Continued)

unique *irr*; interpreting unique, or multiple, *irr*(s) resulting from even-aged to uneven-aged forest management as rates of return must still be consistent with those described in Examples 1 through 4. Is this interpretation consistent with the ideals of sustainably managing even-aged to uneven-aged forests? Unfortunately there is no single accepted definition of sustainable forest management, and I will not attempt to address that issue here;[8] however, common themes do emerge from these definitions which include maintaining the forest-based ecosystem, use of forest-based ecosystem goods and services that does not impact future generations, multiple stakeholders, and scale. I will start by examining if the *irr*(s) calculated (i) for single and multiple rotation even-aged management systems, (ii) for a representative individual tree or stand, and (iii) for a regulated uneven-aged management system are consistent with the common themes of sustainable forest management. I will end this section by examining the reinvestment assumption.

Equation (A9.3) has often been used to calculate the *irr* of a single harvest even-aged management regime where R_t and C_t are defined as cash flows resulting from using forest-based ecosystem goods and services. In addition, equation (A9.3) reveals that flows of any forest-based ecosystem goods or services before $t = 0$ and after $t = T$ have a value of 0 and are not included in the calculation of the *irr*. In addition, Bentley and Teeguarden (1965), Johansson and Löfgren (1985), Uys (1990), and Hseu and Buongiorno (1997) note that formulation of equation (A9.3) implies land is so abundant that it is obtained at zero cost in time 0 and has no value after the harvest in time T, its opportunity cost is zero.[9] The *irr* represents the annual value growth of the net cash flow resulting from a single harvest management regime and nothing else. Interpreting the resulting *irr* as a rate of return in this context must be consistent with Examples 1 through 4. In addition, this *irr* cannot be compared with *irr*s derived from other mutually exclusive patterns of forest-based ecosystem goods and services flows. There is nothing explicit in the formulation of equation (A9.3) or in interpreting the resulting *irr* that states whether the use of the forest-based ecosystem goods and services is consistent with – at the minimum – maintaining the forest-based ecosystem or use of forest-based ecosystem goods and services that does not impact future generations

8 For example Kant (2003) discusses the limitations of a traditional discounting approach to analyze sustainable forest management based on evolutionary and institutional economic schools of thought.

9 A variable for the purchase and sale of land can be included in equation (A9.3)

$$\sum_{t=0}^{T} R_t (1 + irr)^{-t} + L_T (1 + irr)^{-T} - \sum_{t=0}^{T} C_t (1 + irr)^{-t} - L_0 = 0$$

where L_T denotes the market value of the land at time T and L_0 denotes the market value of the land a time 0. If $R_T, C_0, L_T, L_0 > 0$ and $irr > 0$, then including the purchase and sale of the land into the *irr* calculation will, almost assuredly, change the magnitude of the *irr*(s), but will not change whether the net cash flow has a unique or multiple *irr*s. The inclusion of the market value of land does not change the interpretation of the *irr* representing the annual value growth of the net cash flow resulting from a single harvest even-aged management regime.

(i.e., socially, economically, or ecologically sustainable). Consequently, no statement concerning the sustainability of this single harvest even-aged management regime can be made.[10]

Assume the pattern of revenues, R_t, and costs, C_t, from using forest-based ecosystem goods and services defined for the period t = 0,…, T is repeatable. Equation (A9.9) derives the *irr* model given this pattern is repeated an infinite number of times

$$\sum_{\theta=0}^{\infty}\left[\sum_{t=0}^{T}R_t(1+irr)^{-t} - \sum_{t=0}^{T}C_t(1+irr)^{-t}\right]\cdot(1+irr)^{-\theta T} = 0$$

$$\frac{\sum_{t=0}^{T}R_t(1+irr)^{-t} - \sum_{t=0}^{T}C_t(1+irr)^{-t}}{1-(1+irr)^{-T}} = 0 \qquad (A9.9)$$

Equation (A9.9) is analogous to the "Faustmann" model first defined by Faustmann (1849), Pressler (1860), and Ohlin (1921) where the numerator defines the single harvest even-aged management regime and the rotation age T is given. As shown by Uys (1990) if one or more *irr*(s) can be found that sets the numerator of equation (A9.9) equal to zero, then the fraction is equal to zero implying that the *irr*(s) of equation (A9.3) are identical to those of equation (A9.9). This is illustrated in equation (A9.10)

$$\sum_{t=0}^{T}R_t(1+irr)^{-t} - \sum_{t=0}^{T}C_t(1+irr)^{-t} = \frac{\sum_{t=0}^{T}R_t(1+irr)^{-t} - \sum_{t=0}^{T}C_t(1+irr)^{-t}}{1-(1+irr)^{-T}} = 0$$

$$(A9.10)$$

Interpreting the resulting *irr* must be consistent with Examples 1 through 4. The *irr* represents the net cash flow of this particular flow of forest-based ecosystem goods and services for the period t = 1,…, T and nothing else. This *irr* cannot be compared with irrs derived from other mutually exclusive patterns of forest-based ecosystem goods and services flows.

Analyzing equations (A9.9) and (A9.10) in terms of sustainable forest management, while the magnitudes of the *irr*(s) will change as T changes reflecting the different value growth of the net cash flows, *irr* completely ignores the opportunity cost of not harvesting the land at time T (i.e., stand rent and land rent – Johansson and Löfgren, 1985). In addition, the *irr* would not reflect a rate of return on the capital asset land (i.e., the forest-based ecosystem). An argument could be made that the Faustmann model requires maintaining the forest-based ecosystem so that it can produce a flow of resources for current and future generations, *ceteris paribus*. While I am not advocating that a Faustmann type model satisfies all the tenets of sustainable forest management, it is a conceptual improvement compared to the

10 The same argument could also be made with respect to calculating the NPV or BCR in this case.

single harvest model; however, as the *irr* cannot distinguish between a single and infinite harvest model there is nothing explicit in the formulation of equations (A9.9) and (A9.10) to state whether the use of the forest-based ecosystem goods and services is socially, economically, or ecologically sustainable.

Grisez and Mendel (1972), Mendel et al. (1973), Godman and Mendel (1978), Herrick and Gansner (1985), Dennis (1987), Davies (1991), Klemperer (1996), Hseu and Buongiorno (1997), Heiligmann (2008), and Webster et al. (2009) have used the *irr* to describe the annual value growth rate of a representative individual tree or stand. This is illustrated by equation (A9.11)

$$irr = \left(\frac{R_T}{C_t}\right)^{\frac{1}{(T-t)}} - 1; \quad 0 \le t < T \tag{A9.11}$$

where C_t denotes the value of the trees at time t and R_T denotes the harvest value of the trees at time T.[11] Equation (A9.11) has also been used in developing regeneration marking guide's take or leave decision associated with a selection silvicultural harvesting system. Unfortunately, the common interpretation of the *irr* given by equation (A9.11) as a financial rate of return is not accurate in this case. The *irr* given in equation (A9.11) describes the expected annual value growth of the trees *ex-ante*. In addition, the *irr* may only be interpreted as an annual growth in value if and only if reinvestment takes place (Price 1993). *Ex-post*, if C_t and R_T occur, then *irr* does describe the annual value growth of the trees. That said, the only accurate statement that can be made concerning the *irr* is that if it is greater than the appropriate opportunity cost of capital, there is a greater likelihood that a landowner's wealth will increase. Whether a landowner's wealth increases (or not) is tied directly to the opportunity cost of capital not the *irr*. Finally, using equation (A9.11) to help develop a marking guide's take or leave decision does not include the economic implications of (i) one or more trees occupying the same growing space and (ii) the current tree occupying the site and not allowing the next tree to take its place (i.e., the opportunity cost of the land or land rent as described by Johansson and Löfgren (1985) and Hseu and Buongiorno (1997)).

Economic analysis of uneven-aged forest management is often more complex than even-aged management. An uneven-aged forest is comprised of uneven-aged stands. Uneven-aged stands are comprised of three or more distinct age classes. For a regulated uneven-aged stand each age class occupies an equivalent amount of area, as measured by basal area class per acre or number of trees per acre by diameter class, over the length of the cutting cycle. Diameter distributions are commonly used to regulate the uneven-aged stands and thus the forest. The structure of an uneven-aged forest is such that cutting in a diameter class will impact the growth in other diameter classes. This is in contrast to an even-age forest where age classes are spatially distinct such that cutting one age class does not impact the growth of any other age class. It is this particular interrelationship among the diameter classes of an uneven-aged

---

11 Equation (A9.11) is derived from equation (A9.3) directly.

forest that make it more interesting and difficult to analyze economically. While Amacher et al. (2009) provide an excellent summary of current research in this area; for purposes of illustration I will use an economic model based on one developed by Buongiorno et al. (1995) to examine a single uneven-aged stand in a steady-state or regulated condition. The model is given in equation (A9.12)

$$\frac{(R_T - C_T)\cdot(1 + irr)^{-T} - G_t}{1 - (1 + irr)^{-T}} = 0 \qquad (A9.12)$$

where R_T denotes the harvest value at the end of the cutting cycle, C_T denotes the cost of harvesting R_T, G_t denotes the value of the growing stock at the beginning of the cutting cycle, and T denotes the length of the cutting cycle. The numerator of equation (A9.12) is the value associated with a single cutting cycle. The denominator is used to show this cutting cycle will be repeated an infinite number of time. The formulation of equation (A9.12) is the same as that given in equations (A9.9) and (A9.10). Thus, if one or more *irr*(s) can be found that sets the numerator of equation (A9.12) equal to zero, then the fraction is equal to zero implying that the *irr*(s) of a single cutting cycle are identical to those of an infinite number of cutting cycles. The economic interpretation of the resulting *irr* must be consistent with Examples 1 through 4. The *irr* represents the net cash flow of this particular steady-state condition and nothing else. This *irr* cannot be compared with irrs derived from other mutually exclusive steady-state conditions. Again, as the *irr* cannot distinguish between a single and infinite cutting cycle model there is nothing explicit in the formulation of equation (A9.12) to state whether the uneven-aged stand is being managed socially, economically, or ecologically sustainably.

The issue surrounding the reinvestment assumption is not new to the forestry literature (e.g., Marty, 1970; Shallau and Wirth, 1980; Klemperer, 1981 and 1996; Mills and Dixon, 1982; Foster and Brooks, 1983; Harpole, 1984; and Davis et al., 2001). The approaches described to modify an *irr* calculation to account for alternative opportunity costs of capital for reinvestment and financing are similar to that used in equation (A9.7). The result is called a composite internal rate of return, a composite rate of return, or a realized rate of return instead of a modified internal rate of return. Klemperer (1981) criticizes these calculations as biased against investments with long lives. This critique is illustrated by equation (A9.13)

$$\frac{\partial mirr}{\partial T} = \frac{-\ln\left(\frac{FV(R)}{PV(C)}\right)\cdot\left(\frac{FV(R)}{PV(C)}\right)^{\frac{1}{T}}}{T^2} < 0 \qquad (A9.13)$$

with FV(R) > PV(C). Nonetheless, interpreting the composite internal rate of return as a rate of return is similar to the *mirr* given previously. In addition, converting the cash flow from a forest management regime into an equivalent one with an indexed negative net cash flow at time 0, PV(C), and an indexed positive net cash flow at time T, FV(R), using appropriate opportunity costs of capital provides no additional information to assess sustainability: The *mirr* calculated using equation (A9.7) only describes an indexed annual value growth of the net cash flow resulting from the single rotation management regime and would not be consistent with the ideals of sustainable forest management.

In economies that are transitioning from central planning to a mixed or free market, such as those in many of the Eastern European countries that were under the influence of the old Soviet Union, the capital markets may not be well developed enough to assess investments in forest management (Brukas et al., 2001). In addition, the inclusion of economic analyses in forest management is improving but in some countries (e.g., Romania) central planning still can have a nontrivial influence (Sandulescu et al., 2007). Brukas et al. (2001) address the issue of what would be the appropriate discount rates to analyze forest management decisions given this context. The primary tool used by these authors to estimate discount rates is the *irr*. The net cash flows the authors used are based on the existing forest structures resulting from past central planning, the *irr*(s) were determined by optimizing a Faustmann model – similar to equation (A9.9) – that included intermediate stand treatments and given constant real prices (Brukas et al., 2001). The estimated irrs reflect only the internal factors related to timber growth resulting from past central planning, a single harvest regime, and a defined rotation age. They do not account for the opportunity cost of the land or stand properly nor any of the external market and socio-political factors affecting Lithuania. As these irrs are calculated based on a model similar to equation (A9.10), they are not consistent with the ideals of sustainable forest management. Given the lack of well developed capital markets, using a discount rate that is a "little less than *irr*" (Brukas et al., 2001, p. 153) as an initial approximation of the opportunity cost of capital might seem reasonable in this circumstance; however, even in these circumstance I would council extreme caution as these *irr*s do not reflect the macro or micro socio-political–economic environment surrounding managing Lithuania's forests.

Conclusions

Portraying the *irr* as a rate of return or an interest or discount rate (i.e., an opportunity cost of capital set by the capital markets) is an inaccurate interpretation (Hirshleifer, 1970; Hagemann, 1990; Luenberger, 1998; Hazen, 2003; Copeland et al., 2005; and Brealy et al., 2008). Brealey et al. (2008, p. 123) provide an excellent summary of a potential cause of this misinterpretation that I will paraphrase: People may often confuse the *irr* and the opportunity cost of capital because both appear as discount rates in the NPV formula. The *irr* is a *profitability measure* which depends *solely* on the amount and timing of the project's cash flow. The opportunity cost of capital is a *standard of profitability* for the project which is used to calculate how much the project is worth in terms of wealth maximization. Misinterpreting the *irr* gives misleading information to decision makers whether they are private or public land managers or stakeholders. Boulding noted this in 1935:

> We will now assume that there is some rate of return, *i* (*irr*), which is characteristic of the investment as a whole. This is of course a rate of interest, or a

rate of discount. But it must be emphasized that it is a rate of interest which the enterprise itself produces, and which, as we shall see later, is bound up with the very structure of the net revenues themselves. *That is to say, it is an internal rate, and while it may be equal to external rates of interest it must not be confused with them.* (Emphasis added)

(Boulding, 1935, p. 478)

Thus defined and interpreted accurately, the *irr* includes only those internal factors of any net cash flow. A market-derived opportunity cost of capital reflects not only the internal factors but also external factors associated with a project.

I would argue that the *irr* as traditionally defined using equations (A9.3), (A9.9), (A9.10), (A9.11), and (A9.12) reflect a single harvest regime which is inconsistent with concepts of ecological, social, and economic sustainability and therefore the ideals of sustainable forest management. This conclusion was reached assuming a market-derived opportunity cost of capital. I would also argue that the *irr* is not consistent with a social discount rate. The literature on the social discount rate is too vast to review here; however, Just et al. (2004) discusses that the social discount rate is a function of a social rate of time preference, changes in production technology, risk and uncertainty, and macro variables such as population growth, unemployment, and growth of per capita consumption. They point out that a social discount rate cannot be determined without referencing a specific economy. The *irr* does not include any of these external factors. Finally, Nocetti et al. (2008) and Hepburn and Koundouri (2007) note that conventional discounting factor, $(1 + i)^{-t}$, using a constant discount rate, i, is not consistent with inter-generational equity and sustainable development as net cash flows in the far future are virtually irrelevant to decisions made today. Thus, these authors argue that the appropriate social discount rate should decrease with time as this is consistent with human decision behavior. This has lead to research examining the implications of hyperbolic discount rates on long term investments such as in sustainable forest management. A hyperbolic discount rate is a function of time such that it is high for current net cash flows and declines asymptotically for future net cash flows. Again the literature on hyperbolic discounting is too vast to review here succinctly;[12] however, the *irr* as defined by equations (A9.3), (A9.9), (A9.10), and (A9.12) is not consistent with these recent advancements.

Using a tool such as the *irr* to analyze forest management actions (e.g., Heiligmann, 2008; Webster et al., 2009; Pickens et al., 2009) because it is commonly used by business (e.g., Graham and Harvey, 2001; Kierulff, 2008; Pickens et al., 2009) is problematic for two reasons. First, the ideals associated with sustainable forest management are not consistent with the use of *irr*. The *irr* only describes the potential net value growth of the defined net cash flow and nothing else; it cannot be compared with *irr*s derived from other mutually exclusive

---

12 I would point readers to Hepburn and Koundouri (2007) for a recent examination of hyperbolic discounting with respect to forest management.

patterns of forest-based ecosystem goods and services flows; and it does not include any social, economic, or ecological external factors generally associated with sustainability. Second, the *irr* is often represented as a rate of return. This interpretation assumes that all revenues and costs are known with certainty (i.e., no risk) and any net cash flow invested and any net cash not withdrawn are reinvested at the *irr*. The implications of these assumptions restrict interpreting the *irr* as a rate of return and are unfortunately not well understood. Thus, the *irr* conveys very limited short-term information as a profitability measure and nothing else.

Sustainable forest management depends on interpreting the results and management implications of tools such as *irr*, NPV, and BCR accurately. While the focus of this discussion was on *irr*, omitting similar discussions with respect to NPV and BCR is not meant to imply they are not misinterpreted within in the context of sustainable forest management. Academic debates concerning how to account for these cash flows appropriately and defining the concept of optimality within the context of sustainability continues. Unfortunately, these discussions are often not mirrored elsewhere. My purpose is to inform and stimulate these discussions among all those involved in sustainable forest management.

References

Alchian, A.A. (1955). The Rate of Interest, Fisher' Rate of Return over Costs and Keynes' Internal Rate of Return. *The American Economic Review*. 45(5), 938–943.

Amacher, G.S., Olikaninen, M., Koskela, E. (2009). *Economics of Forest Resources*. Cambridge, MA: Massachusetts Institute of Technology Press.

Aufmann, R.N., Barker, V.C., Nation, R.D. (1993). *Precalculus*, 2nd ed. Boston, MA: Houghton Mifflin Company.

Bailey, R.L. (1986). Rotation Age and Establishment Density for Planted Slash and Loblolly Pine. *Southern Journal of Applied Forestry* 10, 166–168.

Battaglia, M., Mummery, D., Smith, A. (2002). Economic Analysis of Site Survey and Productivity Modelling for the Selection of Plantation Areas. *Forest Ecology and Management* 162, 185–195.

Beaves, R.B. (1988). Net Present Value and Rate of Return: Implicit and Explicit Reinvestment Assumptions. *The Engineering Economist* 33(4), 275–302.

Bengston, D.N. (1984). Economic Impacts of Structural Particleboard Research. *Forest Science* 10(3), 685–697.

Bentley, W.R., Teeguarden, D.E. (1965). Financial Maturity: A Theoretical Review. *Forest Science* 11(1), 76–87.

Bettinger, P., Boston, K., Siry J.P., Grebner, D.L. (2009). *Forest Management and Planning*. New York, NY: Elsevier Academic Press.

Bidard, C. (1999). Fixed Capital and Internal Rate of Return. *Journal of Mathematical Economics* 31, 523–541.

Brealey, R.A., Myers, S.C., Allen, F. (2008). *Principles of Corporate Finance*, 9th ed. New York, NY: McGraw-Hill, Inc.

Boulding, K.E. (1935). The Theory of a Single Investment. *The Quarterly Journal of Economics* 49(3), 475–494.

Bullard, S.H., Gunter, J.E., Doolittle, M.L., Arano. K.G. (2002). Discount Rates for Nonindustrial Private Forest Landowners in Mississippi: How High a Hurdle? *Southern Journal of Applied Forestry* 26(1), 26–31.

Buongiorno, J., Gilles, J.K. (2003). *Decision Methods for Forest Resource Management.* Boston, MA: Academic Press.

Buongiorno, J., Peyron, J. L., Houllier, F., Bruciamacchie, M. (1995). Growth and management of mixed-species, uneven-aged forests in the French Jura: Implications for economic returns and tree diversity. *Forest Science.* 41(3): 397–429.

Brukas, V., Thorsen, B.J., Helles, F., Tarp, P. (2001). Discount rate and Harvest policy: Implications for Baltic forestry. *Forest Policy and Economics* 2(2), 143–156.

Cary, D., Dunn. M. (1997). Adjustment of Modified Internal Rate of Return for Scale and Time Span Differences. *Proceedings of the Academy of Accounting and Financial Studies* 2(2), 57–63.

Clutter, J.L., Fortson, J.C., Pienaar, L.V., Brister, G.H., Bailey, R.L. (1983). *Timber Management: A Quantitative Approach.* New York, NY: John Wiley & Sons.

Copeland, T.E., Weston, J.F., Shastri. K. (2005). *Financial Theory and Corporate Policy,* 4th ed. Reading, MA: Pearson Addison-Wesley Publishing Company.

Dangerfield, Jr. C.W., Edwards, M.B. (1991). Economic Comparison of Natural and Planted Regeneration of Loblolly Pine. *Southern Journal of Applied Forestry* 15(3), 125–127.

Davies, K. (1991). Forest Investment Considerations for Planning Thinning and Harvesting. *Northern Journal of Applied Forestry* 8(3), 129–131.

Davis, L.S., Johnson, K.N., Bettinger, P.S., Howard, T.E. (2001). *Forest Management: To Sustain Ecological, Economic, and Social Values,* 4th Ed. Long Grove, IL: Waveland Press.

de Faro, C. (1973). A sufficient condition for a unique nonnegative internal rate of return: A comment. *Journal of Financial and Quantitative Analysis* 8(4), 683–684.

Dennis, D.F. (1983). Tax incentives for Reforestation in Public Law 96–451. *Journal of Forestry* 293–295.

Dennis, D.F. (1987). Rates of value change on uncut forest stands in New Hampshire. *Northern Journal of Applied Forestry* 4(2), 64–66.

Dudley, C.L. Jr. (1972). A note on reinvestment assumptions in choosing between net present value and internal rate of return. *The Journal of Finance* 27(4), 907–915.

Duerr, W.A. (2003). *Introduction to Forest Resource Economics.* New York, NY: McGraw-Hill.

Faustmann, M. (1849). Berechnung des Wertes Welchen Waldboden sowie noch nitch haubare Holzbestände für die Waldwirtschaft besitzen. *Allgemeine Forst- und Jagd-Zeitung.* Vol. 15. Reprinted as Faustmann, Martin. (1995). Calculation of the Value which Forest Land and Immature Stands Possess for Forestry. *Journal of Forest Economics* 1(1):7–44.

Fight, R.D., Bolon, N.A., Cahill, J.M. (1993). Financial Analysis of Pruning Douglas-Fir and Ponderosa Pine in the Pacific Northwest. *Western Journal of Applied Forestry* 10(1), 58–61.

Fisher, I. (1930). *The Theory of Interest.* New York, NY: Macmillan.

Floyd. D. (2002). *Forest Sustainability: The History, the Challenge, the Promise.* The Forest History Society.

Foster, B.B., & Brooks, G.N. (1983). Rates of Return: Internal or Composite. *Journal of Forestry* 81(10), 669–670.

Fortson, J.C., & Field, R.C. (1979). Capital Budgeting Techniques for Forestry: A Review. *Southern Journal of Applied Forestry* 141–143.

Fox, T.R., Allen, H.L., Albaugh, T.J., Rubilar, R., & Carlson, C.A. (2007). Tree Nutrition and Forest Fertilization of Pine Plantations in the Southern United States. *Southern Journal of Applied Forestry* 31(1), 5–11.

Gansner, D.A., & Larsen, D.N. (1969). Pitfalls of Using Internal Rate of Return to Rank Investments in Forestry. United States Dept. of Agriculture Forest Service Research Northeast Research Station, Note NE-106.

van Gardingen, P.R., McLeish, M.J., Phillips, P.D., Fadilah, D., Tyrie, G., & Yasman, I. (2003). Financial and Ecological Analysis of Management Options for Logged-Over Dipterocarp Forests in Indonesian Borneo. *Forest Ecology and Management* 183, 1–29.

Godman, R.M., & Mendel. J.J. (1978). Economic Values for Growth and Grade Changes of Sugar Maple in the Lake States. USDA Forest Service North Central Research Station Research Paper NC-155.

Graham, J.R., & Harvey C.R. (2001) The theory and practice of corporate finance: evidence from the field. *Journal of Financial Economics* 60:187–243.

Gregory, G.R. (1987). *Forest Resource Economics*. New York, NY: John Wiley & Sons.

Grisez, T.J., & Mendel, J.J. (1972). The Rate of Value Increase for Black Cherry, Red Maple, and White Ash. USDA Forest Service Northeastern Research Station Research Paper NE-231.

Hagemann, H. (1990). Internal Rate of Return. In *The New Palgrave Capital Theory*, J. Eatwell, M. Migate, and P. Newman, eds. New York, NY: W.W. Norton & Company.

Hajdasinski, M.M. (1996). Adjusting the modified internal rate of return. *The Engineering Economist* 41(4), 173–186.

Hajdasinski, M.M. (1997). NPV-Compatibility, Project Ranking, and Related Issues. *The Engineering Economist* 42(4), 325–339.

Hansen, B.G. (1986). Selecting a Tax Rate for Use in Analyzing Forest Industry Capital Investments. *Northern Journal of Applied Forestry* 3, 101–103.

Harou, P. (1985). On a social discount rate for forestry, *Canadian Journal of Forest Research*, 15:927–934.

Harpole, G.B. (1984). Internal Rate of Return May be Used to Define Initial Equity for Composite Rate-of-Return Analyses. *Forest Science* 30(4), 1096–1102.

Hartman, J.C., & Schafrick, I.C. (2004). The Relevant Internal Rate of Return. The *Engineering Economist* 49(2), 139–158.

Hatcher, R.L., Johnson, L.A., & Hopper G.M. (1993). Economic Potential of Black Walnut on Small Acreage Tracts. *Southern Journal of Applied Forestry* 17(2), 64–68.

Hazen, G.B. (2003). A New Perspective on Multiple Internal Rates of Return. *The Engineering Economist* 48(1), 31–51.

Heiligmann, R. B. (2008). Here's How To … Increase Financial Returns from your Woodland. *Forestry Source* 13(7), 13–14.

Henderson, J.M., & Quandt, R.E. (1980). *Microeconomic Theory: A Mathematical Approach*, 3rd ed. New York, NY: McGraw-Hill.

Hepburn, C.J. & Koundouri, P. (2007). Recent advances in discounting: Implications for forest economics, *Journal of Forest Economics*, 2–3(13):169–189.

Herrick, O.W., & Gansner, D.A. (1985). Forest-Tree Value Growth Rates. *Northern Journal Applied Forestry* 2(1), 11–13.

Hirshleifer, J. (1970). *Investment, Interest and Capital*. Upper Saddle River, NJ: Prentice-Hall, Inc.

Hseu, J., & Buongiorno. J. (1997). Financial Performance of Maple-Birch Stands in Wisconsin: Value Growth Rate Versus Equivalent Annual Income. *Northern Journal of Applied Forestry* 14(2), 59–66.

Hu, S-C., & Burns, P.Y. (1988). Profits from Growing Virginia Pine Christmas Trees in Louisiana. *Southern Journal of Applied Forestry* 12(2), 122–124.

Jagger, P., & Pender, J., (2003). The Role of Trees for Sustainable Management of Less-Favored Lands: The Case of Eucalyptus in Ethiopia. *Forest Policy and Economics* 5, 83–95.

Johansson, P~O., & Löfgren, K~G. (1985). *The Economics of Forestry and Natural Resources*. Oxford, UK: Basil Blackwell Ltd.

Johnstone, D. (2008). What Does an IRR (or Two) Mean? *Journal of Economic Education* 39(1), 78–87.

Just, R.E., Hueth, D.L., & Schmitz, A. (2004). *The Welfare Economics of Public Policy: A Practical Approach to Project and Policy Evaluation*. Northampton MA: Edward Elgar.

Kadam, K.L., Wooley, R.J., Aden, A., Nguyen, Q.A., Yancey, M.A., & Ferraro, F.M. (2000). Softwood Forest Thinnings as a Biomass Source for Ethanol Production: A Feasibility Study for California. *Biotechnological Progress* 16, 947–957.

Kant, S. (2003). Extending the Boundaries of Forest Economics, *Forest Policy and Economics*. 5:39–56.

Kierulff, H. (2008). MIRR: A better measure. *Business Horizons* 51,321–329.

Keynes, J.M. (1936). *The General Theory of Employment, Interest and Money*. London, UK: Macmillan, (available on the web at http://ebooks.adelaide.edu.au/k/keynes/john_maynard/k44g/ accessed on 16 June 2009).

Klemperer, W.D. (1981). Interpreting the Realizable Rate of Return. *Journal of Forestry* 79(9), 616–617.

Klemperer, W.D. (1996). *Forest Resource Economics and Finance*. New York, NY: McGraw-Hill, Inc.

Kurtz, W.B., Garrett, H.E., & Williams, R.A. (1981). Young Stands of Scarlet Oak in Missouri can be Thinned Profitably. *Southern Journal of Applied Forestry* 5(1), 12–16.

Legault, I., Ruel, J-C., Pouliot, J-M., & Beauregard, R. (2007). Analyse Financière de Scénarios Sylvicoles Visant la Production de Bois D'œuvre de Bouleaux Jaune et à Papier. *The Forestry Chronicles* 83(6):840–851.

Leuschner, W.A. (1984). *Introduction to Forest Resources Management*. New York, NY: John Wiley & Sons,.

Levin, S.A. (2002). Descartes' Rule of Signs – How hard can it be? Available at http://sepwww.stanford.edu/oldsep/stew/descartes.pdf accessed on 8 July 2009.

Lin, S.A.Y. (1976). The modified internal rate of return and investment criterion. *The Engineering Economist* 21(4):237–247.

Liu, J.P., & Wu, R.Y. (1990). Rate of return and optimal investment in an imperfect capital market. *American Economists* 34(2), 65–71.

Lorie, J.H., & Savage, L.J. (1955). Three Problems in Rationing Capital. *The Journal of Business* 28(4):229–239.

Luenberger, D.G. (1998). *Investment Science*. New York, NY: Oxford University Press.

Marty, R. (1970). The Composite Internal Rate of Return. *Forest Science* 16(3), 276–279.

Mehta, N.B., & Leuschner, W.A. (1997). Financial and Economic analyses of Agroforestry Systems and a Commercial Timber Plantation in the La Amistad Biosphere Reserve, Costa Rica. *Agroforestry Systems* 37, 175–185.

Mendel, J.J., Grisez, T.J., & Trimble, Jr. G.R. (1973). The Rate of Value Increase for Sugar Maple. USDA Forest Service Northeastern Research Station Research Paper NE-250.

Mills, T.J., & Dixon, G.E. (1982). Ranking Independent Timber Investments by Alternative Investment Criteria. US Dept. Agr, Pacific Southwest Forest and Range Experiment Station, Research Paper PSW-166.

Newnan. D.G. (1983). *Engineering Economic Analysis*, 2nd ed. San Jose, CA: Engineering Press, Inc.

Nocetti, D., Jouini, E., & Napp, C. (2008). Properties of the social discount rate in a Benthamite framework with heterogeneous degrees of impatience. *Management Science* 54(10):1822–1826.

Norstrøm C.J. (1972). A Sufficien Condition for a Unique Nonnegative Internal Rate of Return. *Journal of Financial and Quantitative Analysis* 7(3), 1835–1839.

Ohlin, B. (1921). Till frågan om skogarnas omloppstid. *Ekonomisk Tidskrift*. Vol. 22. Reprinted as Ohlin B. 1995. Concerning the Question of the Rotation Period in Forestry. *Journal of Forest Economics* 1(1):89–114.

Pearce, P.H. (1990). *Introduction to Forestry Economics*. Vancouver, BC: University of British Columbia Press.

Pickens, J., Johnson, D.L., Orr, B.D., Reed, D.D., Webster, C.E., & Schmierer, J.M. (2009). Expected Rates of Value Growth for Individual Sugar Maple Crop Trees in the Great Lakes Region: A Reply. *Northern Journal of Applied Forestry*. 26(4):145–147.

Pressler, M.R. (1860). Aus der Holzzuwachlehre (zweiter Artikel). *Allgemeine Forst- und Jagd-Zeitung*. Vol. 36. Reprinted as Pressler, M.R. 1995. For the Comprehension of Net Revenue Silviculture and the Management Objectives Derived Thereof. *Journal of Forest Economics* 1(1):45–88.

Price, C. (1993). *Time, Discounting and Value*. Oxford, UK: Blackwell Publisher.

Rideout, D.B., & Hesseln, H. (2001). *Principles of Forest & Environmental Economics*, 2nd ed. Resource & Environmental Management, LLC., Ft. Collins CO.

Samuelson, P.A. (1976). Economics of Forestry in an Evolving Society. Reprinted as Samuelson, P.A., 1995. Economic of Forestry in an Evolving Society. *Journal of Forest Economics* 1(1), 115–149.

Sandulescu, E., Wagner, J.E., Pailler, S., Floyd, D.W., & Davis, C.J. (2007). Policy analysis of a government-sanctioned management plan for a community-owned forest in Romania. *Forest Policy and Economics* 10(1–2), 14–24.

Schallau, C.H., & Wirth, M.E. (1980). Reinvestment Rate and the Analysis of Forestry Enterprises. *Journal of Forestry* 78(12), 740–742.

Shaffer, R.M., McNeel, J.F., Overboe, P.D., & O'Rourke, J. (1987). On-Board Log Truck Scales: Application to Southern Timber Harvesting. *Southern Journal of Applied Forestry* 11(2), 112–116.

Silberberg, E., & Suen.W. (2001). *The Structure of Economics: A Mathematical Analysis*, 3rd ed. New York, NY: McGraw-Hill, Inc.

Siry, J.P., Robison, D.J., & Cubbage, F.W. (2004). Economic Returns Model for Silvicultural Investments in Youn Hardwood Stands. *Southern Journal of Applied Forestry* 28(4):179–184.

Terreaux, J.P.H., & Peyron, J.L. (1997). A critical View of Classical Rotation Optimization. In Moiseev, N.A., von Gadow,K., Krott, M. ed. Plannin and Decision-Making for Forest Management in the Market Economy. IUFRO Internation Conference at Pushkino, Moscow Region Russia September 25–29, 1996.

Venn, T.J. (2005). Financial and economic performance of long-rotation hardwood plantation investments in Queensland, Australia. *Forest Policy and Economic* 7, 437–454.

Wagner, J.E. (2009). Expected Rates of Value Growth for Individual Sugar Maple Crop Trees in the Great Lakes Region: A Comment. *Northern Journal of Applied Forestry*. 26(4):141–144.

Wagner, J.E., Nowak, C.A., & Casalmir, L.M. (2003). Financial Analysis of Diameter-Limit Cut Stands in Northern Hardwoods. *Small-scale Forest Economics, Management and Policy* 2(3), 357–376.

Webster, C., Reed, D., Orr, B., Schmierer, J., & Pickens, J. (2009). Expected Rates of Value Growth for Individual Sugar Maple Crop Trees in the Great Lakes Region. *Northern Journal of Applied Forestry*. 26(4):133–140.

Webster, H.H. (1965). Profit Criteria and Timber Management. *Journal of Forestry* 63(4), 260–266.

Uys, H.J.E. (1990). A New Form of Internal Rate of Return. *South African Forestry Journal* 154, 24–26.

Zinkhan, F.C., & Cubbage, F.W. (2003). Chapter 6 – Financial Analysis of Timber Investments. In Sills, E.O. and Abt, K.L. (eds) *Forest in a Market Economy*. Dordrecht Netherlands: Kluwer Academic Publishers.

Appendix 10

Calculus of the even-aged forest rotation problem

The financially optimal multiple rotation even-aged model developed by Faustmann (1849), Pressler (1860), and Ohlin (1921) is given by equation (A10.1)

$$
\begin{aligned}
LEV_t &= \frac{PQ(t)\cdot(1+r)^{-t} - C_0}{1 - (1+r)^{-t}} \\
&= \frac{PQ(t) - C_0\cdot(1+r)^t}{(1+r)^t - 1}
\end{aligned}
\tag{A10.1}
$$

A continuous interest rate and compounding factor as compared to a discrete interest rate and compounding factor are often used for computational ease. Chiang (1984) derives the continuous compounding factor from the discrete compounding factor given in equation (A10.2):

$$
\lim_{m \to \infty} \left(1 + \frac{r}{m}\right)^{t \cdot m} = e^{\hat{r}t}
\tag{A10.2}
$$

Based on equation (A10.2) a relationship can be derived between a discrete and continuous interest rate and a discrete and continuous compounding factor. These are given in equations (A10.3a) and (A10.3b), respectively

$$
\hat{r} = \ln(1 + r)
\tag{A10.3a}
$$
$$
(1 + r)^t = e^{\hat{r}t}
\tag{A10.3b}
$$

Equation (A10.3b) can be substituted into equation (A10.1). For the purposes of deriving the financially optimal multiple rotation even-aged searching rule described in equation (9.8a), I will drop the distinction between a discrete and continuous interest rate and use the notation e^{rt}. Equation (A10.1) can be re-written as

$$
\begin{aligned}
LEV_t &= \frac{PQ(t)\cdot e^{-rt} - C_0}{1 - e^{-rt}} \\
&= \frac{PQ(t) - C_0\cdot e^{rt}}{e^{rt} - 1}
\end{aligned}
$$

The searching rule given in equation (9.8a) is determined by taking the first derivative of LEV_t with respect to time, t, and setting this derivative equal to zero. This is given in equation (A10.4)

$$\frac{dLEV_t}{dt} = \frac{(1 - e^{-rt}) \cdot \left(P\dfrac{dQ(t)}{dt} e^{-rt} - re^{-rt} PQ(t) \right) - (PQ(t)e^{-rt} - C_0) \cdot (re^{-rt})}{[(1 - e^{-rt})]^2} = 0$$

(A10.4)

Equation (A10.4) can be simplified using the following steps given that T represents the rotation age that sets this expression equal to zero

$$\frac{(1 - e^{-rT}) \cdot \left(P\dfrac{dQ(T)}{dT} e^{-rT} - re^{-rT} PQ(T) \right) - (PQ(T)e^{-rT} - C_0) \cdot (re^{-rT})}{[(1 - e^{-rT})]^2} = 0$$

$$(1 - e^{-rT}) \cdot \left(P\frac{dQ(T)}{dT} e^{-rT} - re^{-rT} PQ(T) \right) - (PQ(T)e^{-rT} - C_0) \cdot (re^{-rT}) = 0$$

$$(e^{-rT}) \cdot \left(P\frac{dQ(T)}{dT} - rPQ(T) \right) - \frac{(PQ(T)e^{-rT} - C_0)}{(1 - e^{-rT})} \cdot (re^{-rT}) = 0$$

$$P\frac{dQ(T)}{dT} - rPQ(T) - r \cdot \frac{(PQ(t)e^{-rT} - C_0)}{(1 - e^{-rT})} = 0$$

$$P\frac{dQ(T)}{dT} = rPQ(T) + r \cdot \frac{(PQ(t)e^{-rT} - C_0)}{(1 - e^{-rT})}$$

$$MRP_T = P\frac{dQ(T)}{dT} = rPQ(T) - r \cdot \frac{(PQ(t)e^{-rT} - C_0)}{(1 - e^{-rT})} = rPQ(T) + rLEV_T = MC_T$$

The last expression is the financially optimal multiple rotation even-aged searching rule described in equation (9.8a). In mathematical terms, equation (A10.4) describes the first order or necessary conditions for an optimum. As the first order conditions could describe the optimum for a maximum or a minimum, an additional set of requirements is needed to ensure a maximum. These are described as second order conditions. The mathematical development of the second order conditions is beyond the scope of this book. I would refer an interested reader to Johansson and Löfgren (1985) and Amacher *et al.* (2009) if they wish to explore this further.

A similar approach can be used to determine the financially optimal single rotation age searching rule given in equation (9.2a). The single rotation age problem can be written as

$$NPV_t = PQ(t) \cdot e^{-rt} - C_0$$

(A10.5)

The first order or necessary conditions for an optimum are

$$\frac{dNPV_t}{dt} = P\frac{dQ(t)}{dt} e^{-rt} - rPQ(t)e^{-rt} = 0$$

(A10.6)

Equation (A10.6) can be simplified using the following steps given that T represents the rotation age that sets this expression equal to zero

$$P\frac{dQ(T)}{dT}e^{-rT} - rPQ(T)e^{-rT} = 0$$

$$P\frac{dQ(T)}{dT} - rPQ(T) = 0$$

$$MRP_T = P\frac{dQ(T)}{dT} = rPQ(T) = MC_T$$

The last expression is the financially optimal single rotation even-aged searching rule described in equation (9.2a). As before, second order conditions are required to insure a maximum. Again, I would refer an interested reader to Johansson and Löfgren (1985) and Amacher *et al.* (2009) if they wish to explore this further.

Appendix 11
The Faustmann–Smith–Samuelson model

The financially optimal multiple rotation age model for even-aged forests developed by Faustmann (1849), Pressler (1860), and Ohlin (1921) – equation (9.7) – describes the present value of a fixed asset to produce a perpetual period income flow. While it is probably the most familiar approach used to analyze an entrepreneur's investment in forest management, it should not be given carte blanche. For example, it is a stand-level model and cannot simply be aggregated to the forest-level (Wear and Newman 1991; Newman and Wear 1993; Yin 1997; Yin and Newman 1997; Yin *et al.* 1998). In addition, it does not allow for developing an aggregate supply curve using Hotelling's lemma (supply $= \dfrac{\partial \Pi(P)}{\partial P}$, see Varian (1992)) as it does not conform to the mathematical conditions of a regular profit function (Wear and Newman 1991; Newman and Wear 1993; Yin 1997; Yin and Newman 1997). To address these concerns, these authors propose a profit function based on the Faustmann–Smith–Samuelson model described by Smith (1968), Samuelson (1976b), and Comolli (1981):

$$\Pi(P_t, r_t, w_t, l_t) = \max\{P_t[Q(t) - r_t I(t)] - w_t K(t) - l_t L(t)\}; \quad t = 1, 2, 3,\dots \quad \text{(A11.1)}$$

where
$\Pi(\bullet)$ = profits;
P_t = stumpage price at time t;
$Q(t)$ = the production system as a function of time;
r_t = the discount rate at time t;
$I(t)$ = the inventory at time t;
w_t = wage for inputs at time t;
$K(t)$ = variable inputs used at time t;
l_t = the market rental rate for land; and
$L(t)$ = land area devoted to forest management at time t.

While a complete discussion comparing and contrasting equations (A11.1) and (9.7) is beyond the scope of this book, I would counsel the reader to think critically when using any analytical model to help develop economic information used in the decision making process.

Notes

1 Introduction

1 While it is true that forests provide both market and nonmarket outputs and services in addition to market outputs such as wood and wood fiber, for the purpose of this argument I will focus only on market outputs such as wood and wood fiber.

2 Wagner *et al.* (2004) examines this issue using hardwood logs.

3 Given the example from the previous paragraph, my assertion does not mean that if the landowner had all the relevant economic information available the choice between action or no action would change. My assertion is that while the observable choice is important, equally, if not more, important is the analysis used to reach that choice.

4 Foerde *et al.* (2006) found that active multi-tasking, such as watching television while studying, is less efficient than focusing only on studying. Thus active multi-tasking comes at a cost of efficiency.

5 Patterson and Xie (1998) did acknowledge that these results are based on a small sample size and "can only be used to develop indications and not to set standards or design values."

6 The treatment of absolute versus relative scarcity is described for those resources, commodities, or services traded in the formal marketplace. Simpson *et al.* (2004) discuss the concept of the "New Scarcity" or the "limitations on the environment to absorb and neutralize the unprecedented waste streams of humanity." While the concept of New Scarcity is important, I have chosen to limit my discussion to the traditional concepts of absolute and relative scarcity.

7 More information on property rights can be found by reading Cole and Grossman (2002), Hanna *et al.* (1995), Bromley (1991), Hardin (1968), and Coase (1960).

8 This can be shown through the use of symbolic logic and the invalid deductive argument form known as the Fallacy of Affirming the Consequence (Copi and Cohen 1998).

9 A discussion of the effects of positive and negative externalities on market equilibrium, also known as a market failure, is in Chapter 7 Market equilibrium and structure.

10 An often heard criticism is that the development of cause-and-effect relationships in economics, or economic theory, is not based on realistic assumptions. Silberberg and Suen (2001) provide a nice concise discussion of this criticism that will not be repeated here.

11 The price of $7.76 per beam represents a two-year weighted average of the market price of various length beams (Patterson *et al.* 2002).

12 Implicit costs can be described as entrepreneurial and social. Entrepreneurial implicit costs are opportunity costs of the time and other resources a firm's owners make available for production with no direct cash outlay (Field and Field 2002; Henderson and Quandt 1980). Social costs are the effects of pollution or negative externalities (Field and Field 2002; Henderson and Quandt 1980).

13 In the terminology of operations research, the least cost and cost effective models describe a primal–dual relationship (Winston 1994).

2 Production systems

1 This is a summary of a more detailed discussion contained in the Chapter 1.
2 The net benefit model will be used in Chapter 6 with respect to the consumer or buyer.
3 This is a summary of a more detailed discussion contained in the Chapter 1.
4 Site index is a species-specific measure of actual or potential forest productivity expressed in terms of the average height of trees included in a specified stand component at a specified index or base age. Site index is often used as an indicator of site quality and productivity. The higher the site index for a given index or base age the better the site quality and productivity.
5 While these are inputs, are they inputs that the manager has direct control over?
6 In geometric terms the volume of a cylinder or cone is a function of diameter and height. Calculating the volume of a tree or log is the same as calculating the volume of a cylinder or cone. Thus, for biometricians whose end goal is to calculate volume (not profit), a measure of diameter (e.g. diameter at breast height) and height (e.g. site index) are sufficient for describing the system.
7 Burkhart *et al.* (1985) examined if including site preparation actions improved the estimation of height and yield, but did not include these variables in the equations because no firm conclusions could be drawn from the analysis due to the study design. In contrast, the mathematical models used to describe a production system for Slash Pine given by Yin *et al.* (1998) had the Architectural Plan for Profit in mind and contain inputs that can be used to help construct the Pillar of Cost.
8 The USDA Forest Service's *Forest Inventory and Analysis National Core Field Guide* gives the grading factors for hardwood and softwood trees (USDA FS FIA 2006).
9 This key component has been taken from the definition of a production function as given by Carlson (1974), Henderson and Quandt (1980) and Silberberg and Suen (2001).
10 Interpreting the concept of technical efficiency is difficult given the production process for the Select Red Oak and the Loblolly Pine Plantation case studies. The idea of technical efficiency has traditionally been defined with respect to the classical definition of a production function (e.g. see Carlson (1974), Henderson and Quandt (1980) and Silberberg and Suen (2001)). The inputs given in a classical production function can all be controlled directly by the manager. Thus, the manager can manipulate the use of the inputs to increase efficiency. In the case of the Select Red Oak and the Loblolly Pine Plantation case studies, the manager cannot manipulate an input such as diameter at breast height directly; therefore, the concept of productivity is more appropriate and is analogous to technical efficiency.
11 Be careful of the over-confidence in the precision that is possible when calculating total product using equations. For example, given equation (2.4), a diameter of 15 inches, and a site index of 67, the board foot volume of a select red oak can be calculated as 150.158896 bd.ft/tree. However, this level of precision is not defensible or reasonable given what is being measured and managed.
12 This definition is based on those provided by Rideout and Hesslen (2001) and Pearce (1994).
13 Appendix 1 provides a calculus-based discussion comparing average and marginal product.
14 While the same information is in Tables 2.8 and 2.9, it may be easier to see on the graphs.
15 A different example that you may be more familiar with is if this semester's grade point average is greater than your cumulative grade point average, then your cumulative grade point average will increase. However, if this semester's grade point average is less than your cumulative grade point average, then your cumulative grade point average will decrease.
16 Sourd (2005) has developed a mathematical model that examines this problem in greater detail. I would recommend reading the Introduction, 2.1 Problem definition, and 2.2 Time windows and breaks.

17 For average product to be increasing, $MP > AP$. Thus, the increased inefficiency is also reflected in marginal product and marginal product does not reflect the Law of Diminishing Returns.

18 While I will not discuss the problems that waste can have on profits, you as a manager should also be aware that waste is an output of the production process and must be included in your analyses.

3 Costs

1 The production system according to Chapter 1 must define input–output combinations that are technically efficient. Appendix 2 develops the relationship between technical efficiency and production cost efficiency.

2 Traditionally, sap was collected in buckets that hung from the taps. This technology has been replaced by vacuum-tubing sap collection systems in commercial operations.

3 The Brix measurement is the ratio of dissolved sugar to water in a liquid. Thus, the higher the Brix measurement the higher the dissolved sugar content.

4 An operating input (e.g. hired labor) is described as an input that is bought or sold as part of normal business operations. A capital input is described as a produced good (e.g. a fishing boat or tackle) that is used to produce an output. Capital inputs usually have a useable life span of one year or greater.

5 The difference between the short-run and the long-run is discussed later in this chapter.

6 The DIRTI-5 are taken from the agricultural farm budgeting literature: American Society of Agricultural Engineers (ASAE) Standards #496, section 6.3.1, and #497, Table 3.

7 Huyler (1982) describes how these fixed equipment costs were annualized.

8 I will revisit how the payments for the Mobile Micromill and support equipment were determined in Chapter 8.

9 Due to the cost of moving and siting the Mobile Micromill, Becker *et al.* (2004) assumed the mill would be relocated at most once per year.

10 If screw-in type fittings have the least variable production costs, does this imply that this fitting is the one that will maximize profit? How do Burdurlu *et al.* (2006) answer this question?

11 If the change in AFC as output changes is constant, then $\dfrac{dAFC}{dQ}$ would be constant. However,

$$\frac{dAFC}{dQ} = \frac{d\left(\dfrac{FC}{Q}\right)}{dQ} = -\frac{FC}{Q^2}$$

which is not a constant.

12 As defined by equation (3.5), AVC is a relationship between "variable cost" and output. With a production process that uses only one input, AVC can be defined as

$$AVC = \frac{TVC}{Q} = \frac{wx}{Q} = \frac{w}{Q/x} = \frac{w}{AP}$$

This relationship does not hold if more than one input is used in the production process. Also note that this ties the concept of AVC to the production system directly.

13 I will assume for purposes of illustrating AVC that the information in Table 3.6b represents the production cost efficient variable costs of a 12,000-tap operation.

14 Economies of scale are due to two factors. First, the spreading out of the TFCs over greater output (i.e. AFC decreases as output increases). Second, while TVC are increasing due

to increased production, output is increasing faster. This is due to the nature of the production system. Thus the ratio of TVC to output (i.e. AVC) is decreasing.

In addition to economies of scale, there are also diseconomies of scale (or decreasing returns to scale) – ATC increases as output production increases – and constant economies of scale (or constant returns to scale) – ATC is constant as output production increases.

15 The fact that TFCs are irrelevant can be shown algebraically

$$\Delta TC = TC_2 - TC_1$$
$$= TVC_2 + TFC - (TVC_1 + TFC)$$
$$= TVC_2 - TVC_1 + TFC - TFC$$
$$= TVC_2 - TVC_1 = \Delta TVC$$

16 As defined by equation (3.7), MC is a relationship between "variable cost" and output. With a production process that uses only one input, MC can be defined as

$$MC = \frac{\Delta TVC}{\Delta Q} = \frac{(wx_2 - wx_1)}{\Delta Q}$$
$$= \frac{w(x_2 - x_1)}{\Delta Q} = \frac{w\Delta x}{\Delta Q}$$
$$= \frac{w}{\Delta Q / \Delta x} = \frac{w}{MP}$$

This relationship does not hold if more than one input is used in the production process. Also note that this ties the concept of MC to the production system directly.

17 Appendix 3 provides a calculus-based discussion comparing average and marginal cost.

18 These last two conditions are not shown in the Maple Syrup Operation case study due to limited data presented by Huyler (2000).

19 What about the land? Are you utilizing all of the land? What about the old house? Could you move your office into one of the other buildings and rent the house to another business?

20 Profit margins will be discussed in more detail in Appendix 6.

21 Information concerning using Lagrangian multipliers to solve constrained optimization problems can be found in Lambert (1985) and Chiang (1984).

4 Revenue

1 While this discussion focuses on output price, it also holds with respect to the prices paid for inputs (i.e. wages).

2 As a hint, compare and contrast the profit model with the least cost/cost effective models to determine what economic information is contained in each.

3 The USDA Forest Service's *Forest Inventory and Analysis National Core Field Guide* gives the grading factors for hardwood and softwood trees (USDA FS FIA 2006). The differences between hardwood grades are due to more than just diameter. Other factors include length of the section being graded, minimum diameter inside bark at top of section being graded, number of clear faces, and cull deductions due to crook and sweep (United States Department of Agriculture, Forest Service 2006).

4 Stumpage price denotes the output price for live standing trees (timber) on the stump.

5 What are the units on average cost? Review Chapter 3 to answer this question.

6 What are the units on marginal cost? Review Chapter 3 to answer this question.

7 Chapter 2 discussed TP in detail and I will not repeat that complete discussion here; however, you are encouraged to review that section of Chapter 2.

8 The focus on this discussion is commercial uses of the tree. Non-commercial uses such as bequest and option values will be discussed in Chapter 12.

9 See note 6 for Chapter 2.

10 Although it is mathematically possible for the average be the same as each observation, this will not be the case for production systems following the Law of Diminishing Returns.

5 Profit

1 Appendix 5 uses calculus to develop the necessary and sufficient conditions for profit maximization using an output and input approach. This appendix also is used to derive the profit searching rule.

2 The assumption we have been using is that the market has been characterized by workable competition – no one buyer or group of buyers and no one seller or group of sellers can influence price. Output price is constant. In other words, the market sets the price and producers have almost no ability to manipulate the price of their output.

3 The definitions of technical efficiency and production cost efficiency are found in Chapter 2 and Chapter 3, respectively.

4 In 2009, a gallon of maple syrup was selling at retail stores for almost $160 per gallon as reported in an 11 March 2009 *New York Times* article entitled "As Maple Syrup Prices Rise, New York Leaders see Opportunity" (http://www.nytimes.com/2009/03/11/dining/11maple.html accessed on 25 September 2009).

5 In essence, the fixed costs listed in Table 5.2a are variable costs when comparing different-sized operations: different levels of the fixed assets of capital and land are required for different-sized operations. This is analogous to the difference between short-run and long-run analysis (see Chapters 7 and 8 in Pindyck and Rubinfeld (1995)).

6 Table 5.2b also illustrates economies of scale (or increasing returns to scale). As the size of the operation increases the total production cost per unit of output decreases. Namely, the average cost decreases as the size of the operation increases and the incremental costs are below the average costs. This was described in Chapter 3.

7 Students who are familiar with calculus are challenged to show that the condition $MC \leq P$ bounds profit between its minimum and maximum points.

8 Information concerning using Lagrangian multipliers to solve constrained optimization problems can be found in Lambert (1985) and Chiang (1984).

6 Supply and demand

1 The reader should review the concept of an economic model as described in Chapter 1.

2 For the purposes of this chapter, the supply curve will describe a representative entrepreneur and the demand curve will describe a representative buyer or group of buyers. Various different market structures and their implications for market equilibriums will be discussed briefly in Chapter 7.

3 Dimension lumber is lumber that is cut and finished to standardized width and depth usually measured in inches; for example, common sizes are 2 × 4, 2 × 6, 2 × 8, and 4 × 4, etc. The actual finished width and depth are 1.5 × 3.5, 1.5 × 5.5, 1.5 × 7.5, and 3.5 × 3.5, etc. This is due to the width of the saw and planing to square the lumber. Board lengths of dimension lumber can vary; however, standard lengths are 6, 8, 10, 12, 14, 16, 18, 20, 22, and 24 feet.

4 Chapter 3 shows the direct relationship between the definition of the production system and the determination of total and variable costs; average total and variable costs; and marginal cost. I would recommend that the reader review this information briefly.

5 For example, the retail price of a $2 \times 4 \times 8$ ranges from 2.29 to 2.98 US dollars per piece depending on quality (http://www.lowes.com/lowes/lkn?action=productList&Ne= 4294947504&category=Dimensional%20Lumber%20Area%2016&N=4294927926 accessed on 11 November 2009). These retail prices equate to approximately 661.81 to 861.22 US dollars per 1,000 board feet of lumber. The illustration below shows approximately how many $2 \times 4 \times 8$s are in 1,000 board feet of lumber.

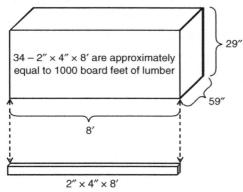

Figure N.1 Illustration of how many $2'' \times 4'' \times 8'$
are in 1,000 board feet of lumber.

6 If a buyer is willing to pay more than $340.36 per MBF of lumber for less than 450.10 MBF of lumber per month and the market price is greater than $340.36 per MBF of lumber, then the sawmill's owners will produce at the technically efficient and production cost efficient output levels for the given market price and sell any excess lumber in the market.

7 The extrapolation was based on a 2nd-degree polynomial of average total costs.

8 The area under the supply curve can be calculated using two mensuration formulae.

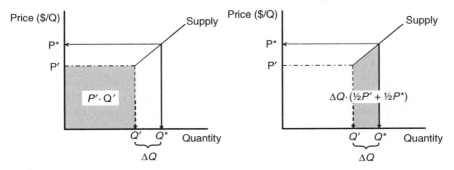

Figure N.2 Illustration of calculating the area under the supply curve.

‡P′ denotes the minimum average variable cost. P* and Q* denote the market equilibrium price and quantity respectively.

The first is calculating the area of the rectangle: $P' \cdot Q'$ which represents the TVC of producing Q'. The second is using the trapezoid rule to calculate the area of the remaining component: $\Delta Q \cdot (\frac{1}{2}P' + \frac{1}{2}P^*)$. The trapezoid rule is an algebraic approximation of the area under the marginal cost curve. For example, the TVC of producing 470.33

MBF of lumber per month is \$160,110.99 (Table 6.2). The approximation of TVC based on the two mensuration formulae is \$160,096.67. It is left to the reader to check my calculations.

9 The reader will notice that this paragraph is almost identical to the first paragraph under the Supply section. This was done on purpose and helps illustrate the parallels between the concepts of supply and demand.

10 Forestland is defined as land that is at least 10 percent stocked by forest trees of any size, including land that formerly had such tree cover and that will be naturally or artificially regenerated. The minimum area for classification of forest land is one acre (Smith *et al.* 2009).

11 We will revisit this same issue in Chapter 7.

12 The calculation of consumer surplus used was based on a linear demand curve. If the demand curve is nonlinear, then readers familiar with calculus will be able to verify that consumer surplus is

$$CS = \left[\int_{P}^{P*} Q(P)\, dP \right] - P* \cdot Q(P*)$$

where $Q(P)$ denotes output quantity is a function of the price.

13 Own-price elasticity is technically a negative number as the demand curve has a negative slope. However, it is often given as the absolute value. If the demand curve is perfectly inelastic, own-price elasticity is equal to 0. If the demand curve is perfectly elastic, own-price elasticity is equal to infinity.

7 Market equilibrium and structure

1 Transactions costs can be described as searching and information costs (e.g. both consumers and producers search for each other, information about products, etc.), bargaining costs (e.g. agreeing upon a price and quantity), and enforcement costs (e.g. redress for any failure to complete agree upon actions by buyer and seller).

2 The default market structure is defined as workable competition (Chapter 1). I will expand upon this in the section entitled "Market structure."

3 Entrepreneur A

Marginal Cost = Supply = $1 + Q$

Entrepreneur B

Marginal Cost = Supply = $6 + 0.25 \cdot Q$

Let market price be 10 $/Q$. Given this information, it is left to the reader to show that market producer surplus of \$72.5 is equal to the sum of the individual producer surplus of each entrepreneur.

4 I would recommend that readers review the section entitled "Efficiency versus equity" in Chapter 1.

5 The concept of workable competition was first described by J.M. Clark in 1940 (Clark 1940). As the conditions required for perfect competition are rarely satisfied, he argued that the goal of governmental policy should be to make competition "workable," not necessarily perfect. His proposal sparked much discussion at the time. My use of the term implies that neither entrepreneurs nor consumers have sufficient market power to significantly alter the price searching rule of $MR = P = Supply = MC$ even though the conditions required for perfect competition are not met.

6 Under purely competitive conditions, marginal cost (MC) curves are equated with supply and the marginal revenue (MR) curves are horizontal, each MR curve can be represented by a specific, exogenously determined price, P_0. Give the MC curve, we can

therefore express the optimal output as $Q^* = Q^*(P_0)$, or the optimal output is a function of price, P_0. The expression $Q^* = Q^*(P_0)$ is a function mapping a real number (price) into a real number (optimal output). However, when the MR curves are downward-sloping under imperfect competition (such as monopoly or monopolistic competition), the optimal output of the entrepreneur with a given MC curve will depend on the specific position of the MR curve; namely, an entrepreneur's MR is a function of the amount produced. In such a case, since the output decision involves a mapping from *curves* to real numbers, this is described as a *functional*: $Q^* = Q^*[MR]$. It is, of course, precisely because of this inability to express the optimal output as a *function* of price that makes it impossible to draw a supply curve for a firm under imperfect competition, as can be for its competitive counterpart (Chaing 1992).

7 As shown by Figure 6.15b, it is not the demand curve but the area $P \cdot Q$ that defines an entrepreneur's total revenue. Using the definition of average revenue (AR) from equation (4.3) (Chapter 4)

$$AR = \frac{TR}{Q} = \frac{P \cdot Q}{Q} = P$$

where *TR* denotes total revenue. The interpretation of AR is total revenue per unit output, P. Consequently

$$TR = P \cdot Q = \left(\frac{P \cdot Q}{Q}\right) \cdot Q = AR \cdot Q$$

or AR multiplied by output (Q) gives total revenue. If the demand curve represents AR and $AR = P$, then the output level a consumer is willing and able to buy, Q, multiplied by AR would give $TR = AR^*Q = P^*Q$. Thus, I would argue that a consumer's demand curve represents an entrepreneur's average revenue curve. Consequently, the entrepreneur's interpretation of market price is as the revenue per unit sold or average revenue.

8 In Chapters 2 and 3, I have developed a relationship between marginal and average product and marginal and average variable cost, respectively; namely, if the average curve was decreasing (increasing), the margin curve was below (above) the average curve. This holds for average revenue and marginal revenue. Let total revenue (TR) be defined as

$$TR = P(Q) \cdot Q$$

where $P(Q)$ denotes that output price is a function of quantity; that is, demand has a negative slope. The first derivative of total revenue with respect to output defines marginal revenue (MR) (equation (4.5))

$$\frac{dTR}{dQ} = MR = \frac{dP(Q) \cdot Q}{dQ}$$

$$= \frac{dP(Q)}{dQ} \cdot Q + P(Q)$$

Because the demand curve has a negative slope, $\frac{dP(Q)}{dQ} < 0$, and the revenue received for all units of the output sold not just the incremental one, $\frac{dP(Q)}{dQ} \cdot Q$, decreases, the marginal revenue is less than the average revenue, $P(Q)$, for any given Q. If $\frac{dP(Q)}{dQ} = 0$ then $MR = P$ and P is not a function of the amount of output sold.

9 I would recommend the reader review the discussion in Chapter 1 on accounting versus economic profit as illustrated by Table 1.1.

10 Based on Figure 7.6a, is there a price below which a monopolists would stop producing?
11 The terms marginal factor cost as well as average factor cost in Figures 7.7a and 7.7b are used to denote that we are examining the cost of a factor of production such as labor. For example, the marginal factor cost of labor would be the hourly wage rate paid ($/hour). In the case of workable competition in the input market, marginal factor cost would equal average factor cost.
12 Another analogy that can be used is that of marginal and average product (Chapter 2). Figures 2.12 and 2.13 show that whenever average product is increasing (decreasing) marginal product is greater (less) than average product. This relationship between margins and averages holds for production systems, costs, and revenues.
13 As the monopsonist is the only buyer of the input, their expenditures bring forth the "supply" of the input (Pindyck and Rubinfeld 1995). The monopsonist's average expenditure per unit of input, or average factor cost, would then define the market supply curve for the input. Given the definition of average factor cost

$$AFC \cdot x_j = \frac{w_j \cdot x_j}{x_j} \cdot x_j = w_j \cdot x_j = TFC_j$$

The monopsonist's total factor cost of labor is illustrated in Figure 7.7a as the area of the rectangle defined by

$$AFC \cdot L^* = W^* \cdot L^*$$

14 Keat and Young (2000) describe this as the three Cs of pricing: cost, customers, and competition.
15 The curious reader may want to compare and contrast equations (7.4a) and (7.4b) with equation (5.6a).
16 The use of the term marginal cost in marginal cost-pricing is not accurate. As Silberberg and Suan (2001: 182) have pointed out "Marginal cost (in the finite sense) is the increase (or derease) in cost resulting from the production of an extra increment of output, which is not the same thing as the 'cost of the last unit.'" Thus, a more accurate description of this should be incremental cost pricing.
17 The reader is referred to Tables 5.4, 5.5a, and 5.5b from the section entitled "To produce or not to produce?" in Chapter 5.
18 Appendix 6 gives a description of price searching given an entrepreneur has market power in their output market.

8 Capital theory: investment analysis

1 The following formula can be used to calculate the monthly payments

$$a = V_0 \frac{\frac{i}{m} \cdot \left(1 + \frac{i}{m}\right)^{t \cdot m}}{\left(1 + \frac{i}{m}\right)^{t \cdot m} - 1}$$

Where a defines the payment, i defines the annual interest rate, t defines the length of the loan, m defines the number of payments per year, and V_0 defines the principle borrowed. Based on the loan information

$$202.65 = 17{,}125 \cdot \frac{\frac{0.0743}{12} \cdot \left(1 + \frac{0.0743}{12}\right)^{10 \cdot 12}}{\left(1 + \frac{0.0743}{12}\right)^{10 \cdot 12} - 1}$$

2 The concept of time preference is not universally accepted as the principle factor of a positive interest rate. Silberberg and Suen (2001) summarize these arguments. They instead describe the concept of dynamic consistency of intertemporal utility functions; namely, they assume that the marginal value of consumption in period k in terms of forgone consumption in period $k + j$ is independent of the date but depends on the consumption levels in the two time periods; that is, the marginal value of consumption depends only on the level of consumption not on which time periods are involved.

3 Could you define an example in which a college student may exhibit a zero rate of time preference? What about loaning a fellow student the money for a 12-ounce can of Coke?

4 Specifically, Silberberg and Suen (2001) qualify this statement by assuming that if a person's utility function is strictly increasing and quasi-concave, then the tradeoff between consumption in two time periods is the rate of production technology change if bound solutions are ruled out. The example they use is of Robinson Crusoe and an edible bush. If bush grows at a constant rate, the tradeoff between consumption in any two consecutive periods is 1 plus the growth rate.

5 In addition, Kennedy *et al.* (1996) propose a novel approach to reduce this particular type of political risk based on the work by the Ronal Coase in his 1960 article entitled "The Problem of Social Cost" for which he received the Nobel Memorial Prize in Economic Sciences in 1991.

6 If the anticipated inflation rate is close to zero, equation (8.1a) is often written as:

$$i = r + \omega$$

However, I would counsel readers strongly *not* to use this simplified version.

7 Silberberg and Suen (2001) and Brealey *et al.* (2008) provide a more complete discussion on this topic.

8 If readers are interested, Brealey *et al.* (2008) provide a very detailed discussion of this topic.

9 A cash flow represents the pattern of revenues and costs over time.

10 While accounting for imperfect capital markets is beyond the scope of this book, interested readers may review the articles by Brazee (2003) and Wagner *et al.* (2003) as examples of analyzing a forestland owner's timber harvest behaviors given imperfect capital markets assuming that landowners maximize wealth.

11 Appendix 7 contains some of the more common formulae.

12 Profit (Architectural Plan for Profit) or income is a flow concept while wealth is a stock concept. Wealth is the capitalization of profit or income.

13 I am indebted to the many forest economists that I have drawn my non-case study examples from and must acknowledge their work: Buongiorno and Gilles (2003), Davis *et al.* (2001), Rideout and Hesseln (2001), Klemperer (1996), Pearce (1990), Gregory (1987), Johansson and Löfgren (1985), Leuschner (1984), and Clutter *et al.* (1983).

14 I will use the real interest rate in defining the future and present value formulae which is consistent with my suggestion of using real cash flows and the real interest rate. However, nominal cash flows and the nominal interest rate can be used as well.

15 Appendix 8 revisits the concept of the interest rate in a little more depth.

16 I will discuss this topic further for all seven investment analysis tools in the section entitled "Capital budget analysis."

17 I would advise not to use the *irr* function contained in many spreadsheet programs unless you understand completely how this function works. There are other functions such as MIRR, but again I would advise that you not use these unless you understand completely how these functions work, what they are calculating, and how to interpret the results.

18 Appendix 9 contains a paper that I wrote detailing the problem with misinterpreting the *irr*.

19 The following discussion resulted from discussions with Dr David Newman, Professor and Chair, Department of Forest and Resources Management, College of Environmental Science and Forestry, Syracuse, New York USA.
20 This is approach is called sensitivity analysis and will be described in more detail in Chapter 10.
21 Chiang (1984) derives the continuous compounding factor from the discrete compounding factor

$$\lim_{m \to \infty} \left(1 + \frac{i}{m}\right)^{t \cdot m} = e^{it}$$

Thus the continuous discounting factor is e^{-it}.

22 However, if $-\kappa > \dfrac{i \cdot t}{\ln t}$, $\forall t > 1$, then $0 < e^{\kappa \ln t} < e^{-it} < 1$.

9 The forest rotation problem

1 Chapter 12 will examine approaches that can be used to estimate values of these goods and services.
2 Why is the emphasis on *their* ownership goals and not what you think their ownership goals should be?
3 Non-timber goods and services are often called ecosystem goods and services. I will discuss these more in depth later in this chapter.
4 To be consistent with Chapter 7, I will use real interest rates, prices, and costs.
5 Planting 1–0 stock means the seedlings are between 8 and 12 months old when planted.
6 The reader may want to question if a 19-year-old loblolly pine tree does in fact describe a sawtimber tree.
7 I have taken some mathematical liberty in deriving the condition given in equation (9.2a). Appendix 10 develops these conditions more completely.
8 The interested reader may want to examine Binkley (1987) for a more in-depth discussion on this topic.
9 Could the model described in equation (9.5) be used to examine sustainable forest management planning and sustainable forest management?
10 This left-hand term has also been called soil expectation value (SEV) or bare land value (BLV).
11 The problem formulation and its solution as presented by Faustmann in 1849 is

$$\begin{aligned}
LEV_t &= P \cdot Q(t) \cdot (1 + r)^{-t} - C_0 + [P \cdot Q(t) \cdot (1 + r)^{-t} - C] \cdot (1 + r)^{-t} \\
&\quad + [P \cdot Q(t) \cdot (1 + r)^{-t} - C] \cdot (1 + r)^{-2t} + \cdots \\
&= [P \cdot Q(t) - C] \cdot (1 + r)^{-t} + [P \cdot Q(t) - C] \cdot (1 + r)^{-2t} \\
&\quad + [P \cdot Q(t) - C] \cdot (1 + r)^{-3t} + \cdots + C_0
\end{aligned}$$

where the initial afforestation or establishment cost, C_0, is isolated from the reforestation cost, C. The assumption is that $C_0 = C$. This expression can be solved giving

$$\begin{aligned}
LEV_t &= \sum_{\theta=0}^{\infty} (PQ(t) - C) \cdot [(1 + r)^{-t}]^{\theta} - C_0 \\
&= \frac{PQ(t) - C}{(1 + r)^t - 1} - C_0 \\
&= \frac{NFV_t}{(1 + r)^t - 1} - C_0
\end{aligned}$$

where NFV_t denotes a net future value at time t. Finally, it can be shown mathematically that

$$LEV_t = \frac{NFV_t}{(1 + r)^t - 1} - C_0 = \frac{NPV_t}{1 - (1 + r)^{-t}}$$

12 Navarro (2003) reviews two variations of the Faustmann Formula given in equation (9.7).
13 The reader may want to question if a 15-year-old loblolly pine tree does in fact describe a sawtimber tree.
14 I have taken some mathematical liberty in deriving the condition given in equation (9.8a). Appendix 10 develops these conditions more completely.
15 Amacher *et al.* (2009) shows that if the market rental rate for bare land is included in the financially optimal single rotation formulation, equation (9.1), then the financially optimal single and multiple rotation ages calculated using equations (9.1) and (9.7) respectively will be the same. Samuelson (1976b) first discussed this in his seminal article.
16 While the Faustmann model is probably the most familiar approach used to analyze an entrepreneur's investment in forest management, it should not be used carte blanche. Appendix 11 provides a brief discussion of an alternative approach.
17 A fully regulated or normal forest has equal acres in each age class up to and including the rotation age (Helms 1998).
18 Navarro (2003) discusses the theoretical and economic implications of the "Faustmann" assumption of afforesting or reforesting given bare land.
19 Hyytiäinen and Tahvonen (2003) describe conditions under which a forest rent rotation age would be shorter than the financially optimal multiple rotation age.
20 It is left to the reader to determine if this statement is accurate for the Loblolly Pine Plantation case study. A hint to this analysis is to find and interpret the solution of $\frac{dFR_T}{dT}$ or the first derivative of FR with respect to the financially optimal single and multiple rotation ages.
21 It is interesting to note for the Loblolly Pine Plantation case study that if the existing plantation is younger than a plantation age of 15, the optimal time to cut the existing plantation will always be at plantation age of 15. If the existing plantation is older than a plantation age of 15, the optimal time to cut the existing plantation will always be to harvest right away. You can verify this by calculating the first order condition of equation (9.12) or $\frac{dFV_t}{dt} = 0$.
22 Of course changing the management regime would also change the amount of time for a loblolly pine plantation to grow 12-inch trees. For example, multiple herbicide and fertilization treatments and one or more thinnings would allow growing 12-inch loblolly pines much faster than 30 to 40 years.
23 The results described in Table 9.6 and Figure 9.8 hold if output price reflects volume production regardless of product class.
24 Even-aged stands can be regenerated also using a shelterwood or seed tree silvicultural system. The end result is still an even-aged stand in which the overstory has been removed completely after sufficient regeneration.
25 The literature abounds with articles on this topic and it would be impossible to give a complete list here; however, I would recommend an interest reader start with the following: Adams and Ek (1974), Buongiorno and Michie (1980), Chang (1982), Hall (1983), Nautiyal (1983), Michie (1985), Rideout (1985), Buongiorno *et al.* (1995), Klemperer (1996), Kant (1999), Buongiorno (2001), Tarp *et al.* (2005), Tahvonen (2007), Zhou *et al.* (2008), and Amacher *et al.* (2009).
26 The entrepreneur's wealth objective presented in equation (9.13) was developed given an existing uneven-aged forest. Klemperer (1996) develops a cash flow diagram and

corresponding economic model given the entrepreneur starts with bare land, regenerates an even-aged forest, and then converts it to an uneven-aged forest. Buongiorno (2001), Tarp *et al.* (2005), and Tahvonen (2007) provide an economic analysis of converting an even-aged to an uneven-aged forest.

27 If the uneven-aged stand is in a steady state condition, equation (9.13) can be defined as

$$USV_{t,G} = \sum_{\theta=0}^{\infty} \{P \cdot [Q(t, G) \cdot (1 + r)^{-t} - G]\} \cdot [(1 + r)^{-t}]^{\theta}$$
$$= \frac{P \cdot [Q(t, G) \cdot (1 + r)^{-t} - G]}{1 - (1 + r)^{-t}}$$
$$= \frac{P \cdot [Q(t, G) - G] \cdot (1 + r)^{-t}}{1 - (1 + r)^{-t}} - P \cdot G$$

28 I would highly recommend that the reader review the discussion on the internal rate of return found in Chapter 8.
29 In addition, Amacher *et al.* (2009) described the case when the wealth maximizing optimal multiple rotation age is infinite and the stand should never be harvested: If the ecosystem good or service values increase faster than the decrease in the harvest revenues due to diminishing returns in the timber production system. Finally, they also examine the case of wealth maximizing optimal multiple rotation ages if there is an existing stand. A discussion of these topics is beyond the scope of this book and I would point the reader to Amacher *et al.* (2009).
30 The following discussion is based on Escobedo *et al.* (In Press).
31 This will be the topic of Chapter 12.

10 Capital theory: risk

1 This relative sensitivity index has the same mathematical form as elasticities described in Chapter 6.
2 While I have focused on NPV and LEV, the same discussion would apply to the investment analysis tools described in Chapter 8.
3 A more complete description of these parameters is given in Chapter 9.
4 The reader can verify this conclusion by examining equation (9.3).
5 Theory of comparative statics in economics refers to examining the direction of change in an outcome (e.g. quantity demanded) when an exogenous variable (e.g. own-price) changes. It is the direction of change (e.g. increase in own-price implies a decrease in the quantity demanded) and not the amount of the change that is predicted (Silberberg and Suen 2001).
6 Technically, changing the interest rate to include different levels of risk and uncertainty could have been included in the discussion of sensitivity analysis. However, it was separated to highlight how time impacts the weight placed on risk.

11 Forest taxes

1 I will discuss these tax incentive programs later in this chapter.
2 This also called an unmodified property tax (Chang 1982).
3 Chang (1982) showed a similar result when both timber and/or land were reassessed annually.
4 A property tax that only includes the value of the forestland is called a land value or site value tax and in some cases a current use tax.
5 This result is consistent with how total fixed costs impact the development of the profit searching rule discussed in Chapter 5.

6 There is a tax called site productivity tax which is also based on the ability of the land to grow timber. The concept of site is based on a site index or land expectation value concept. This is generally an annual tax.
7 Equation (11.5) has the form of the present value of a perpetual every-period series described Appendix 6.
8 An interested reader may want to compare and contrast equations (11.5) and (9.10).
9 A fully regulated or normal forest has equal acres in each age class up to and including the rotation age (Helms 1998).
10 A yield tax is often called a severance tax.

12 Estimating nonmarket values

1 The concepts of ecosystem processes, goods, and services were defined in Chapter 9.
2 For example, a 2009 special issue of the *Journal of Forest Economics*, Vol. 15 Nos. 1–2, is devoted to illustrating the use and limitations of estimating the nonmarket values of forest-based ecosystem goods and services in decision making related to the management of forest resources in Europe.
3 Indirect use values can also be described as positive externalities that ecosystems provide. A positive externality is defined as an activity by one agent causes a gain of welfare to another agent and the affected agent is not indifferent to the change in welfare and the gain in welfare is uncompensated.
4 There is an additional concept termed quasi-option which is the value attached to the preservation of an option in order to stress the crucial role played by irreversibility, the expected value of information, and learning (Arrow and Fisher 1974; Conrad 1980; Basili and Fontini 2005). A discussion of quasi-option value is beyond the scope of this book.
5 I have limited the discussion of nonuse values to existence and bequest values as these are probably the most recognizable and thus relevant. However, the discussion and identification of nonuse values is not limited to existence and bequest values. For example, intrinsic and cognitive values are often included as nonuse values; but, the discussion of these concepts is beyond the scope of this book.
6 An alternative approach to determine an individual's expressed relative importance of an ecosystem good or services is to estimate their willingness to accept. A comparison of willingness to pay and willingness to accept will be discussed later in this chapter.
7 Willingness to pay is synonymous with the concept of demand and the demand curve (Chapter 6).
8 Could contributions to organizations like Greenpeace be used as a proxy for nonuse existence value? While it is an observable choice that an individual could make, an individual could give to Greenpeace for many different reasons (e.g. a counterbalance to exploitive uses or a political statement). It would be very hard to determine if this contribution was to save the whales (existence value) without asking individuals this question directly. Asking an individual to value the existence of a species or ecosystem would be a stated preference approach.
9 While travel cost and hedonic models are probably the most recognized approaches used in the context described here, other approaches such as damage cost avoided, replacement cost, and opportunity cost methods are also available. The first three are based on the premise that if an individual incurs costs to avoid damages caused by the loss of an ecosystem service or replace the lost ecosystem services then those services must be worth at least what was paid to replace. For example, water treatment plants to replace the loss of water quality services of the ecosystem or wetlands mitigation banking. The opportunity cost can be used to estimate nonconsumptive and nonuse values of a resource. This approach examines the value of the resource in its next best use. For example, determining the net present value associated with the cash flows

from the use values of two different land utilizations such as timber production versus a nature reserve. If a public decision maker value chooses the nature reserve with a smaller positive net present value the nonconsumptive and nonuse values are greater than that for timber production (Newman 2010). I will not discuss these approaches beyond this note.

10 I would recommend that the reader keep Figure 12.2 firmly in mind during this discussion.

11 There are many different functional forms that a hedonic price model may take. In addition to the linear–linear model shown in equation (12.2), the other common forms are

Linear–log

$$P_j = \ln(\beta_0) + \sum_{k=1}^{K} \beta_k \ln(X_{jk}) + \ln(\varepsilon_j)$$

Log–linear

$$\ln(P_j) = \beta_0 + \sum_{k=1}^{K} \beta_k X_{jk} + \varepsilon_j$$

Log–log

$$\ln(P_j) = \ln(\beta_0) + \sum_{k=1}^{K} \beta_k \ln(X_{jk}) + \ln(\varepsilon_j)$$

where "ln" denotes the natural log. It should be noted that hedonic theory does not define which mathematical form should be used.

12 The first derivative of the hedonic price function with respect to a given attribute, holding the level of all other attributes constant, defines the implicit price function of that attribute; that is, $\dfrac{\partial P_j}{\partial X_k}$. For example using equation (12.1) the implicit price function of the kth attribute, X_k is

$$\frac{\partial P_j}{\partial X_k} = \frac{\partial(\beta_0 + \beta_1 X_{j1} + \beta_2 X_{j2} + \beta_3 X_{j3} + \cdots + \beta_K X_{jK} + \varepsilon_j)}{\partial X_k} = \beta_k$$

The implicit price function using a linear–linear form is constant and can be interpreted as a price. This may not be the case if the form of the hedonic price function is linear–log, log–linear, or log–log. It is up to the reader to determine if this statement is accurate. Consequently, interpreting the resulting implicit price function accurately is dependent directly on the hedonic price function used to generate it.

13 Other examples are described in Holmes *et al.* (1990) and Bare and Smith (1999) who used a hedonic approach to examine the impact of various timber sale attributes on timber sale prices in Connecticut and the Pacific Northwestern United States, respectively.

14 A lump-sum sale is the sale of standing timber for a fixed dollar amount agreed upon in advance by the landowner and the buyer. The dollar amount agreed upon is not a function of the volume of timber actually cut.

15 Klemperer (1996) noted this similarity by described a statistically estimated ZTCM as a hedonic travel cost model.

16 If an item is purchased and used for the site in question and nowhere else, then perhaps the cost might be included; however, as most equipment is used at multiple sites and there is no good way to allocate the cost to the site being valued, the equipment costs are generally excluded.

17 A map of the DEC administrative regions can be found at http://www.dec.ny.gov/about/50230.html accessed on 12 September 2010.

18 The concept of WTP as described is derived from Hicksian compensating variation; that is, the maximum amount of money an individual would be willing and able to spend that would leave them at the same level of happiness, welfare, or utility after obtaining the good or service than before obtaining the good or service.

19 What exists is seen as a reference point (i.e. part of an individual's endowment).

20 The concept of WTA as described is derived from Hicksian equivalent variation; that is, the minimum amount of money an individual would require in compensation such that would leave them at the same level of happiness, welfare, or utility after giving up the good or service that before giving up the good or service.

21 I would encourage an interested reader to review that following articles on this topic: Miceli and Minkler (1995), Loomis *et al.* (1998), Morrison (2000), Venkatachalam (2004), Plott and Zeiler (2005), Tom *et al.* (2007), and Grutters *et al.* (2008).

22 A public good is defined traditionally as a nonrival, nonexclusive, uncongested good. The example often given is national defense. A private good is defined traditionally as a rival, exclusive, congested good. Examples of private goods are bicycles, fly fishing rods, cars, or clothes. Public or private goods are distinguished by their definitions, *not* by the type of organization or institution making it available.

23 Michael Ahlheim and Wolfgang Buchholz in an online article entitled "WTP or WTA – Is that the Question? Reflections on the difference between 'Willingness to Pay' and Willingness to Accept'" (http://www.wiwi.uni-regensburg.de/buchholz/forschung/buchholz/WTP_%20WTA.pdf accessed on 16 September 2010) argue that whether WTA is greater than WTP is moot as the economic concept of welfare derived from WTA and WTP are ordinal not cardinal measurements (see Varian 1992 or Silberberg and Suen 2001). Thus, the magnitude of difference between the measurements for an individual is irrelevant. In an ordinal world, the only fact that counts is the sign of a welfare change; that is, if the compensating variation value is negative (positive) this represents the maximum (minimum) WTP (WTA) and if the equivalent variation is negative (positive) this represents the maximum (minimum) WTP (WTA). The problem of comparing absolute values arises only if we want to add up individual WTPs or WTAs according to the Hicks–Kaldor criterion (i.e. an outcome is more efficient if those that are made better off could in theory compensate those that are made worse off), which means leaving the grounds of ordinal utility theory. Aggregation of individual values, which is crucial for cost–benefit analysis, implies the use of arbitrary political value judgments. Using aggregate WTPs or WTAs thus implies leaving the world of pure economic science and entering the world of applied policy. Unfortunately, in most cases of nonmarket ecosystem goods and services we are in the world of applied policy and WTP and WTA provide metrics to assess welfare. Thus, aggregating estimated WTP or WTA as part of project evaluation should be made dependent on the political and socio-economic circumstances of the specific environmental change to be valued.

24 Kilgore *et al.* (2008) posed an interesting question with respect to forest stewardship or sustainably managing the forested ecosystem by family forest landowners. They examined the probability of these landowners' willingness to pay to be certified through programs such as Sustainable Forestry Initiative or Forest Stewardship Council versus willingness to accept payments from a governmental organization to practice good forest stewardship. The results indicated as the cost (willingness to pay) for certification increased, the probability that family forest landowners would become certified decreased. As the payments (willingness to accept) for enrolling in stewardship programs increased, so did the probability that family forest landowners would enroll. In this case, markets existed and the researchers examined participation probabilities based on the levels of willingness to pay and accept.

25 Probably the most cited reference with respect to survey methodology is Dillman (1978). A newer version of the book is Dillman *et al.* (2009). I would point the

interested reader to these sources whether developing a survey for a stated or revealed preference study.

26 Reliability is a measure of accuracy that predicts how often one can repeat applications and obtain the same results. Validity is a measure of accuracy concerned with identifying the presence of systematic error and eliminating it.

27 The willingness to pay estimates derived from revealed preference techniques result in Marshallian demand as are the demand curves described in Chapters 6 and 7. Marshallian demand includes both income and substitution effects associated with changes in prices (Silberberg and Suen 2001). The consumer surplus resulting from revealed preference techniques reflects this economic information. A Marshallian demand is sometimes called a money demand. The willingness to pay estimates derived from contingent valuation result in Hicksian demand. Hicksian demand examines the pure substitution effects associated with changes in prices (Silberberg and Suen 2001). The consumer surplus resulting from contingent valuation (a stated preference technique) reflects this economic information. A Hicksian demand is sometimes called an income-compensated demand as real income is held constant. As a result, the Hicksian demand curve is everywhere below (above) the Marshallian demand curve given a decrease (increase) in price. If the percent of income spent on the good is small – and thus the income effect is small – and the there are many substitutes – and thus the substitution effect is small and the Hicksian demand curve is more elastic – then the Hicksian and Marshallian demand curves will be similar. While it is not theoretically accurate to drop the distinction between Marshallian and Hicksian willingness to pay or demand, for purposes of illustration I will assume that revealed and stated preference methods provide comparable measures of demand.

28 It should be noted that there is no nonarbitrary way of allocating truly joint costs (Carlson 1974; Blocher *et al.* 2002). Thus, any cost allocation based on these arbitrary procedures must be a policy decision.

Appendix 2: technical efficiency versus production cost efficiency

1 Information concerning using Lagrangian multipliers to solve constrained optimization problems can be found in Lambert (1985) and Chiang (1984).

Appendix 4: profit and least cost models

1 Information concerning using Lagrangian multipliers to solve constrained optimization problems can be found in Lambert (1985) and Chiang (1984).

Appendix 8: sustainability and the interest rate

1 Chiang (1984) derives the continuous compounding factor from the discrete compounding factor

$$\lim_{m \to \infty} \left(1 + \frac{i}{m}\right)^{t \cdot m} = e^{it}$$

Thus the continuous discounting factor is e^{-it}.

2 However, if $-\kappa > \frac{i \cdot t}{\ln t}$, $\forall t > 1$, then $0 < e^{\kappa \ln t} < e^{-it} < 1$.

Bibliography

Adams, D.M. and Ek, A.R. (1974) Optimizing the Management of Uneven-aged Forest Stands, *Canadian Journal of Forest Research*, 4(3):274–287.

Adams, D.M. and Haynes, R. (1980) The 1980 Softwood Timber Assessment Market Model: Structure, Projections and Policy Pimulations, *Forest Science Monograph* Number 26.

Alexander, S.J., Pliz, D., Weber, N.S., Brown, E., and Rockwell, V.A. (2002) Mushrooms, Trees, and Money: Value Estimates of Commercial Mushrooms and Timber in the Pacific Northwest, *Environmental Management*, 30(1):129–141.

Amacher, G.S., Olikaninen, M., and Koskela, E. (2009) *Economics of Forest Resources*, Cambridge, MA: Massachusetts Institute of Technology Press.

Amateis, R.L. and Burkhart, H.E. (1985) Site Index Curves for Loblolly Pine Plantations on Cutover, Site-Prepared Lands, *Southern Journal of Applied Forestry*, 9(3): 166–169.

Arano, K.G. and Munn, I. (2006) Evaluating Forest Management Intensity: A Comparison Among Major Forest Landowner Types, *Forest Policy and Economics*, 9(3):237–248.

Arrow, K., Solow, R., Portney, P.R., Leamer, E.E., Radner, R., and Schuman, H. (1993) Report of the NOAA Panel on Contingent Valuation, Federal Register, 58, 10, 4602–14.

Arrow, K.J. and Fisher, A.C. (1974) Environmental Preservation, Uncertainty, and Irreversibility, *The Quarterly Journal of Economics*, 88(2):312–319.

Bare, B.B. and Smith R.L. (1999) Estimating Stumpage Values from Transaction Evidence Using Multiple Regression, *Journal of Forestry*, 97(7):32–39.

Basili, M. and Fontini, F. (2005) Quasi-option Value Under Ambiguity, *Economics Bulletin*, 4(3):1–10.

Baumol, W.J. (1968) On the Social Rate of Discount, *The American Economic Review*, 58(4):788–802.

Baye, M.R. (2009) *Managerial Economics and Business Strategy*, 6th edn., New York, NY: McGraw-Hill Irwin.

Becker, D.R., Hjerpe, E.E., and Lowell, E.C. (2004) Economic Assessment of Using a Mobile Micromill® for Processing Small-Diameter Ponderosa Pine, Gen. Tech. Rpt. PNW-GTR-623 Portland, OR: US Dept. of Agriculture, Forest Service, Pacific Northwest Research Station.

Belin, D.L., Kittredge, D.B., Stevens. T.H., Dennis, D.C., Schweik, C.M., and Morzuch, B.J. (2005) Assessing Private Forest Owner Attitudes Toward Ecosystem-Based Management, *Journal of Forestry*, 103(1):28–35.

Bentley, W.R. and Teeguarden, D.E. (1965) Financial Maturity: A Theoretical Review, *Forest Science*, 11(1):76–87.

Bettinger, P., Boston, K., Siry, J.P., and Grebner, D.L. (2009) *Forest Management and Planning*, New York, New York: Elsevier Academic Press.

Bierman, H.S. and Fernandez, L. (1993) *Game Theory with Economic Applications*, Redding, MA: Addison-Wesley Publishing Company, Inc.

Binkley, C.S. (1987) When is the Optimal Economic Rotation Longer Than the Rotation of Maximum Sustained Yield? *Journal of Environmental Economics and Management*, 14(2):152–158.

Blocher, E.J., Chen, K.H., and Lin, T.W. (2002) *Cost Management: A Strategic Emphasis*, 2nd edn., New York, NY: McGraw-Hill Irwin.

Bockstael, N.E., McConnell, K.E., and Strand, I.E. (1989) A Random Utility Model for Sportfishing: Some Preliminary Results for Florida, Marine Resource Economics, 6:245–260.

Boone, R.S., Kozlik, C.J., Bois, B.J., and Wengert, E.M. (1988) Dry Kiln Schedules for Commercial Woods – Temperate and Tropical, Gen. Tech. Rpt. FPL-GTR-57 Madison, WI: US Dept. of Agriculture, Forest Service, Forest Products Laboratory.

Boulding, K.E. (1935) The Theory of a Single Investment. *The Quarterly Journal of Economics*, 49(3):475–494.

Bowes, M. and Krutilla, J. (1985) Multiple-Use Management of Public Forest lands. In *Handbook of Natural Resource and Energy Economics*, Vol. II A.V. Kneese and L.L. Sweeney (eds.), Amsterdam: North Holland.

Bowker, J.M., Newman, D.H., Warren, R.J., and Henderson D.W. (2003) Estimating the Economic Value of Lethal versus Nonlethal Deer Control in Suburban Communities, *Society and Natural Resources*, 16(2):143–158.

Boyd, J.W. and Banzhaf, H.S. (2005) Ecosystem Services and Government: The Need for a New Way of Judging Nature's Value, *Resources*, 158:16–19.

Boyd, J.W. and Banzhaf, S. (2007) What are the Ecosystem Services? The Need for Standardized Environmental Accounting Units, *Ecological Economics*, 63: 616–626.

Boyd, J.W. and Krupnick, A. (2009) The Definition and Choice of Environmental Commodities for Nonmarket Valuation. RFF Discussion paper DP 09-35. September, p. 57.

Brazee, R.J. (2003) The Volvo Theorem: From Myth to Behavior Model. In F. Helles, N. Strange and L. Wichmann (eds.), *Recent Accomplishments in Applied Forest Economics Research*, New York, New York: Kluwer Academic Publishers.

Brealey, R.A., Myers, S.C., and Allen, F. (2008) *Principles of Corporate Finance*, 9th edn., New York, NY: McGraw-Hill, Inc.

Brickley, J., Smith, Jr, C.W., and Zimmerman, J.L. (2007) *Managerial Economics and Organizational Architecture*, 4th edn., New York, NY: McGraw-Hill, Inc.

Bromley, D.W. (1991) *Environmental Economy: Property Rights and Public Policy*, Oxford: Blackwell.

——(1992) Property Rights as Authority Systems: The Role of Rules in Resource Management, In Nemetz, P.N. (ed.) (1992), *Emerging Issues in Forest Policy*. Vancouver, BC: UBC Press, pp. 471–496.

Brown, T.C. (1984) The Concept of Value in Resource Allocation, *Land Economics*, 60(3):231–246.

Brown, T.C., Bergstrom, J.C., and Loomis, J.B. (2007) Defining, Valuing and Providing Ecosystem Goods and Services, *Natural Resources Journal*, 47(2):329–376.

Brown, W.G. and Nawas, F. (1973) Impact of Aggregation on the Estimation of Outdoor Recreation Demand Functions, *American Journal of Agricultural Economics*, 55(2):246–249.

Bullard, S.H., Gunter, J.E., Doolittle, M.L., and Arano. K.G. (2002) Discount Rates for Nonindustrial Private Forest Landowners in Mississippi: How High a Hurdle? *Southern Journal of Applied Forestry*, 26(1):26–31.

Buongiorno, J. (2001) Quantifying the Implications of Transformation From Even to Uneven-Aged Forest Stands, *Forest Ecology and Management*, 151:121–132.

Buongiorno, J. and Gilles, J.K. (2003) *Decision Methods for Forest Resource Management*. Academic Press.

Buongiorno, J. and Michie, B. (1980) A Matrix Model for Uneven-Aged Forest Management, *Forest Science*, 26:609–625.

Buongiorno, J., Peyron, J.L., Houllier, F., and Bruciamacchie, M. (1995) Growth and Management of Mixed-Species, Uneven-Aged Forests in the French Jura: Implications for Economic Returns and Tree Diversity, *Forest Science*, 41(3):397–429.

Burdurlu, E., Ciritcioğlu, H.H., Bakir, K., and Özdemir, M. (2006) Analysis of the Most Suitable Fitting Type for the Assembly Of Knockdown Panel Furniture, *Forest Products Journal*, 56(1):46–52.

Bureau of Land Resources (2008) *Stumpage Price Report No. 73*, Division of Lands and Forests, Summer 2008, New York State Department of Environmental Conservation, Albany, New York.

Burkhart, H.E., Cloeren, D.C., and Amateis, R.L. (1985) Yield Relationships in Unthinned Loblolly Pine Plantations on Cutover, Site-Prepared Lands, *Southern Journal of Applied Forestry*, 9(2):84–91.

Burns, R.M. and Honkala, B.H. (eds.) (1990) *Silvics of North America Volume 2, Hardwoods*, Agriculture Handbook 654, Washington DC: United States Department of Agriculture – Forest Service.

Butler, B.J. (2008) Family Forest Owners of the United States, 2006, General Technical Report NRS-27 Newtown Square PA, US Department of Agriculture – Forest Service, Northern Research Station.

——(2010) The Average American Family Forest Owner, The Consultant – The Annual Journal of the Association of Consulting Foresters (http://www.acf-foresters.org/Content/NavigationMenu/Media/ConsultantMagazine/default.htm accessed on 8 December 2009).

Butler, B.J. and Leatherberry, E.C. (2004) American's Family Forest Owners, *Journal of Forestry*, 102(7):4–9.

Butler, B.J., Tyrrell, M., Feinberg, G., VanManen, S., Wiseman, L., and Wallinger, S. (2007) Understanding and Researching Family forest Owners: Lessons from Social Marketing Research, *Journal of Forestry*, 105(7):348–357.

Calish, S., Flight, R.D., and Teeguarden. D.E. (1978) How Do Nontimber Values Affect Douglas-fir Rotations? *Journal of Forestry*, 76:217–221.

Canadian Farm Business Management Council (CFBMC) (2000) *Report of the Economics of Maple Syrup Production*, 75 Albert St, Suite 903 Ottawa, Ontario, K1P 5E7.

Canham, H.O. (1986) Comparable Valuation of Timber and Recreation for Forest Planning, *Journal of Environmental Management*, 23:335–339.

Carlson, S. (1974) *A Study on the Pure Theory of Production*, Clifton, NJ: Augustus M. Kelley Publishers.

Carson, R.T., Mitchell, R.C., Hanemann, M., Kopp, R.J., Presser, S., and Ruud, P.A. (2003) Contingent Valuation and Lost Passive Use: Damages from the Exxon Valdez Oil Spill, *Environmental and Resource Economics*, 25:257–286

Casalmir, L.M.P. (2000) *Economic Optimal Rotation Age Determination under Sustainable Forestry*. MS Thesis State University of New York – College of Environmental Science and Forestry, Syracuse, NY.

Chang, S.J. (1982) An Economic Analysis of Forest Taxation's Impact on Optimal Rotation Age, *Land Economics*, 58(3):310–323.

Chang, S.J., Cooper, C., and Guddanti, S. (2005) Effects of the Log's Rotational Orientation and the Depth of the Opening Cut on the Value of Lumber Produced in Sawing Hardwood Logs, *Forest Products Journal*, 55(10):49–55.

Chapeskie, D. and Koelling, M.R. (2006) Chapter 11 Economics of Maple Syrup Production. In *North American Maple Syrup Producers Manual*, 2nd edn., Heiligmann, R.B., Koelling, M.R., and Perkins, T.D. (ed.) (2006) Ohio State University Extension, The University of Ohio, Wooster, Ohio.

Chiang, A.C. (1984) *Fundamental Methods of Mathematical Economics*, 3rd edn., New York, NY: McGraw-Hill, Inc.

——(1992) *Elements of Dynamic Optimization*, New York, NY: McGraw-Hill, Inc.

Chichilnisky, G. (1997) What is Sustainable Development? *Land Economics*, 73(4):467–491.

Ciriacy-Wantrup, S.V. (1947) Capital Returns from Soil Conservation Practices, *Journal of Farms Economics*, 29(4):1181–1196.

Clark, J.M. (1940) Toward a Concept of Workable Competition, *The American Economic Review*, 39(2):241–256.

Clawson, M. and Knetsch, J.L. (1966) *Economics of Outdoor Recreation*, Resources for the Future, Baltimore, ME: Johns Hopkins Press.

Clutter, J.L., Fortson, J.C., Pienaar, L.V., Brister, G.H., and Bailey, R.L. (1983) *Timber Management: A Quantitative Approach*, Hoboken, New Jersey: John Wiley & Sons.

Coase, R.H. (1960) The Problem of Social Cost, *Journal of Law and Economics*, 3:1–44.

Cole, D.H. and Grossman, P.Z. (2002) The Meaning of Property Rights: Law Versus Economics? *Land Economics*, 78(3):317–330.

Comolli, P.M. (1981) Principles and Policies in Forestry Economics, *The Bell Journal of Economics*, 12(1):300–309.

Conrad, J.M. (1980) Quasi-option Value and the Expected Value of Information, *The Quarterly Journal of Economics*, 94(4):813–820.

Copeland, T.E., Weston, J.F., and Shastri, K. (2005) *Financial Theory and Corporate Policy*, 4th edn., Pearson Addison-Wesley Publishing Company.

Copi, I.M. and Cohen, C. (1998) *Introduction to Logic*, 10th edn., Englewood Cliffs, NY: Prentice-Hall, Inc.

Cropper, M. and Laibson, D. (1999) The Implications of Hyperbolic Discounting for Project Evaluation. In *Discounting and Intergenerational Equity*, P.R. Portney and J.P Weyant (eds.), Washington DC: Resources for the Future.

Davies, K. (1991) Forest Investment Considerations for Planning Thinning and Harvesting, *Northern Journal of Applied Forestry* 8(3):129–131.

Davis, L.S., Johnson, K.N., Bettinger, P.S., and Howard, P.S. (2001) *Forest Management: To Sustain Ecological, Economic, and Social Values*, 4th edn., Long Grove, Illinois: Waveland Press.

Davis, R.K. (1963) The Value of Outdoor Recreation: An Economic Study of the Maine Woods, Ph.D. Dissertation, Harvard University, Cambridge, MA.

Dennis, D.F. (1987) Rates of Value Change on Uncut Forest Stands in New Hampshire, *Northern Journal of Applied Forestry*, 4(2):64–66.

Dillman, D.A. (1978) *Mail and Telephone Surveys – The Total Design Method*, New York, New York: John Wiley and Sons.

Dillman, D.A., Smyth, J.D., and Christian, L.M. (2009) *Internet, Mail, and Mixed-Mode Surveys: The Tailored Design Method*, 3rd edn., New York, New York: John Wiley and Sons.

Dobbs, I.M. (1993) Individual Travel Cost Method: Estimation and Benefit Assessment with Discrete and Possibly Grouped Dependent Variable, *American Journal of Agricultural Economics*, 75(1):84–94.

Dubois, M.R., McNabb, K., and Straka, T.J. (1997) Costs and Cost Trends for Forestry Practices in the South, *Forest Landowner*, March/April:7–13.

Duerr, W.A. (1960) *Fundamentals of Forestry Economics*, New York, New York: McGraw-Hill, Inc.

Ellefson, P.V., Cheng, A.S., and Moulton, R.J. (1995) Regulation of Private Forestry Practices by State Governments, Minnesota Agricultural Experiment Station, Station Bulletin 605-1995, St. Paul, MN: University of Minnesota.

Erickson, J.D., Chapman, D., Fahey, T.J., and Christ, M.J. (1999) Non-renewability in Forest Rotations: Implications for Economic and Ecosystem Sustainability, *Ecological Economics*, 31:91–106.

Escobedo, F., Kroeger, S.T., and Wagner, J.E. Urban Forests and pollution mitigation: Analyzing Ecosystem services and disservices, *Environmental Pollution*, In Press.

Faustmann, M. (1849) Berechnung des Wertes Welchen Waldboden sowie noch nitch haubare Holzbestände für die Waldwirtschaft besitzen. *Allgemeine Forst- und Jagd-Zeitung*. Vol. 15. Reprinted as Faustmann, M. (1995) Calculation of the Value which Forest Land and Immature Stands Possess for Forestry. *Journal of Forest Economics*, 1(1):7–44.

Federal Inter-Agency River Basin Committee (FIRBC) (1958) Proposed Practices for Economic Analysis of River Basin Projects (Known as the "GreenBook"), Federal Inter-Agency River Basin Committee, Washington, DC, Subcommittee on Benefits and Costs.

Feldstein, M.S. (1964a) The Social Time Preference Discount Rate in Cost Benefit Analysis, *The Economic Journal*, 74(294):360–379.

——(1964b) Net Social Benefit Calculation and the Public Investment Decision, *Oxford Economic Papers*, 16(1):114–131.

Feng, Y., Fullerton, D., and Gan, L. (2005) Vehicle Choice, Miles Driven and Pollution Policies. National Bureau of Economic Research, Working Paper #11553, Cambridge, MA (http://www.nber.org/papers/w11553 accessed on 29 December 2009).

Field, B.C. and Field, M.K. (2002) *Environmental Economics: An Introduction*, 3rd edn., New York, NY: McGraw-Hill, Inc.

Fisher, I. (1930) *The Theory of Interest*, New York, NY: Macmillan.

Foerde, K., Knowlton, B.J., and Poldrack, R.A. (2006) Modulation of Competing Memory Systems by Distraction, *Proceedings of the National Academy of Sciences*, 103(31):11778–83.

Friedman, M. (1953) *Essays in Positive Economics*, Chicago, IL: The University of Chicago Press.

Gevorkiantz, S.R. and Scholz, H.F. (1948) Timber Yields and Possible Returns from the Mixed-Oak Farmwoods of Southwestern Wisconsin, Publication No. 521, US Dept. of Agriculture, Forest Service Lake States Forest Experiment Station, In cooperation with Wisconsin Department of Conservation and University of Wisconsin.

Gibbons, R. (1992) *Game Theory for Applied Economists*, Princeton, NJ: Princeton University Press.

Godman, R.M. and Mendel, J.J. (1978) Economic Values for Growth and Grade Changes of Sugar Maple in the Lake States, USDA Forest Service North Central Research Station Research Paper NC-155.

Gowdy, J. and Erickson, J.D. (2005) The Approach of Ecological Economics, *Cambridge Journal of Economics*, 29:207–222.

Graham, J.R. and Harvey, C.R. (2001) The Theory and Practice of Finance: Evidence from the Field, *Journal of Financial Economics*, 60:187–243.

Graham, W.G., Goebel, P.C., Heiligmann, R.B., and Bumgardner, M.S. (2006) Maple Syrup in Ohio and The Impact of Ohio State University (OSU) Extension Program, *Journal of Forestry*, 104(7):94–100.

Gregory, G.R. (1972) *Forest Resource Economics*, New York, NY: John Wiley and Sons.

—— (1987) *Forest Resource Economics*, Hoboken, New Jersey: John Wiley & Sons.

Grisez, T.J. and Mendel, J.J. (1972) The Rate of Value Increase for Black Cherry, Red Maple, and White Ash, USDA Forest Service Northeastern Research Station Research Paper NE-231.

Grutters, J.P.C., Kessels, A.G.H., Dirksen, C.D., van Helvoort-Postulart, D., Anteunis, L.J.C., and Joore, M.A. (2008) Willingness to Accept versus Willingness to Pay in a Discrete Choice Experiment, *Value in Health*, 11(7):1110–1119.

Hagan, J.M., Irland, L.C., and Whitman, A.A. (2005) Changing Timberland Ownership in the Northern Forest and Implications for Biodiversity, Report # MCCS-FCP-2005-1 Brunswick, ME: Manomet Center for Conservation Science.

Hahn, J.T. and Hansen, M.H. (1991) Cubic and Board Foot Volume Models for the Central States, *Northern Journal of Applied Forestry*, 8(2):47–57.

Hall, D.O. (1983) Financial Maturity for Even-Aged and All-Aged Stands, *Forest Science*, 29(4):833–836.

Halstead, J.M., Luloff, A.E., and Stevens, T.H. (1992) Protest Bidders in Contingent Valuation, *Northeast Journal of Agricultural and Resource Economics*, 21(2): 160–169.

Hancock Timberland Investor (2000) *Impact of Carbon Credits on Forestland Values*, Hancock Timber Resource Group 4th Quarter 2000 (http://www.htrg.com/research_archives_2000.htm accessed on 21 May 2010).

—— (2006) *Housing Starts, Lumber Prices, and Timberland Investment Performance*, Hancock Timber Resource Group 3rd Quarter 2006 (http://www.htrg.com/research_archives_2006.htm accessed on 21 February 2010).

Haney, H.L., Hoover, W.L., Siegel, W.C., and Green, J.L. (2001) *The Forest Landowners' Guide to the Federal Income Tax*, United States Forest Service Agricultural Handbook No. 718 (http://www.fs.fed.us/publications/2001/01jun19-Forest_Tax_Guide31201.pdf accessed on 5 August 2010).

Hanna, S., Folke, C., and Mäler, K.-G. (1995) Property Rights and Environmental Resources, In *Property Rights and the Environment: Social and Ecological Issues*, S. Hanna and Munasinghem, M. (ed.) (1995) The Beijer International Institute of Ecological Economics and The World Bank, The International Bank for Reconstruction and Development, Washington, D.C.

Hardin, G. (1968) The Tragedy of the Commons, *Science*, 162(3859):1243–1248.

Harou, P. (1985) On a Social Discount Rate for Forestry, *Canadian Journal of Forest Research*, 15:927–934.

Hartman, R. (1976) The Harvesting Decision when a Standing Forest has Value, *Economic Inquiry*, 14:52–55.

Haynes, R.W. and Monserud, R.A. (2002) A Basis for Understanding Compatibility Among Wood Production and Other Forest Values. United States Department of Agriculture Forest Service Pacific Northwest Research Station, PNW-GTR-529

Heiligmann, R.B. (2008) Here's How To ... Increase Financial Returns from your Woodland, *Forestry Source*, 13(7):13–14.

Heiligmann, R.B., Koelling, M.R., and Perkins, T.D. (ed.) (2006) *North American Maple Syrup Producers Manual*, 2nd edn., Ohio State University Extension, The University of Ohio, Wooster, Ohio.

Helms, J.A. (ed.) (1998) *The Dictionary of Forestry*, Bethesda, Maryland: The Society of American Foresters.

Henderson, J.M. and Quandt, R.E. (1980) *Microeconomic Theory: A Mathematical Approach*, 3rd edn., New York, NY: McGraw-Hill, Inc.

Hepburn, C.J. and Koundouri, P. (2007) Recent Advances in Discounting: Implications for Forest Economics, *Journal of Forest Economics*, 2–3(13):169–189.

Herrick, O.W. and Gansner, D.A. (1985) Forest-Tree Value Growth Rates, *Northern Journal Applied Forestry*, 2(1):11–13.

Hesseln, H., Loomis, J.B., González-Cabán, A., and Alexander S. (2003) Wildfire Effects on Hiking and Biking Demand in New Mexico: A Travel Cost Study, *Journal of Environmental Management*, 69:359–368.

Hesseln, H., Loomis, J.B., and González-Cabán, A. (2004) Comparing the Economic Effects of Fire on Hiking Demand in Montana and Colorado, *Journal of Forest Economics*, 10:21–35.

Heyne, P., Boettke, P., and Prychitko, D. (2006) *The Economic Way of Thinking*, 11th edn., Upper Saddle River, NJ: Pearson Prentice-Hall.

—— (2010) *The Economic Way of Thinking*, 12th edn., Upper Saddle River, NJ: Pearson Prentice-Hall.

Hines, S.J., Heath, S.L., and Birdsey, R.A. (2010) *Annotated Bibliography of Scientific Literature on Managing Forest for Carbon Benefits*, General Technical Report GTR-NRS-57, Newtown Square, PA: US Dept. of Agriculture, Forest Service, Northeastern Research Station.

Hirshleifer, J. (1970) *Investment, Interest and Capital*, Englewood Cliff, NJ: Prentice-Hall, Inc.

Holmes, T.P., Bentley, W.R., Broderick, S.H., and Hobson, T. (1990) Hardwood Stumpage Price Trends and Characteristics in Connecticut, *Northern Journal of Applied Forestry*, 7(1):13–16.

Hool, J.N. (1966) A Dynamic Programming-Markov Chain Approach to Forest Production Control, *Forest Science Monograph no. 12*.

Horowitz, J.K. and McConnell, K.E. (2002) A Review of WTA/WTP studies, *Journal of Environmental Economics and Management*, 44:426–447.

Hseu, J. and Buongiorno, J. (1997) Financial Performance of Maple-Birch Stands in Wisconsin: Value Growth Rate Versus Equivalent Annual Income, *Northern Journal of Applied Forestry*, 14(2):59–66.

Huyler, N.K. (1982) Economics of Maple Sap and Syrup Production, In Clayton, M.C. Jr, Donnelly, J.R., Gabreil, W.J., Garrett, L.D., Gregory, R.A., Huyler, N.K., Jenkins, W.L., Sendak, P.E., Walters, R.S., and Yawney, H.W. (1982) *Sugar* Maple Research: Sap Production, Processing And Marketing of Maple Syrup, General Technical Report NE-72, Broomall, PA: US Dept. of Agriculture, Forest Service, Northeastern Forest Experiment Station.

——(2000) *Cost of Maple Sap Production for Various Size Tubing Operations*, Research Paper NE-RP-712, Newtown Square, PA: US Dept. of Agriculture, Forest Service, Northeastern Research Station.

Hyytiäinen, K. and Tahvonen, O. (2003) Maximum Sustained Yield, Forest Rent or Faustmann: Does it Really Matter?, *Scandinavian Journal of Forest Research*, 18(5):457–469.

Jensen, O.W. (1982) Opportunity Costs: Their Place in the Theory and Practice of Production, *Managerial and Decision Economics*, 3(1):48–51.

Johansson, P-O. and Löfgren, K.-G. (1985) *The Economics of Forestry and Natural Resources*, Basil Blackwell Ltd.

Johnstone, D. (2008) What Does an IRR (or Two) Mean?, *Journal of Economic Education*, 39(1):78–87.

Jorgensen, B.S. and Syme, G.J. (2000) Protest Responses and Willingness to Pay: Attitude Toward Paying for Stormwater Pollution Abatement, *Ecological Economics*, 33:251–265.

Kant, S. (1999) Sustainable Management of Uneven-Aged Private Forests: A Case Study from Ontario, Canada, *Ecological Economics*, 30:131–146.

—— (2003) Extending the Boundaries of Forest Economics, *Forest Policy and Economics*, 5:39–56.

Kaoru, Y., Smith, V.K., and Liu, J.L. (1995) Using Random Utility Models to Estimate the Recreational Value of Estuarine Resources, *American Journal of Agricultural Economics*, 77(1):141–151.

Karush, W. (1939) Minima of Functions of Several Variables with Inequalities as Side Conditions. M.S. Thesis, Department of Mathematics, University of Chicago.

Kay, J. (1996) *The Business of Economics*, Oxford University Press Inc.

Keat, P.G. and Young, P.K.Y. (2000) *Managerial Economics Economic Tools for Today's Decision Makers*, 3rd edn., Upper Saddle River, NJ: Prentice-Hall, Inc.

Kemkes, R.J., Farley, J., and Koliba, C.J. (2009) Determining When Payments are an Effective Policy Approach to Ecosystem Service Provisions, *Ecological Economics*, In Press (doi:10.1016/j.ecolecon.2009.11.032).

Kennedy, E.T., Costa, R., and Smathers Jr, W.M. (1996) Economic Incentives: New Directions for Red-Cockaded Woodpecker Habitat Conservation, *Journal of Forestry*, 94(4):22–26.

Keynes, J.M. (1936) *The General Theory of Employment, Interest and Money*. London: Macmillan (available on the web at http://ebooks.adelaide.edu.au/k/keynes/john_maynard/k44g/ accessed on 16 June 2009).

Kierulff, H. (2008) MIRR: A Better Measure. *Business Horizons*, 51:321–329.

Kilgore, M.A., Snyder, S., Taff, S., and Schertz, J. (2008) Family Forest Stewardship: Do Owners Need a Financial Incentive, *Journal of Forestry*, 106(7):357–362.

Klemperer, W.D. (1996) *Forest Resource Economics and Finance*, New York, NY: McGraw-Hill, Inc.

Kline, J.D., Mazzotta, M.J., and Patterson, T.M. (2009) Toward a Rational Exuberance for Ecosystem Services Markets. *Journal of Forestry*,107(4):204–212.

Koskela, E. and Ollikainen, M. (2003) Optimal Forest Taxation under Private and Social Amenity Valuation, *Forest Science*, 49(4):596–607.

Kreps, D.M. (1990) *Game Theory and Economic Modelling*, Oxford, UK: Oxford University Press.

Kroeger, T. and Casey, F. (2007) An Assessment of Market-Based Approaches to Providing Ecosystem Services on Agricultural Lands, *Ecological Economics*, 64:321–332.

Kuhn, H.W. and Tucker, A.W. (1951) Nonlinear Programming In Jerzy Neyman (ed.), *Proceedings of the Second Berkeley Symposium*, University of California Press, Berkeley, pp. 481–492.

Laibson, D. (1997) Golden Eggs and Hyperbolic Discounting, *The Quarterly Journal of Economics*, 112(2):443–447.

Lambert, P.J. (1985) *Advanced Mathematics for Economists: Static and Dynamic Optimization*, Oxford UK: Basil Blackwell Ltd.

Lembersky, M.R. and Johnson, K.N. (1975) Optimal Policies for Managed Stands: An Infinite Horizon Markov Decision Process Approach, *Forest Science*, 21(2):109–122.

Leuschner, W.A. (1984) *Introduction to Forest Resources Management*, New York, New York: John Wiley & Sons.

Li, C.Z. and Löfgren, K.G. (2000) Renewable Resources and Economic Sustainability: A Dynamic Analysis with Heterogenous Time Preferences, *Journal of Environmental Economics and Management*, 40:236–250.

Lichtkoppler, F.R. and Kuehn, D. (2003) New York's Great Lakes Charter Fishing Industry in 2002, Ohio Sea Grant College Program OHSU-TS-039 Columbus OH: Sea Grant Great Lakes Network, The Ohio State University.

Longenecker, J.G., Moore, C.W., Petty, J.W., and Palich, L.E. (2006) *Small Business Management: An Entrepreneurial Emphasis*, 3rd edn., Mason, Ohio: Thomson South-Western.

Loomis, J.B. and Walsh, R.G. (1997) *Recreation Economic Decisions: Comparing Benefits and Costs*. State College, PA: Venture Publishing, Inc.

Loomis, J.B., Peterson, G., Champ, P., Brown, T., and Lucero, B. (1998) Paired Comparison Estimates of Willingness to Accept versus Contingent Valuation Estimates of Willingness to Pay, *Journal of Economic Behavior and Organization*, 35:501–515.

Loomis, J., Tadjion, O., Watson, P., Wilson, J., Davies, S., and Thilmany, D. (2009) A Hybrid Individual-Zonal Travel Cost Model for Estimating the Consumer Surplus of Golfing in Colorado, *Journal of Sports Economics*, 10(2):155–167.

Luenberger, D.G. (1998) *Investment Science*, Oxford, England: Oxford University Press.

McConnell, C.R. and Brue, S.L. (2005) *Economics*, 16th edn., New York, NY: Irwin/ McGraw-Hill, Inc.

McGuigan, J.R. and Moyer, R.C. (1993) *Managerial Economics*, 6th edn., Minneapolis, MN: West Publishing Company.

McIntyre, R.K., Jack, S.B., McCall, B.B., and Mitchell, R.J. (2010) Financial Feasibility of Selection-Based Multiple-Value Management on Private Lands in the South: A Heuristic Case Study Approach, *Journal of Forestry*, 108(3):230–237.

McKenzie, R.B. and Lee, D.R. (2006) *Microeconomics for MBAs: The Economic Way of Thinking for Managers*, Cambridge, United Kingdom: Cambridge University Press.

Maital, S. (1994) *Executive Economics: Ten Essential Tools for Managers*, New York, NY: The Free Press.

Marckers, H., Heiligmann, R.B., and Koelling, M.R. (2006) Chapter 8 Syrup Filtrationk Grading, Packaging, and Storage, In *North American Maple Syrup Producers Manual*, 2nd edn., Heiligmann, R.B., Koelling, M.R., and Perkins, T.D. (eds.) (2006) Ohio State University Extension, The University of Ohio, Wooster, Ohio.

Mendel, J.J., Grisez, T.J., and Trimble, Jr G.R. (1973) *The Rate of Value Increase for Sugar Maple*, USDA Forest Service Northeastern Research Station Research Paper NE-250.

Mentzer, J.T., DeWitt, W., Keebler, J.S., Min, S., Nix, N.W., Smith, C.D., and Zacharia, Z.G. (2001) Defining Supply Chain Management, *Journal of Business Logistics*, 22(2):1–25.

Meyerhoff, J. and Liebe, U. (2006) Protest Beliefs in Contingent Valuation: Explaining Their Motivation, *Ecological Economics*, 57:583–594.

Miceli, T.J. and Minkler, A.P. (1995) Willingness-To-Accept Versus Willingness-To-Pay Measures of Value: Implications for Rent Control, Eminent Domain, and Zoning, *Public Finance Review*, 23(2): 255–270.

Michie, B.R. (1985) Uneven-Aged Stand Management and the Value of Forest Land, *Forest Science*, 31(1):116–121.

Mikesell, R.F. (1977) *The Rate of Discount for Evaluating Public Projects*, Washington D.C.: American Enterprise Institute for Public Policy Research.

Möhring, B. (2001) The German Struggle Between the 'Bodenreinertragslehre' (Land Rent Theory) and 'Waldreinertragslehre' (Theory of the Highest Revenue) Belongs to the Past – but What is Left?, *Forest Policy and Economics*, 2:195–201.

Montgomery, C.A., Brown, Jr G.M., and Adams, D.M. (1994) The Marginal Cost of Species Preservation: The Northern Spotted Owl, *Journal of Environmental Economics and Management*, 26:111–128.

Morrison, G.C. (2000) WTP and WTA in Repeated Trial Experiments: Learning or Leading?, *Journal of Economic Psychology*, 21:57–72.

Munasinghe, M. (1993) *Environmental Economics and Sustainable Development*, World Bank Environment Paper No. 3, The World Bank, Washington, D.C.

Murray, B.C. (1995a) Measuring Ologopsony Power with Shadow Prices: U.S. Market for Pulpwood and Sawlogs, *The Review of Economics and Statistics*, 77(3):486–498.

——— (1995b) Oligopsony, Vertical Integration, and Output Substitution: Welfare Effects in U.S. Pulpwood Markets, *Land Economics*, 71(2):193–206.

Mutanen, A. and Toppinen, A. (2005) Finnish Sawlog Market under Forest Taxation Reform, *Silva Fennica*, 39(1):117–130.

Navarro, G.A. (2003) Re-examining the Theories Supporting the So-called Faustmann Formula. In F. Helles, N. Strange, and L. Wichmann (eds.) *Recent Accomplishments in Applied Forest Economics Research*, Dordrecht, The Netherlands: Kluwer Academic Publishers.

Nautiyal, J.C. (1983) Towards a Method of Uneven-Aged Forest Management Based on the Theory of Financial Maturity, *Forest Science*, 29(1):47–58.

New England Agricultural Statistics (2008) Maple Syrup 2008, NEAS is a Field Office of the U.S.D.A. National Agricultural Statistics Service (http://www.nass.usda.gov/Statistics_by_State/New_England_includes/Publications/0605mpl.pdf accessed on 5 September 2008).

Newman, D.H. (1988) *The Optimal Forest Rotation: A Discussion and Annotated Bibliography*, General Technical Report SE-48, Ashville, NC: U.S.D.A. Forest Service Southeastern Forest Experiment Station.

—— (2002) Forestry's Golden Rule and the Development of the Optimal Forest Rotation Literature, *Journal of Forest Economics*, 8:5–27.

—— (2010) Forestry Foundations – Total Economic Value, *The New York Forester*, 66(3):5–7.

Newman, D.H. and Wear, D.N. (1993) Production Economics of Private Forestry: A Comparison of Industrial and Nonindustrial Forest Owners, *American Journal of Agricultural Economics*, 75(3):674–684.

New York Department of Environmental Conservation – NYS DEC. (1997) *New York Statewide Angler Survey 1996 Report 1: Angler Effort and Expenditures*, New York State Department of Environmental Conservation, Division of Fish and Wildlife, Albany NY.

Nguyen, D. (1979) Environmental Services and the Optimal Rotation Problem in Forest Management, *Journal of Environmental Management*, 8:127–236.

Nyland, R. (1996) *Silviculture: Concepts and Applications*, New York, NY: McGraw-Hill, Inc.

Ohlin, B. (1921) Till frågan om skogarnas omloppstid, *Ekonomisk Tidskrift*. Vol. 22. Reprinted as Ohlin, B. (1995) Concerning the Question of the Rotation Period in Forestry, *Journal of Forest Economics*, 1(1):89–114.

Pagiola, S., von Ritter, K., and Bishop, J. (2004) How Much is an Ecosystem Worth? Assessing the Economic Value of Conservation, The International Bank for

Reconstruction and Development, Washington D.C.: World Bank (http://www-wds. worldbank.org/external/default/main?pagePK=64193027&piPK=64187937&theSitePK= 523679&menuPK=64187510&searchMenuPK=64187283&siteName=WDS&entityID= 000012009_20041207120119 accessed on 24 August 2010).

Patterson, D.W. and Xie, X. (1998) Inside-out beams From Small-Diameter Appalachian Hardwood Logs, *Forest Products Journal*, 48(1):76–80.

Patterson, D.W., Kluender, R.A., and Granskog, J.E. (2002) Economic Feasibility of Producing Inside-Out Beams From Small-Diameter Logs, *Forest Products Journal*, 52(1):23–26.

Patterson, T.M. and Coelho, D. L. (2009) Ecosystem Services: Foundations, Opportunities, and Challenges for Forest Products Sector, *Forest Ecology and Management*, 257(8):1637–1646.

Pearce, D.W. (ed.) (1994) *The MIT Dictionary of Modern Economics*, 4th edn., Cambridge, MA: The MIT Press.

Pearce, D.W., Groom, B., Hepburn, C., and Koundouri. C. (2003) Valuing the Future: Recent Advances in Social Discounting, *World Economics*, 4(2):121–141.

Pearce, P.H. (1990) *Introduction to Forestry Economics*. Vancouver, CA: University of British Columbia Press.

Penfield, P. (2007) 3 Avenues to Cost Reduction, *Supply Chain Management Review*, 11(1):30–36.

Peters, C.M., Gentry, A.H., and Mendelsohn, R.O. (1989) Valuation of an Amazonian Rainforest, *Nature*, 339:655–656.

Pickens, J.B., Johnson, D.L., Orr, B.D., Reed, D.D., Webster, C.E., and Schmierer, J.M. (2009) Expected Rates of Value Growth for Individual Sugar Maple Crop Trees in the Great Lakes Region: A Reply, *Northern Journal of Applied Forestry*, 25(4):1345–147.

Pindyck, R.S. and Rubinfeld, D.L. (1995) *Microeconomics*, 3rd edn., Englewood Cliffs, NJ: Prentice-Hall.

Plott, C.R. and Zeiler, K. (2005) The Willingness to Pay–Willingness to Accept Gap, the "Endowment Effect," Subject Misconceptions, and Experimental Procedures for Eliciting Valuations, *The American Economic Review*, 95(3).

Png, I. and Leman, D. (2007) *Managerial Economics*, Malden, MA: Blackwell Publishing.

Poudyal, N.C., Hodges, D.G., Tonn, B., and Cho, S.-H. (2009) Valuing Diversity and Spatial Pattern of Open Space Plots in Urban Neighborhoods, *Forest Policy and Economics*, 11:194–201.

Pressler, M.R. (1860) Aus der Holzzuwachlehre (zweiter Artikel), *Allgemeine Forst- und Jagd-Zeitung*, Vol. 36. Reprinted as Pressler, M.R. (1995) For the Comprehension of Net Revenue Silviculture and the Management Objectives Derived Thereof, *Journal of Forest Economics*, 1(1):45–88.

Prestemon, J.P., Turner, J.A., Buongiaorno, J., Zhu, S., and Li, R. (2008) Some Timber Product Market and Trade Implications of an Invasive Defoliator: The Case of Asian Lymantria in the United States, *Journal of Forestry*, 106(8):409–415.

Price, C. (1993) *Time, Discounting and Value*, Oxford, England: Blackwell Publisher.

—— (2004) Hyperbole, Hypocrisy and Discounting that Slowly Fades Away. In *Scandinavian Forest Economics* No. 40, 2004 Heikki Pajuoja and Heimo (eds.). Proceedings of the Biennial Meeting of the Scandinavian Society of Forest Economics held in Vantaa, Finland 12–15 May, 2004.

Puttock, G.D., Prescott, D.M., and Meilke, K.D. (1990) Stumpage Prices in Southwestern Ontario: A Hedonic Function Approach, *Forest Science*, 36(4):1119–1132.

Rideout, D. (1985) Managerial Finance for Silvicultural Systems, *Canadian Journal of Forest Research*, 15:163–166.

Rideout, D.B. and Hesseln, H. (2001) *Principles of Forest & Environmental Economics*, 2nd edn., Ft. Collins, CO: Resource & Environmental Management, LLC.

Rosen, S. (1974) Hedonic Prices and Implicit Markets: Product Differentiation in Pure Competition, *Journal of Political Economy*, 82(1):34–55.

Rosenthal, D.H. and Brown, T.C. (1985) Comparability of Market Prices and Consumer Surplus for Resource Allocation Decisions, *Journal of Forestry*, 83(2):105–109.

Ruddell, S., Sampson, R., Smith, M., Giffen, R., Cathcart, J., Hagan, J., Sosland, D., Godbee, J., Heissenbuttel, Lovett, S., Helms, J., Price, W., and Simpson, R. (2007) The Role for Sustainably Managed Forests in Climate Change Mitigation. *Journal of Forestry*, 105(6):314–319.

Samuelson, P.A. (1976a) *Economics*, 10th edn., New York, NY: McGraw-Hill, Inc.

—— (1976b) Economics of Forestry in an Evolving Society, *Economic Inquiry*, 14(4):466–492. Reprinted as Samuelson, P.A., (1995) Economics of Forestry in an Evolving Society. *Journal of Forest Economics*, 1(1):115–149.

Sander, H., Polasky, S., and Haight, R.G. (2010) The Value of Urban Tree Cover: A Hedonic Property Price Model in Ramsey and Dakota Counties, Minnesota, USA, *Ecological Economics*, 69:1646–1656.

Sendak, P.E. and Bennink, J. (1985) The Cost of Maple Sugaring in Vermont, Research Paper NE-565, Broomail, PA: US Dept. of Agriculture, Forest Service, Northeastern Research Station.

Sharma, M. and Oderwald, R.G. (2001) Dimensionally Compatible Volume and Taper Equations. *Canadian Journal of Forest Research*, 31:797–803.

Siegel, W.C., Haney, Jr. H.L., and Greene, J.L. (2009) *Estate Planning for Forest Landowners: What will become of your timberland?*, General Technical Report SRS-112, Asheville, NC, US Dept. of Agriculture, Forest Service, Southern Research Station (available at http://www.srs.fs.fed.us/pubs/31987 accessed on 5 August 2010).

Silberberg, E. and Suen, W. (2001) *The Structure of Economics: A Mathematical Analysis*, 3rd edn., New York, NY: Irwin/McGraw-Hill, Inc.

Simpson, R.D., Toman, M.A., and Ayres, R.U. (2004) Scarcity and Growth in the New Millennium: Summary, Resources for the Future Discussion Paper 04-01, Washington DC: Resources for the Future.

Smith, A. (1776) *An Inquiry into the Nature and Causes of the Wealth of Nations*, Edited by Edwin Cannan (1976) Chicago: University of Chicago Press.

Smith, E.D., Szidarovszky, F., Karnavas, W.J., and Bahill, A.T. (2008) Sensitivity Analysis, a Powerful System Validation Technique, *The Open Cybernetics and Systemics Journal*, 2:39–56.

Smith, V.L. (1968) Economics of Production From Natural Resources, *The American Economic Review*, 58(3):409–431.

Smith, W.B. (technical coordinator), Miles, P.D. (data coordinator), Perry, C.H. (map coordinator), and Pugh, S.A. (data CD coordinator) (2009) *Forest Resources of the United States 2007*, General Technical Report WO-78, US Department of Agriculture – Forest Service Washington Office.

Smith, W.R. (1988) The Fallacy of Preferred Species, *Southern Journal Applied Forestry*, 12(2):79–84.

Sourd, F. (2005) Optimal timing of a sequence of tasks with general completion costs, *European Journal of Operational Research*, 165:82–96.

Spelter, H. and Alderman, M. (2005) Profile 2005: Softwood sawmills in the United States and Canada. Research Paper FPL-RP-630. Madison, WI: US Department of Agriculture, Forest Service, Forest Products Laboratory.

Stednick, J.D. (1996) Monitoring the Effects of Timber Harvest on Annual Water Yield, *Journal of Hydrology*, 176:79–95.

Stier, J.C. (2003) What is my Timber Worth? And Why?, Forestry Facts University of Wisconsin Extension, University of Wisconsin – Madison, School of Natural Resources, Department of Forest Ecology and Management, Madison Wisconsin.

Stowe, B., Wilmot, T., Cook, G.L., Perkins, T., and Heiligmann, R.B. (2006) Chapter 7 Maple Syrup Production, In *North American Maple Syrup Producers Manual*, 2nd edn., Heiligmann, R.B., Koelling, M.R., and Perkins, T.D. (eds.) (2006) Ohio State University Extension, The University of Ohio, Wooster, Ohio.

Strange, N., Brodie, J., Meilby, H., and Helles, F. (1999) Optimal Control of Multiple-Use Products: The Case of Timber, Forage and Water Protection, *Natural Resources Modeling*, 12:335–354.

Tahvonen, O. (2007) Optimal Choice Between Even-Aged and Uneven-Aged Forest Management Systems. Working papers of the Finnish Forest Research Institute, No. 60. (http://www.metla.fi/julkaisut/workingpapers/2007/mwp060.pdf accessed on 16 June 2010).

Tarp, P., Buongiorno, J., Helles, F., Larsen, J.B., and Strange, N. (2005) Economics of Converting an Even-Aged *Fagus sylvatica* Stand to an Uneven-Aged Stand Using Target Diameter Harvesting, *Scandinavian Journal of Forest Research*, 20:63–74.

Tasissa, G., Burkhart, H.E., and Amateis, R.L. (1997) Volume and Taper Equations for Thinned and Unthinned Loblolly Pine Trees in Cutover, Site-Prepared Plantations. *Southern Journal of Applied Forestry*, 21(3):146–152.

Thomas, G.B. Jr (1951) *Calculus and Analytic Geometry*, Reading, MA: Addison-Wesley Publishing Company, Inc.

Tom, S.M., Fox, C.R., Trepel, C., and Poldrack., R.A. (2007) The Neural Basis of Loss Aversion in Decision-Making Under Risk, *Science*, 315:515–518.

United States Department of Agriculture, Forest Service. (2006) *Forest Inventory and Analysis National Core Field Guide, Volume I: Field Data Collection Procedures for Phase 2 Plots Version 3.1*, Newtown Square, PA. (http://www.fs.fed.us/ne/fia/datacol-lection/manualver3_1/index.html accessed on 27 July 2007).

Uys, H.J.E. (1990) A New Form of Internal Rate of Return, *South African Forestry Journal*, 154:24–26.

Varian, H.R. (1992) *Microeconomic Analysis*, 3rd edn., New York, NY: W. W. Norton and Company.

Vaughan, W.J. and Russell, C.S. (1982) Valuing a Fishing Day: An Application of a Systematic Varying Parameter Model, *Land Economics*, 58(4):450–463.

Venkatachalam, L. (2004) The Contingent Valuation Method: A Review, *Environmental Impact Assessment Review*, 24:89–124.

Volk, T.A. and Luzadis, V.A. (2008) Willow biomass production for bioenergy, biofuels and bioproducts in New York. In *Renewable Energy from Forest Resources in the United States*, B. Solomon and V.A. Luzadis (eds.), Oxon, UK: Routledge Press.

Wagner, J.E. (2009) Expected Rates of Value Growth for Individual Sugar Maple Crop Trees in the Great Lakes Region: A Comment, *Northern Journal of Applied Forestry*, 25(4):141–145.

Wagner, J.E. and Choi, J. (1999) Economic Assessment of Recreation, Wildlife, and Timber Resources on Niagara Mohawk Power Company's Moose River Land Holdings.

In Porter, W. and Gibbs, J.P. (project directors) (1999) *Ecological Assessment of Niagara Mohawk Power Corporation Lands*, College of Environmental Science and Forestry, State University of New York, Syracuse New York.

Wagner, J.E. and Holmes, T.P. (1999) Estimating Economic Gains for Landowners Due to Time-Dependent Changes in Biotechnology, *Forest Science*, 45(2):163–170.

Wagner, J.E. and Sendak, P.E. (2005) The Annual Increase of Northeastern Regional Timber Stumpage Prices: 1961–2002, *Forest Products Journal*, 55(2):36–45.

Wagner, J.E., Nowak, C.A., and Casalmir, L.M. (2003) Financial Analysis of Diameter-Limit Cut Stands in Northern Hardwoods, *Small-scale Forest Economics, Management and Policy*, 2(3):357–376.

Wagner, J.E., Smalley, B., and Luppold, W. (2004) Factors Affecting the Merchandising of Hardwood Logs in the Southern Tier of New York, *Forest Products Journal*, 54(11):98–102.

Wear, D.N. and Newman, D.H. (1991) The Structure of Forestry Production: Short-Run and Long-Run Results, *Forest Science*, 37(2):540–551.

Webster, C.E., Reed, D.D., Orr, B.D., Schmierer, J.M., and Pickens, J.B. (2009) Expected Rates of Value Growth for Individual Sugar Maple Crop Trees in the Great Lakes Region, *Northern Journal of Applied Forestry*, 25(4):133–140.

Whitehead, J.C., Pattanayak, S.K., Van Houtven, G.L., and Gelso, B.R. (2008) Combining Revealed and Stated Preference Data to Estimate the Nonmarket Value of Ecological Services: An Assessment of the State of the Science, *Journal of Economic Surveys*, 22(5):872–908.

Willis, K.G. and Garrod, G.D. (1991) An individual Travel-Cost Method of Evaluating Forest Recreation, *Journal of Agricultural Economics*, 42(1):33–42.

Winston, W.L. (1994) *Operations Research: Applications and Algorithms*, 3rd edn., Belmont, CA: Duxbury Press.

Wunder, S. and Wertz-Kanounnikoff, S. (2009) Payments for Ecosystem Services: A New Way of Conserving Biodiveristy in Forests, *Journal of Sustainable Forestry*, 28(3):576–596.

Yin, R. (1997) An Alternative Approach to Forest Investment Assessment, *Canadian Journal of Forest Research*, 27(12):2072–2078.

Yin, R. and Newman, D.H. (1997) Long-Run Timber Supply and the Economics of Timber Production, *Forest Science*, 43(1):113–120.

Yin, R., Pienaar, L.V., and Aronow, M.E. (1998) The Productivity and Profitability of Fiber Farming, *Journal of Forestry*, 96(11):13–18.

Zhou, M., Liang, J.J., and Buongiorno, J. (2008) Adaptive versus Fixed Policies for Economic or Ecological Objectives in Forest Management, *Forest Ecology and Management*, 254:178–187.

Zudak, L.S. (1970) Productivity, Labor Demand and Cost in a Continuous Production Facility, *The Journal of Industrial Economics*, 18(3):255–274.

Index

ROUTLEDGE
Revivals

Made in United States
North Haven, CT
26 January 2022

15291216R00226